Form and Function in the Honey Bee

Lesley J Goodman
28 August 1931–6 March 1998

Dr Lesley Goodman was inspired by teachers at her High School to take up a career as a research scientist, and in 1950 she went to Girton College, Cambridge to study zoology. After Cambridge, she moved to Liverpool University where she completed her PhD research into landing mechanisms in flying insects. In 1960 she became a Lecturer and later a Reader in Zoology at Queen Mary, University of London, where she remained until her official retirement in 1996. Her enthusiasm for the study of invertebrate, especially insect, sensory physiology and behaviour stimulated interest in many of her students.

Dr Goodman's research interests were chiefly in the field of neurobiology and insect physiology, in particular honey bee flight and visual systems. Little was known about the way in which the insect thoracic flight motor machinery was influenced by commands from the brain and she addressed this problem in the bee, mapping the route of visual input to certain flight motor neurons.

Interaction with bee scientists and beekeepers stimulated her desire to produce a book that would be readable (and affordable) for both non-scientists and students, and so in 1996 Dr Goodman began work on Form and Function in the Honey Bee. It was a book she particularly hoped would assist students in this field, and the project was still evolving when lung cancer eventually overwhelmed her — when she died in 1998, it was still far from complete. Before her death she set up the L J Goodman Insect Physiology Research Trust to ensure the book was completed and published. The book has been completed posthumously by Prof. Richard J Cooter, Chair of the L J Goodman Insect Physiology Trust, and Dr Pamela Munn, Deputy Director of IBRA.

Lesley's aim was to describe some of the topics in bee biology that would target a broad audience of people wanting to know more about how bees function, and the subsequent contributors and the publishers have adhered faithfully to her plan.

Form and Function in the Honey Bee

Lesley Goodman

Completed and edited by Richard J Cooter and Pamela A Munn

IBRA

INTERNATIONAL BEE RESEARCH ASSOCIATION

Jointly published by:
The International Bee Research Association,
a Company Limited by Guarantee, 1, Agincourt Street, Monmouth, NP25 3DZ (UK) &
Northern Bee Books, Scout Bottom Farm, Mytholmroyd, Hebden Bridge HX7 SJS (UK).

Obtainable from:
www.ibra.org.uk & www.northernbeebooks.co.uk

IBRA is a not-for-profit organization that exists to increase people's awareness of the vital
role of bees in agriculture and the natural environment. It aims to promote the study and
conservation of bees, and provide useful and reliable information to beekeepers and bee
scientists all over the world.

First published: January 2003
Reprinted: 2012
Reprinted: 2022

British Library Cataloguing-in-Publication Data
A cataloguing record for this book is available from the British Library

Designed at The Design Stage (Digital) Ltd, Cardiff Bay, UK www.design-stage.co.uk
Artwork: DM Design and Print

IBRA Proof Editor - Stuart A. Roberts

ISBN 978-1-913811-11-2 Softcover
ISBN 978-1-913811-12-9 Hardcover

Dedication

This book is dedicated with love and gratitude to my parents,

Harry and Agnes Goodman, who encouraged me in everything

I undertook, to Kathleen Ludbrook, Margaret Edorall and

Eulalia Higgins, teachers who awoke in me a love of science,

to all my research students who later became my friends and

colleagues, and who made it all a glorious adventure,

and finally to my husband, Leslie Heald, who has supported

me through a very difficult time.

Lesley J Goodman

Contents

Foreword

Dr Lesley Goodman was Reader in Zoology at Queen Mary College, University of London, until her official retirement in 1996. During the previous thirty years, her enthusiasm for the study of invertebrate, especially insect, sensory physiology and behaviour stimulated interest in many of her students, some of whom went on to study for PhDs under her supervision. I was one of Lesley's first PhD students and I know from personal experience that she took a great interest not only in her students' studies but also in their careers; she became a good friend and mentor. Vision and visually-controlled behaviour in insects, especially honey bees, was a particular passion of hers. In 1983 she was elected to the Council of the International Bee Research Association (IBRA); her interaction with bee scientists and bee keepers stimulated her desire to produce a book that would be readable — and affordable — for both non-scientists and undergraduate scientists. Her aim was to describe some of the topics in bee biology that were of particular fascination for her and which she knew would be of interest to others. One of her ideas was to commission the paintings that illustrate topics in the book. She planned to fund its production through IBRA, and the income from sales of the book would be used by IBRA to support its activities.

The book project was still evolving when lung cancer eventually overwhelmed her, and when she died, in 1998, it was still far from complete. Before her death she set up the L J Goodman Insect Physiology Research Trust, and tasked the Trustees with finishing the project after her death. It therefore fell to me, as Trust Chairperson, to ensure that the book was completed and published. The subsequent delay in its production has been largely due to my inability to find enough time to devote to the project — I believe that she would have understood.

We have adhered faithfully to Lesley's plan for the book, and I hope that we have satisfied her stipulation that it should target a broad audience of people wanting to know more about how bees function. Any errors that remain are mine.

Please note that unless stated otherwise, all mentions of bees or honey bees refer to the European honey bee, *Apis mellifera*. *The New Flora of the British Isles* by C Stace (Cambridge University Press, 1995) was used as the reference for botanical names referred to in the book, and the CAB-I *Thesaurus* for other biological terms. The spelling of the singular and plural of sensilla(e) is as given in *Henderson's Dictionary of Biological Terms* edited by E Lawrence (12th edition).

Richard J Cooter

formerly Professor of Applied Entomology and Head of the Agricultural Resources Management Department, Natural Resources Institute, University of Greenwich

Acknowledgements

Many people have assisted Lesley, and then myself, in writing and completing this book; a number of them merit particular thanks. Haidee Price Jones, Laboratory Services Manager, and Keith Pell, Electron Microscopist of the Schools of Biological Sciences and Basic Medical Sciences, and John Cowley of the Chemistry Department, Queen Mary, University of London, assisted Lesley in many ways. Haidee facilitated the use of light and electron microscopy equipment and John maintained the honey bee colonies. Keith identified, dissected and prepared the honey bee tissue and took the light and electron micrographs for the book. He also elucidated the ultrastructure of the Nasonov gland. The diagrams and beautiful paintings that further illustrate the book were produced by Michael J Roberts. Thanks to The Design Stage (Digital) Ltd, Cardiff Bay: to Mark Roberts for the original page design and especially to John Dixon, who played an integral part in the book's production and undertook the difficult task of designing the layout which brought the text and illustrations to life in a way that satisfied Lesley's particular requirements. Kim Gilmore helped Lesley with literature searches, drafting chapter 8 and with editing and early proofreading of the first completed chapters. I was assisted by Lynn Dicks in organizing material for writing chapter 7 which had not been started before Lesley died. Penelope Walker, who joined the continually evolving project at a late stage, has done a very thorough job in final checking, proofreading and indexing the book. The whole project has been skilfully and efficiently managed by Pamela Munn, Deputy Director of IBRA, without whose dedication we might have failed to finish our task.

Lesley left the dedication that is printed on page iii.

Richard J Cooter

1. The antennal sense organs:

smelling, tasting, touching and hearing in the bee

FIG. 1.1 *The left antenna of a worker bee: basal scape, pedicel and flagellum comprising 10 annuli. Note the hairy head capsule of the bee. The compound eye is also covered with hairs (mh) although these hairs are innervated and have a mechanoreceptive function. The points of articulation between the scape and the head capsule, and the pedicel and the scape are indicated by arrows.*

The **antennae**, paired, mobile appendages arising from the head, perform many functions in the life of the bee. Their surface is packed with sensory receptors, **sensillae**, present in enormous numbers (figs 1.1, 1.2). Most of the receptors for smelling and many of those for tasting are situated on the antennae, which means that they play a crucial role in the chemical communication between bees that regulates the life of the colony. Outside the hive the sense of smell is one of the most important factors in the location of forage.

The antennae bear other types of sense organ, including those that detect the airborne sounds made by the dancing bee during the waggle dance — sounds that communicate information directing followers to a good forage site. Tactile receptors on the surface of the antennae are involved in many behaviours, for example when a bee solicits for food; in the determination of the thickness and smoothness of the wax walls of the comb; and, according to some authors, detecting the microsculpturing on the surface of petals. Receptors that can detect the displacement of the antennae during flight are present and are thought to play some part in flight control. In addition, receptors for detecting temperature and humidity changes, and carbon dioxide have been reported.

Each antenna consists of three segments, a basal **scape**, a **pedicel** and a **flagellum**. The flagellum is subdivided into a number of annuli, 10 in the female, 11 in the male. The annuli are commonly referred to as 'segments' in the literature, although this is not strictly correct since they lack intrinsic musculature. The antenna has two joints: the scape is attached to the head capsule by a socket joint allowing rotational movement of the whole antenna; the flagellum moves together with the pedicel and there is a hinge joint between the pedicel and the scape allowing these two segments to move up and down relative to the scape.

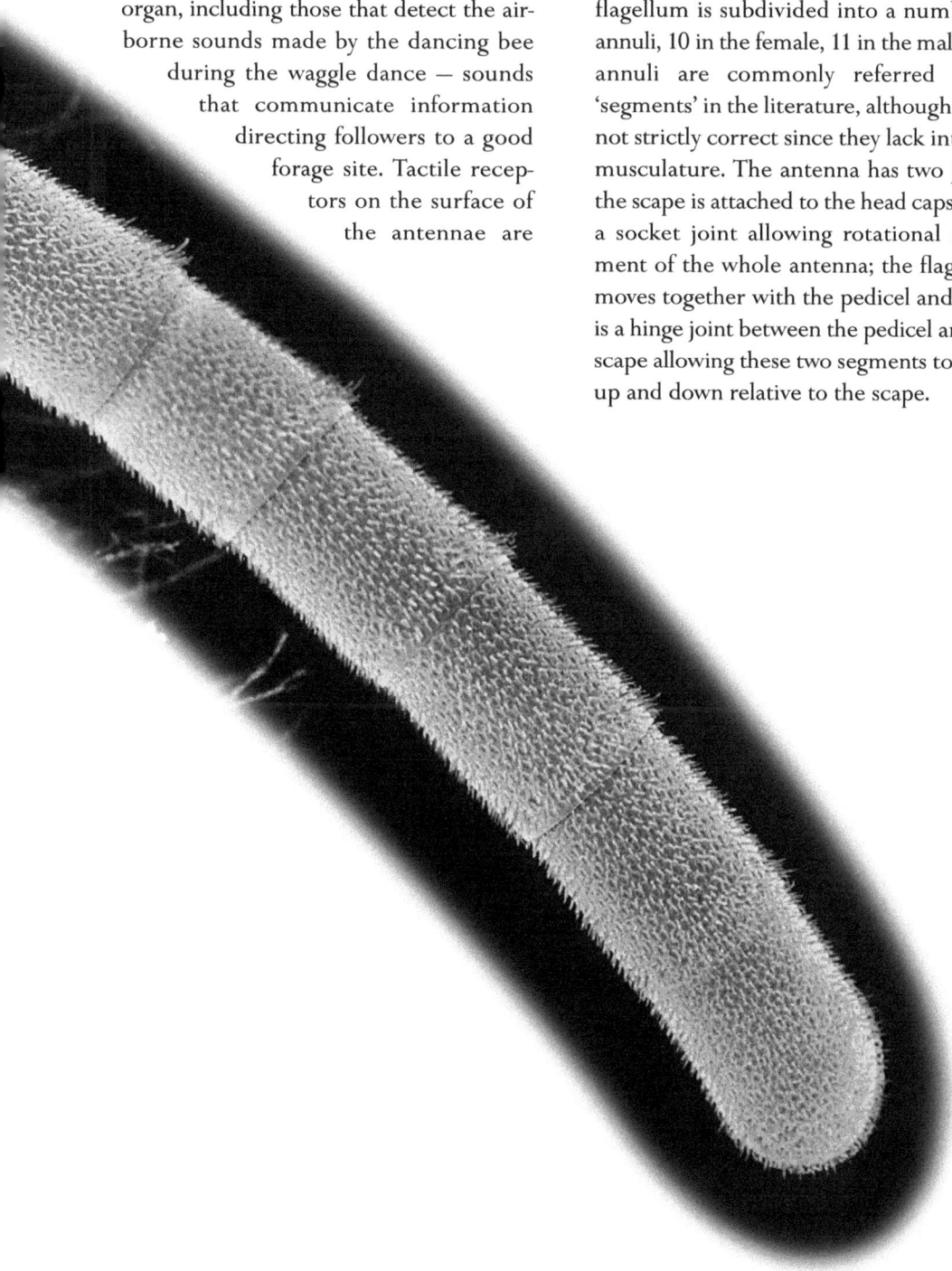

FIG. 1.2 *A small section of two of the flagellar annuli showing representative antennal sensillae. Different types of trichoid sensilla (ts1, ts2, ts3, ts4), basiconic sensilla (bs), placoid sensilla (pls), coeloconic sensilla (arrow 1), campaniform sensilla (cf), coelocapitular sensilla (arrow 2).*

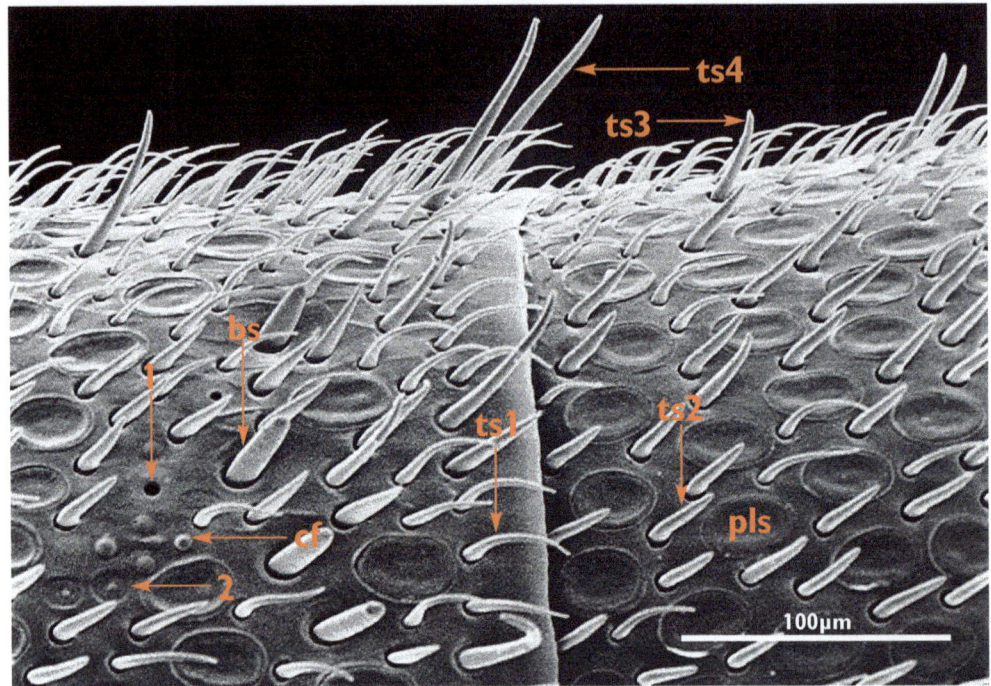

The sense of smell in bees

Odours emanating from biological organisms, both plants and animals, are complex blends of chemicals. Even pheromones, the chemicals secreted by animals to communicate with other members of their own species, are often blends of two or more compounds. The Nasonov pheromone, secreted in a variety of circumstances in which aggregation of workers occurs, contains at least seven components. Olfactory receptors on the antennae respond to all seven. The queen pheromone contains many more components. The bee lives in a rich and complex odour environment, both within the hive and outside, and has to cope with distinguishing the blend of chemicals that represents a nestmate, a pheromone signal or a forage source among a wide variety of plants. Colour and shape of a flower are the important cues for a bee at a distance from forage, but at close quarters odour becomes the dominant cue. On a warm day a flowering plant may give off as many as 40–50 or even more volatile compounds; if its nectar reward is rich, the bee has to discriminate this plant from its equally odour-rich neighbours (fig. 1.5). In fact, the bee usually uses a proportion of these volatiles to represent the plant's odour 'signature' but this still leaves it with a formidable task as it moves among a mixture of plants. The bee's sensitivity to odours has been shown to be as good as man's in cases where thresholds have been determined for each species. For example, in both man and bee thresholds are similar for citral, jasmine oil and proprionic acid[1,2] at concentrations between 10^{-10} and 10^{-11} molar. The bee's capacity to discriminate between hundreds of different odours is also similar to that of man. Furthermore, compounds with different structures but with similar odours for man are also confused by the bee; however, compounds with nearly identical structures but with very different odours for man are again well discriminated by bees[1]. The bee is remarkable in the rapidity with which it learns odours. Most odours can be learnt to the level of a 90% correct choice after a single visit to a rewarded odour source[1,3]. The importance of odours in the life of the bee is emphasized by the fact that the majority of receptors on the antenna are used for smelling. Which are the receptors for smell and how do they work?

Smelling in humans

It is easier to understand the arrangement of the olfactory structures in insects if first we consider the process of smelling in humans. In order to sample the odours in the environment, we have to expose our delicate olfactory receptor cells to the air but at the same time we have to protect them and prevent them from dehydrating. This is achieved by associating them with the respiratory system (fig. 1.3). The olfactory receptor cells are gathered together in the upper part of the nasal cavity forming the olfactory mucosa. A stream of air is constantly being drawn into the nasal cavity, where it is filtered, warmed and kept moist. Just inside the nasal cavity are hairs which act as a coarse filter trapping large particles entering with the air. The nasal cavity is lined with a mucous membrane comprised of goblet cells that secrete mucus, and ciliated epithelial cells that beat continuously, moving the mucus

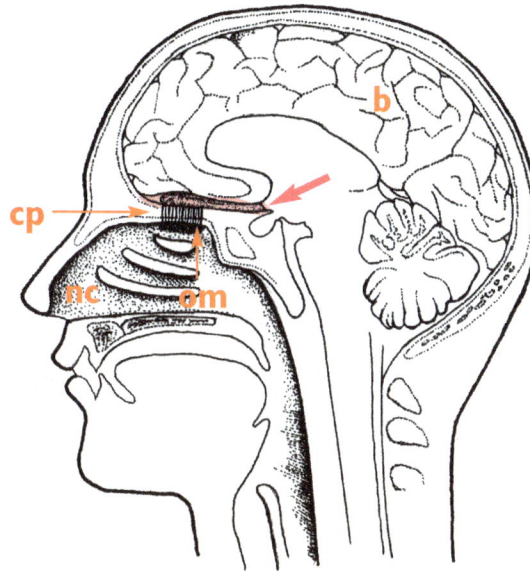

FIG. 1.3 *Longitudinal section through the human head showing the olfactory mucosa (om) projecting into the nasal cavity (nc). The olfactory cells of the mucosa send their axons through the pores of the cribriform plate (cp) into the olfactory bulb (arrow). This structure is the primary processing centre for olfactory information, which it then relays to specific areas of the brain (b).*

over the surface of the nasal cavity and down into the pharynx. The nasal cavity is plentifully supplied with blood vessels which function to warm the incoming air while the mucus keeps the air moist and protects the delicate cilia. Normally only 2–5% of the inspired air passes over the olfactory mucosa but, when we sniff, eddies of air are wafted over it. We have around 30 million olfactory receptor cells in each nostril and each cell has a neuron leading into the brain[4]. Each cell has several cilia hanging down into the mucous layer, extending its surface area by several hundred fold (fig. 1.4). The protein molecules that form the acceptor sites for the olfactory molecules are borne on these

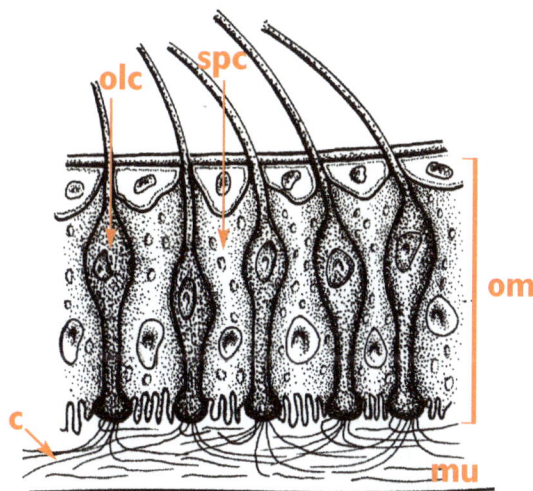

FIG. 1.4 *A section through the olfactory mucosa (om) showing the individual olfactory cells (olc) with their neurons extending towards the olfactory bulb. Their outer ends are ciliated (c) and lie in a layer of mucus (mu) secreted by the glands of the mucosal layer. The cilia bear the receptor sites and, since a single cell may have as many as 1000 cilia, the receptive area of the mucosa, around 5 cm^2, is effectively increased to around 600 cm^2; supporting cells (spc).*

cilia. In order to interact with the acceptor sites, the olfactory molecules in the inspired air have to cross the mucous layer. Odorants with a high water solubility can dissolve in and traverse the watery mucous layer to the acceptor sites. However, many odorants are insoluble in water and must be transported across the mucous layer. Small proteins that show a high affinity for a wide variety of odorant molecules, odorant-binding proteins (OBPs), are secreted as a fine spray or aerosol in the nasal cavity. It is believed that these OBPs may play a part in olfaction, transporting hydrophobic odorant molecules across the mucus to the vicinity of the acceptor sites where stimulation occurs[5].

How do the odorant molecules stimulate the receptor cells? The molecules bind to the protein acceptor sites located in the membrane of the cilia. This interaction initiates a cascade of chemical reactions within the cell that ultimately causes a brief opening of ion pores in the membrane, resulting in a flow of ions between the outside and inside of the cell. This alters the electrical charge across the surface of the cell membrane causing the cell to depolarize, which, in turn, results in the initiation of nerve impulses in the neuron leading from the receptor cell into the brain. The odorant molecules are then removed from the acceptor binding site by degrading enzymes. The OBPs may also be involved in terminating the stimulus, and the site is then ready to interact with another molecule.

Most people can differentiate up to 2000 different scents: how are they able to do this? It is unlikely that there are as many different acceptor sites on the olfactory cilia as there are odours since we are able to smell newly synthesized odours as soon as they are created. It is thought that the factors governing the interaction of an odorant molecule with an acceptor site are a set of structural elements belonging to that molecule[6]; these include its three-dimensional shape, the functional groups present on the molecule and its physicochemical characteristics. Some acceptor sites are very specific

in their requirements, especially those interacting with pheromones (this is particularly true of insects); others are less specific and will interact with a structurally similar class of compounds. Some receptor cells may have more than one type of acceptor site on their cilia which increases the number of compounds to which they can respond. Cells can be inhibited by odours as well as stimulated by them, so that a certain odour may excite some receptor cells while inhibiting others[7]. Where the stimulus comprises a blend of odours, which is normally the case in everyday life, complex effects are produced in the receptor cells as excitatory and inhibitory effects compete.

When odorant molecules are wafted up to the olfactory mucosa they dissolve in, or are carried across, the mucous layer and bind to those acceptor sites that have the appropriate configuration, thus stimulating the cells bearing those sites. This results in a pattern of firing across many of the millions of olfactory cell neurons entering the first relay station of the brain. A specific spatio-temporal pattern of firing develops that is characteristic for the particular odour, i.e. a different constellation of cells will be activated by a different odour and the firing pattern that develops will encode that odour. These spatio-temporal firing patterns can be discriminated in the higher centres of the brain[7,8].

Smelling in bees

The basic mechanisms of olfaction are the same in the bee as they are in humans: odorant molecules interact with appropriately structured acceptor sites resulting in a pattern of excitation across olfactory neurons running into the brain. Indeed, so similar are the mechanisms believed to be, that many of our current ideas on olfaction are based on work carried out on insects[7]. Comparative studies have shown that there is a remarkable similarity of design in the macro- and microarchitecture of olfactory circuits in the animal kingdom. Molluscs, insects and vertebrates all possess numerous olfactory receptor cells whose neurons converge on a neuropile composed of discrete groupings of cells, the **glomeruli**. The neurons from this first-order relay station in turn project to a second-order relay station associated with memory function (the piriform cortex of vertebrates and the mushroom bodies of insects)[7].

At the periphery, the same requirements are placed on the insect as on the human: a necessity to expose an enormous number of very delicate receptor cells to airborne odorant molecules. However, this poses a major problem for insects because their surface area to volume ratio is large, which means that dehydration is a danger. Insects have minimized the risk of dehydration by covering the body with a layer of **cuticle** which, in addition to acting as an external skeleton, also helps to prevent loss of water from the body. Waterproofing is completed by covering the cuticle with a thin layer of wax. The cuticle extends, in a thinner form, into the orifices of the body, including the mouth, rectum and the air passages of the respiratory system. However, the respiratory system of insects (see chapter 6) does not lend itself to association with the olfactory system as air is piped right to the cells and organs of the body in cuticle-lined tubes which lack the protective mucous layer. Respiratory movements are often irregular in insects and the apertures into the system are kept closed as far as possible to prevent water loss. Since these apertures are often closed or partially closed, the entry of odorant molecules would be greatly restricted. Moreover, the apertures are located along the sides of the body and not at the front end where they need to be to explore the environment into which the insect is moving.

Insects have overcome this problem by placing the majority of their olfactory receptors on the antennae, which protrude forward from the body and which, in flight particularly, have the odour-bearing

FIG. 1.5 *Bees visiting oilseed rape (Brassica napus), for nectar and pollen. Many of the plant volatiles emitted by oilseed rape have been identified, and studies have shown that a small number of key volatiles cue rape plant recognition by the bee[34].*

FIG. 1.6 *Examples of olfactory receptors from insect antennae.*
a *Large trichoid sensilla covered in minute pores (arrows). The dendrites (d) of the olfactory receptor cells extend up into the shaft and lie adjacent to the pores. Only two dendrites are shown.*
b *Small trichoid sensilla.*
c *Coeloconic sensilla, small, flask-shaped depression in the cuticle containing a short peg covered in pores.*
d *Placoid sensilla, an oval-shaped area of the cuticle with the outer edge bearing many small ridges with rows of pores in the troughs between them. The dendrites (d) of the sensory cells run around the rim underlying the pores.*

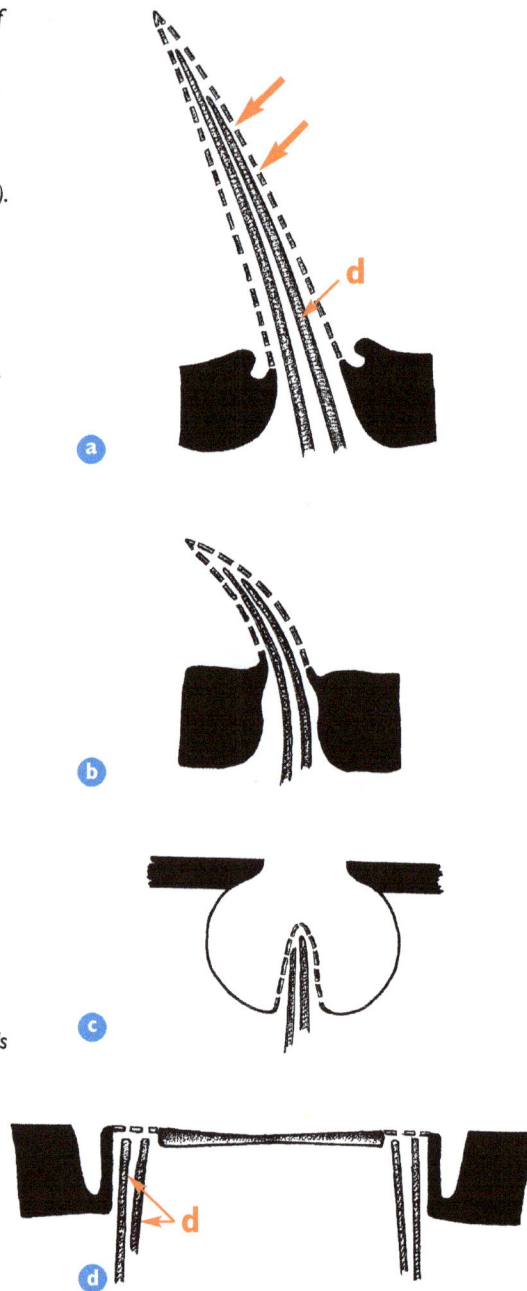

these structures are pierced by many very fine pores, around 10 nm in diameter: there may be up to 1000 pores on one hair (figs 1.6a, b; 1.7b)[9]. The fine dendritic extensions of the olfactory receptor cells either extend up into the hairs and pegs or lie beneath the pores of the plates (figs 1.6, 1.7). The olfactory molecules pass through these fine pores to reach the acceptor binding sites on the dendritic membranes of the receptor cells. An antenna bears many thousands of these cuticular modifications or sensillae, usually including several different types. A sensilla commonly has the dendritic branches of 3–5 receptor cells in its lumen but in some sensillae there can be up to 50 olfactory cells. Although the dendritic branches are bathed in a fluid, the receptor lymph, the restricted aperture of the pores prevents excessive water loss from the sensillae. In this way, the insect exposes a large surface area of receptor membrane to airborne olfactory molecules while keeping the membrane protected and moist. The antennal geometry and sensilla array in some insects, notably male moths, is highly specialized for the most efficient filtering of odorant molecules from the air.

The pore apertures occupy a very small part of the surface area of a sensory hair or plate but olfactory molecules are trapped all over the surface of the sensilla in the thin layer of lipoprotein that covers the cuticular surface (fig. 1.7c). The trapped molecules migrate in the lipoprotein layer to the pore apertures and once inside, they may pass to the receptor membrane via tubules running from the pore aperture to the surface of the membrane (fig. 1.7c). In cases where no pore tubules are present, they cross the receptor lymph directly to the receptor membrane[10]. In insects, it has been shown that OBPs are present in the lymph surrounding the dendrites and that they do bind to and transport odorant molecules to the acceptor sites on the dendrites. The dendritic endings of the receptor cells bear many protein acceptor sites in their membrane, and an odorant molecule interacts with a

airstream drawn over them. Nonetheless, there is still the difficulty of exposing the receptor cells to the airstream with minimal water loss. Instead of the receptor cells being located in one patch, like the olfactory mucosa, the cells here are split into small groups, each group associated with a modification of the cuticular surface. These modifications take many forms, for example, cuticular hairs, pegs, pegs sunk into pits and plate-like structures (figs 1.6, 1.7, 1.8, 1.9). The thin cuticular walls of site if it has the appropriate configuration. As in humans, this interaction indirectly opens ion gates in the membrane and the consequent flow of ions into the cell results in a depolarization of the receptor membrane. This then leads to the initiation of nerve impulses in the receptor cell neuron which are transmitted to the glomeruli of the antennal lobes within the brain. In the bee, as in man, the odorant molecules are believed to be removed from the receptor sites after stimulation either by enzymatic degradation of the molecule or by its binding to an OBP.

Olfactory receptors among the sensillae on the antenna

The surface of the antenna is covered with a wide variety of hair-like structures of varying length and thickness: stout and slim pegs, pegs sunk beneath the surface and oval plates flush with the surface. Which of these structures function as olfactory sensillae? Assigning a function to each of them is proving a daunting task because their very small size and close proximity makes recording the signals from the individual receptors very difficult. Function thus has to be inferred from structure alone for the majority of the receptors.

The receptor organs on the surface of the bee antenna have been classified into seven different types (figs1.8, 1.9)[11,12,13]. There are many **trichoid** sensillae, around 3000 on the worker antenna. These are hair-like structures that can be subdivided into as many as five different types, ranging in form from short, slender, flexible hairs to long, stiff, thick-walled sensillae. The latter are sometimes known as sensillae **chaeticae**. In addition, there are four different types of hair-like structures present, the **setae**, that are not innervated. It is almost impossible to distinguish these from the trichoid sensillae on the basis of external structure alone. The **basiconic** sensillae are peg or cone-like structures. The numerous, plate-like structures flush with the antennal surface are known as **placoid** sensillae. The **coeloconic** sensillae and **ampullaceae** consist

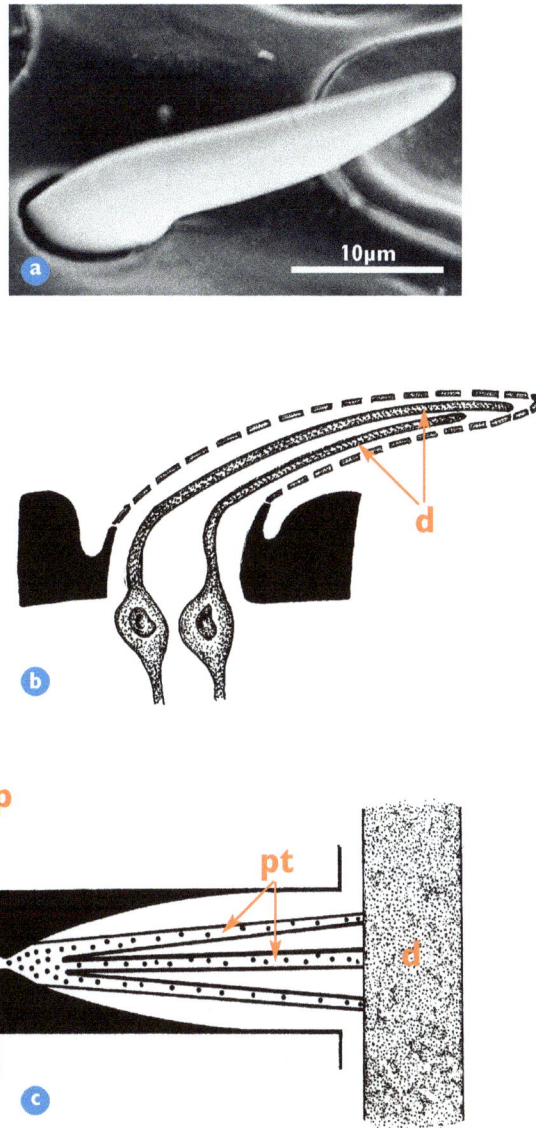

FIG. 1.7 **a** *An S-shaped trichoid sensilla (see also fig. 1.2, ts2) from the antennal flagellum of the worker bee.* **b** *A diagrammatic section through the sensilla, showing pores through the cuticle, together with the dendrites (d) of two of the 5–10 olfactory cells innervating the sensilla.* **c** *Section through the aperture of one of the pores showing the pore tubules (pt) through which the olfactory molecules (black dots) pass to the surface of a dendrite (d). The receptor sites are located on the surface of the dendrite. A thin layer of lipoprotein (lp) traps molecules over the entire surface of the sensilla. The molecules travel within the layer to the pore entrances.*

of fluted pegs or cones set in flask-shaped depressions in the cuticle. They differ only in the depth of the depression in the cuticle, the sensillae ampullaceae being deeper and having a narrower opening to the surface. The **coelocapitular** sensillae consist of circular, shallow depressions in the cuticle, each containing a mushroom-shaped peg that lies just below the antennal surface. Also present are a small number of **campaniform** sensillae, small, dome-shaped structures surrounded by a ring of slightly depressed cuticle. Deciding whether a sensilla is best described as a trichoid or chaetical hair or whether the depression containing a peg is shallow or deep is very much subject to individual interpretation. In addition to these external structures, Johnston's organ, which functions as a rather specialized type of ear, is located inside the pedicel.

The size of the placoid sensilla has permitted electrophysiological recordings which have shown that they are olfactory receptors. One type of trichoid sensilla is believed to be olfactory in function on the basis of structure, since its surface is covered with fine pores. A few of the basiconic sensillae present are also reported to be multiporous.

Placoid olfactory sensillae

These structures occupy a considerable amount of the surface area of the antenna. There are around 2700 present in the worker bee, with a smaller number (1600) in the queen. The flagellum surface of the drone is twice as large as that of the worker and it carries between 15 000 and 16 000[13] placoid sensillae (fig. 1.11 a, b, c, d). In the worker, they occur on the eight distal-most annuli of the flagellum (fig. 1.1), packed on the side facing away from the head which is forwardly-pointing during flight. Each sensilla consists of an oval area of thin cuticle surrounded by a slightly depressed ring-shaped margin in which can be seen radially-arranged ridges of cuticle. Under very high magnification in the electron microscope, fine pores are visible in the thinner cuticle between the ridges, about 20 per

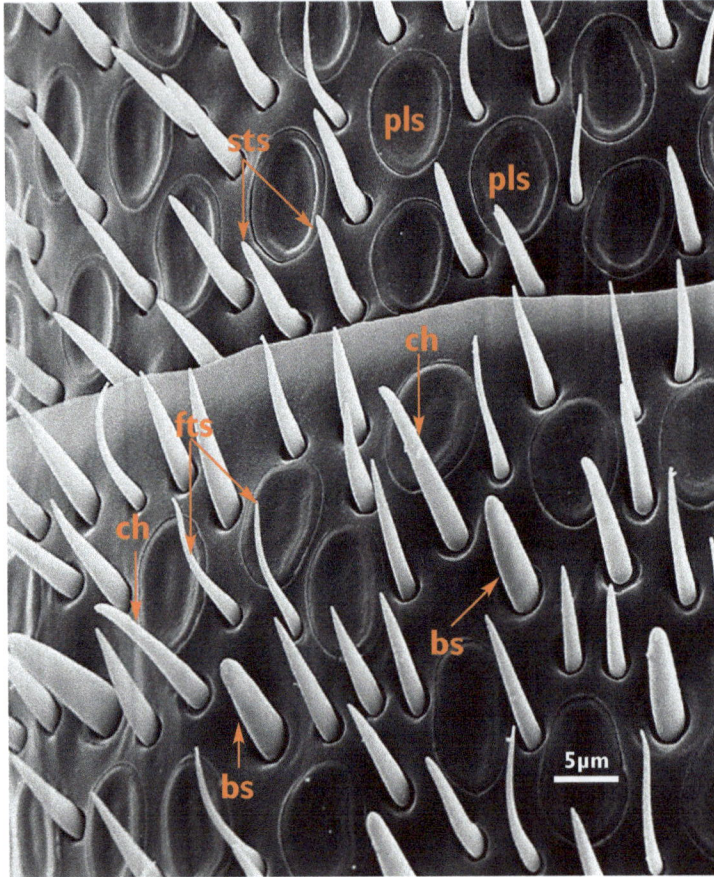

FIG. 1.8 *Two annuli from the flagellum of a worker bee showing a variety of sensilla types: fine trichoid sensillae (fts), S-shaped trichoid sensillae (sts), stout peg-like trichoid sensillae sometimes known as sensillae chaeticae (ch), basiconic sensillae (bs), placoid sensillae (pls).*

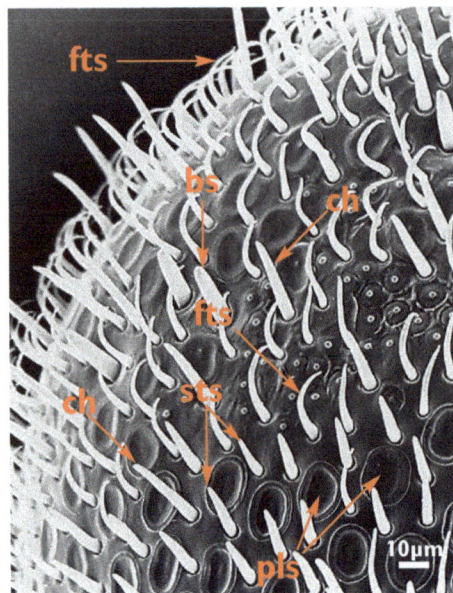

FIG. 1.9 *Tip of the flagellum of the worker bee antenna showing a variety of sensillae: fine trichoid sensillae (fts), S-shaped trichoid sensillae (sts), stout peg-like trichoid sensillae or sensillae chaeticae (ch), placoid sensillae (pls), basiconic sensillae (bs).*

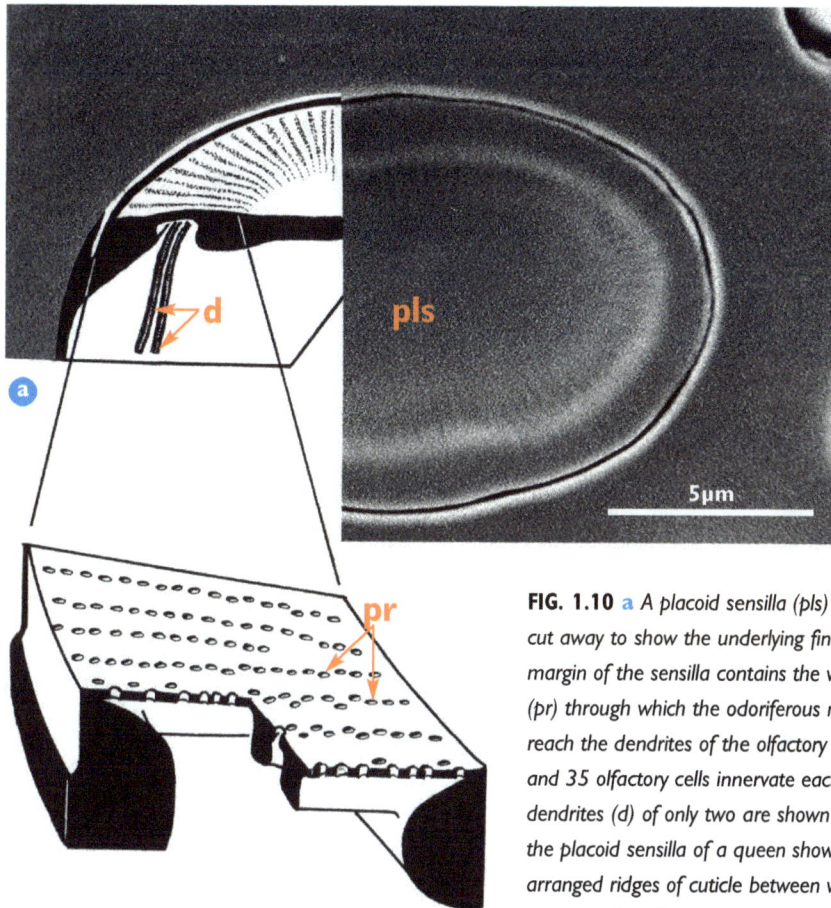

FIG. 1.10 a *A placoid sensilla (pls) from the worker cut away to show the underlying fine structure. The margin of the sensilla contains the very fine pores (pr) through which the odoriferous molecules pass to reach the dendrites of the olfactory cells. Between 5 and 35 olfactory cells innervate each sensilla; the dendrites (d) of only two are shown here.* **b** *Part of the placoid sensilla of a queen showing the radially arranged ridges of cuticle between which the fine pores are found.*

strip of cuticle. There are some 120–150 ridges of cuticle around each margin so that each sensilla has some 2400–3000 pores (fig. 1.10)[11]. The dendritic branches of the receptor cells lie around the margin of the pore plate, underneath the many pores which allow olfactory molecules access to the acceptor sites on the dendrites. Surprisingly perhaps, the placoid sensillae are not all innervated by the same number of neurons: between 5 and 35 neurons innervate each sensilla[14]. In the worker, some 48 000 neurons in total run from the placoid sensillae to the antennal centres in the brain.

We do know quite a lot about the responses of these sensillae in the bee since they are large enough to permit the entry of an electrode. By this means we can record the nerve impulses generated in the receptor cells following olfactory stimuli. Individual cells in various placoid organs, particularly in drones, have been found to respond to

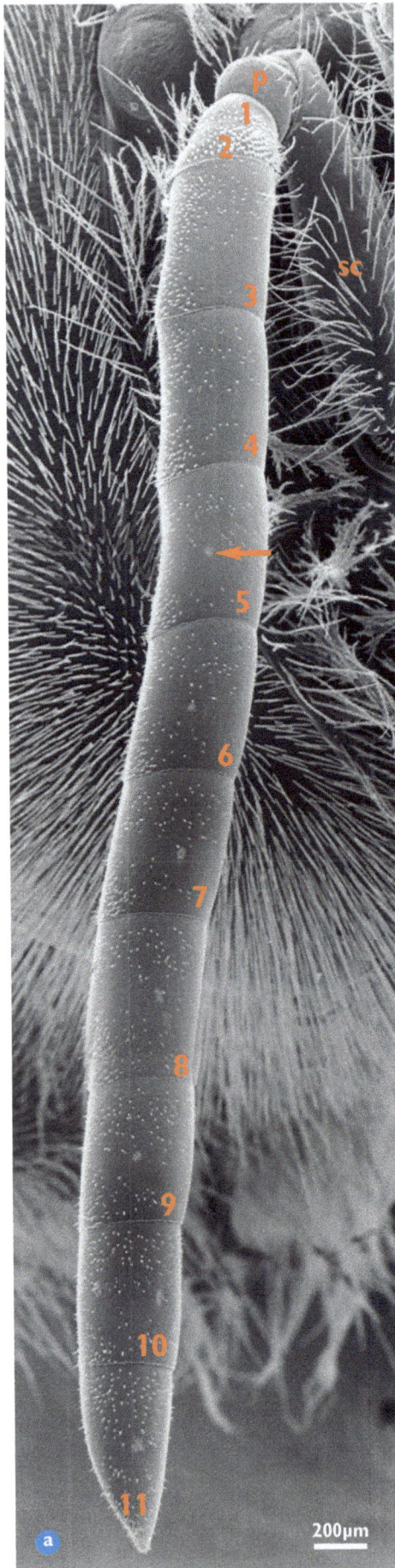

FIG. 1.11 **a** The right antenna of a drone bee. The flagellum has 11 annuli or segments in contrast to the 10 of the worker and the queen. Annuli (1,2,3… 11), scape (sc), pedicel (p). The light patches (*arrow*) situated centrally in annuli 5–11, are groups of coelocapitular and coeloconic sensillae. See also figures 1.14 and 1.15. **b** The drone antenna is solidly packed with placoid sensillae but other types of sensillae are fewer in number. *Arrow* indicates the patch of coelocapitular and coeloconic sensillae. **c** The placoid sensillae (pls) are packed close together over the forward-facing region of the flagellum, with no other sensillae between them over quite large sections of each annulus. **d** Small basiconic sensillae (bs) can be found between the placoid sensillae near the margins of the annuli and at the flagellar tip.

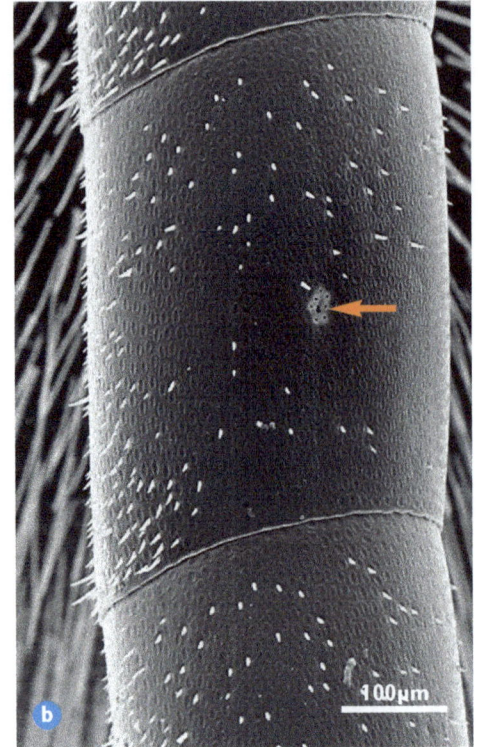

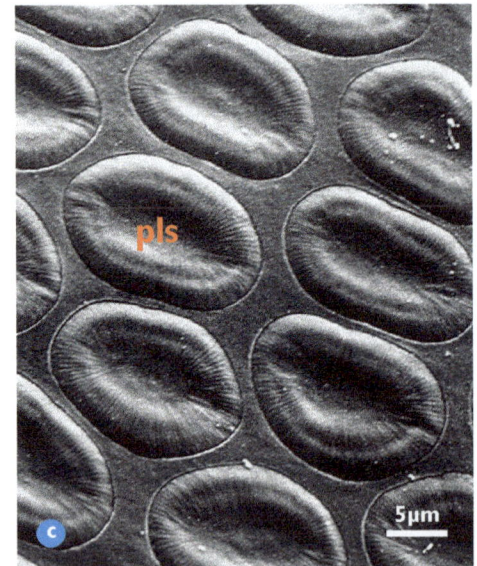

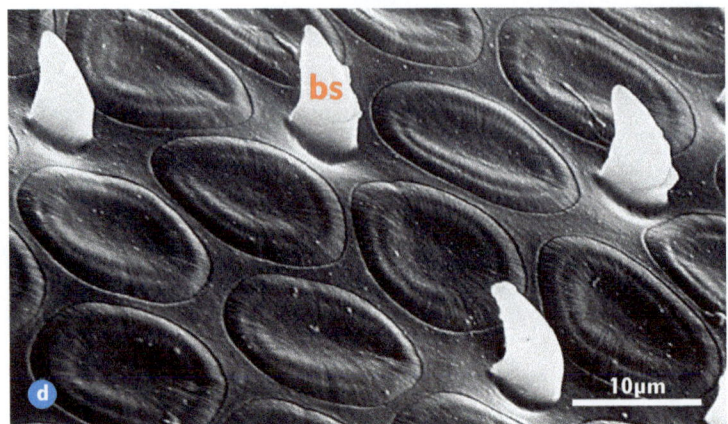

components of the queen pheromone, to the various components of the Nasonov and alarm pheromones, and to a very wide range of plant and floral odours[14,15,16]. No receptor cells have been found that respond only to one compound, but there are some that have a very narrow response spectrum, i.e. the cell gives a large response to one particular compound but also gives some response to a few other compounds. One such group of cells are those that respond to the Nasonov gland components, with the peak response being given to geraniol. Another group gives a peak response to the alarm pheromone component, isopentyl acetate, and smaller responses to related substances. The narrowest response spectrum reported is that of certain drone placoid cells which respond to 9-hydroxy-(*E*)-2-decenoic acid (9-HDA) and 9-oxo-(*E*)-2-decenoic acid (9-ODA), the major components of the queen pheromone.

Most of the placoid receptor cells examined have been found to respond to a number of compounds[14]. Cells with a wide response spectrum are known as generalist cells, and in the bee such cells may be stimulated by a range of plant volatile compounds and possibly some pheromone components in addition. The response spectra of the cells differ, although they may overlap to some extent. The bee uses a broad variety of host plants and, even though it may use only a few of the many volatiles given off by a plant as that plant's signature, it has to discriminate those components from the rich odour environment. Possession of an enormous number of placoid cells with differing response spectra can result in a distinctive pattern of firing developing across the thousands of neurons running centrally when a bee smells a plant. This pattern represents the blend of odours that form the odour signature of that plant to the bee. The complex pattern arising from the involvement of so many receptor cells makes possible discrimination between plants and even the discrimination of subtle changes in the signature odours of a particular

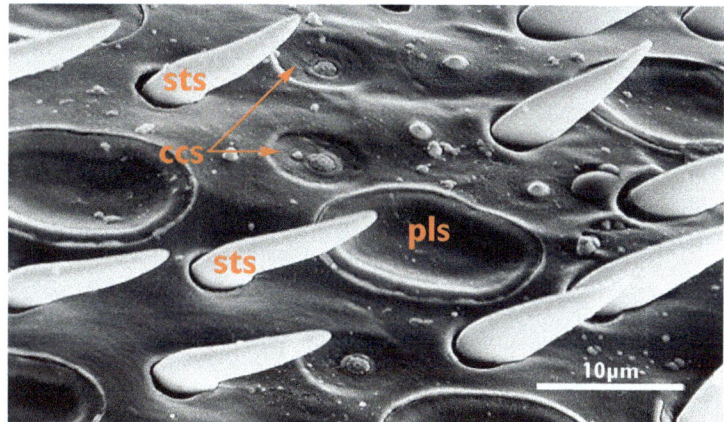

FIG. 1.12 *S-shaped olfactory trichoid sensillae from the worker antennae (sts). These sensillae are covered in fine pores. Also present are coelocapitular sensillae (ccs) and placoid sensillae (pls).*

plant that come with age or even with the time of day. Work on sunflowers has shown that recognition of the floral extract by free-flying bees is mediated by about 10% of the total extract and that even the small differences associated with different genotypes may be detected by visiting bees[17].

In contrast, in male insects like the drone, the neurons from cells with a narrow response spectrum for a mating pheromone run to a discrete part of the ipsilateral antennal lobe within the brain. In the bee, the drone possesses four large glomerular complexes within the antennal lobes that are not found in the worker and the queen. Stimulation of this region of the brain signals the presence of the queen pheromone to the drone[18].

Trichoid and basiconic olfactory sensillae

Many of the trichoid sensillae are covered in the hundreds of minute pores characteristic of olfactory receptors[11] and are assumed to have this function, although as yet it has not been confirmed by electrophysiological recording. There are around 2000 of the olfactory trichoid sensillae, thin-walled pegs, slightly S-shaped, distributed fairly evenly over the last eight annuli of the flagellum (fig. 1.12). The drone antenna bears only 350–400 of these sensillae. Each sensilla is innervated by 5–10 neurons.

The presence of many pores on some of the basiconic sensillae has also been reported[11]. These much wider, thin-walled pegs are innervated by 15–20 neurons and have a more restricted distribution: there are only 150 on each antenna, the majority of them being located on the outer annulus of the flagellum (figs 1.8, 1.9). We can only speculate as to the olfactory function of these trichoid and basiconic sensillae.

In addition to smells associated with foraging, the bee must also discriminate between the many pheromones used to regulate the activities of the colony. It must also discriminate the complex of components that make up the hive or nest smell and lead to the recognition of nestmates and identification of intruders. It is tempting to speculate that the S-

shaped trichoid sensillae and perhaps some of the basiconic sensillae may be involved with the perception of this type of 'smell' although there is currently no evidence to support this suggestion.

Tasting in bees

As well as olfactory receptors the bee's antenna bears receptors for taste. It is not unusual for an insect to have taste receptors (known also as contact chemoreceptors) on its antennae; most insects have them at several different sites on the body. In structure they are very similar to the olfactory sensory receptors, often taking the form of hairs or pegs. The main difference is that they are not covered in minute pores but have one aperture at the apex of the peg filled with a viscous fluid (fig. 1.13). Unlike smelling, where airborne molecules travel from a distant source to the receptors, in tasting the stimulus source has to be in contact with the taste receptors and the chemicals present must come in contact with the fluid-filled aperture. Commonly 3–5 receptor cells innervate the sensilla, their dendritic branches, bearing the acceptor sites for the stimulating molecules, projecting up the shaft of the hair or peg to the apex. The stimulation process is similar to that of the olfactory receptors, the molecules of the gustatory stimulus dissolving in or being transported through the fluid to the acceptor sites on the dendritic membrane. If the compound entering can react with suitable sites on one of the receptor cell membranes then, as usual, this will lead to the opening of ionic gates and the eventual depolarization of the receptor cell. This in turn leads to the initiation of an impulse in the receptor cell neuron. The bee's ability to taste will be discussed in detail in chapter 5. However, it should be remembered that tasting in the bee does not only involve sampling potential food but also detecting pheromones. Some of the pheromones used to regulate colony activity are relatively involatile and are adsorbed onto the body surface of the bee. Other bees detect these pheromones, which include those

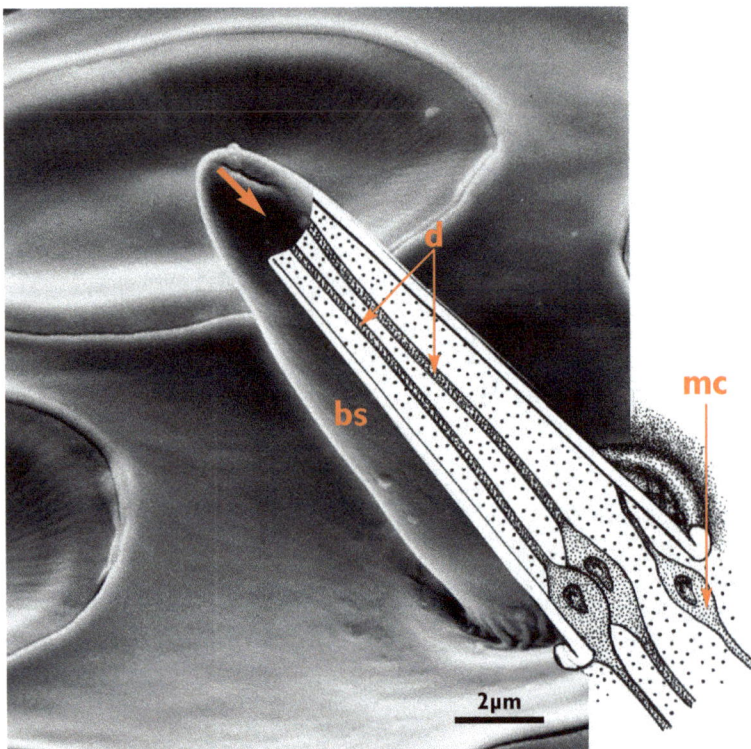

FIG. 1.13 *Contact chemoreceptors, or taste receptors, have apertures (arrow) at the apex of the peg. The sensory dendrites (d) reach up the shaft terminating just beneath the aperture. In many cases, contact chemoreceptors, such as this basiconic sensilla (bs) have a mechanoreceptor cell (mc) terminating at the base of the shaft. This cell is also stimulated if the sensilla is moved.*

responsible for caste regulation and brood regulation, by using their contact chemo-receptors[19]. Many of the antennal taste sensillae are used for this purpose as the bee palpates other bees, although other antennal taste sensillae are sensitive to the same four taste modalities as man, namely sweet, salt, sour and bitter.

Which of the antennal receptors are sensitive to taste?

While we know from behavioural experiments that the antennae bear taste receptors it is difficult without electrophysiological recordings from individual sensillae to say which of the hairs and pegs are responsible for this sense. Light microscope studies suggest that some of the long, bluntly-rounded trichoid sensillae, also known as sensillae chaeticae (fig. 1.8, ch), have single apertures at the apex and these are possible candidates for taste sensillae. They have a widespread distribution over the last eight annuli of the flagellum. In addition, some of the basiconic sensillae, particularly those encountered at the forward or distal margins of the flagellar annuli (fig. 1.8, bs), have apertures at the apex of the peg, very characteristic of contact chemoreceptors. These sensillae are more restricted in number than those of the trichoid sensillae. We cannot tell at this point which receptors respond to food stimuli and which respond to contact with glandular secretions.

One of the receptor cells innervating a taste receptor peg often terminates at the base of the peg where it is stimulated, not by chemical compounds but by the movement of the shaft of the peg relative to the surface of the cuticle (fig. 1.13). Taste receptors are normally associated with parts of the body that are concerned with exploring or manipulating food. The mechanoreceptive cell at the base of the peg signals contact with objects and gives positional information about the receptors in relation to objects that they are exploring.

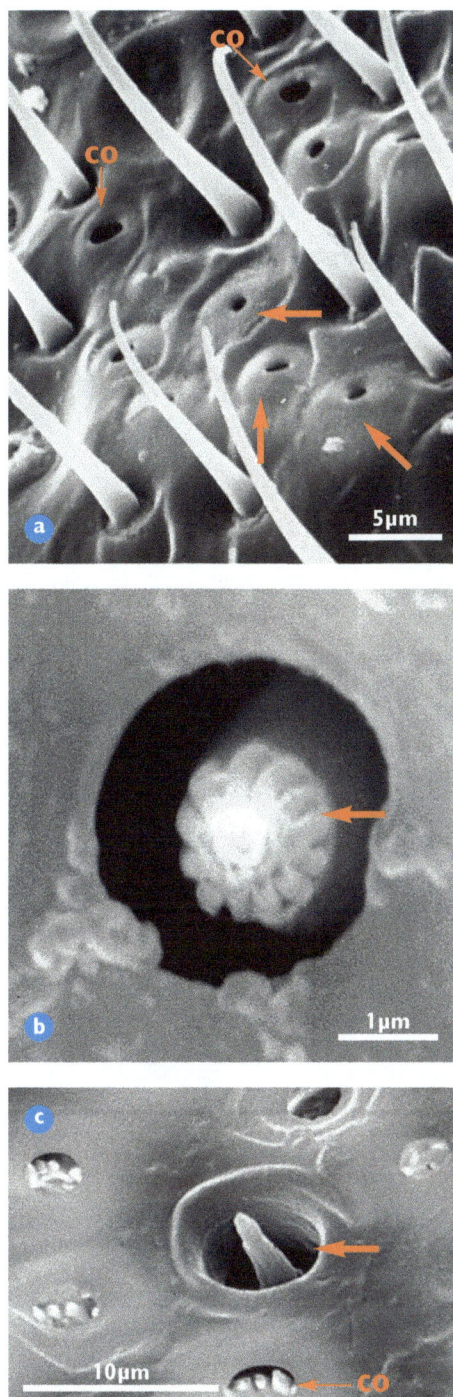

FIG. 1.14 a *A small patch of coeloconic sensillae (co) and ampullaceae (arrows) on the tip of the worker antenna.* **b** *Coeloconic sensilla from a drone antenna with a fluted peg (arrow) set in the floor of a pit in the cuticle. These receptors have been proposed as humidity and temperature receptors.* **c** *A few small, peg-like sensillae (arrow) of unknown function are found among the patch of coeloconic sensillae (co) on the annuli of the drone.*

FIG. 1.15 a *A small patch of coelocapitular sensillae (arrow) on the tip of the worker antenna.*

b *The coelocapitular sense organ forms a shallow depression in the cuticle. A small peg with a mushroom-shaped head (arrow) lies just below the surface of the cuticle.*

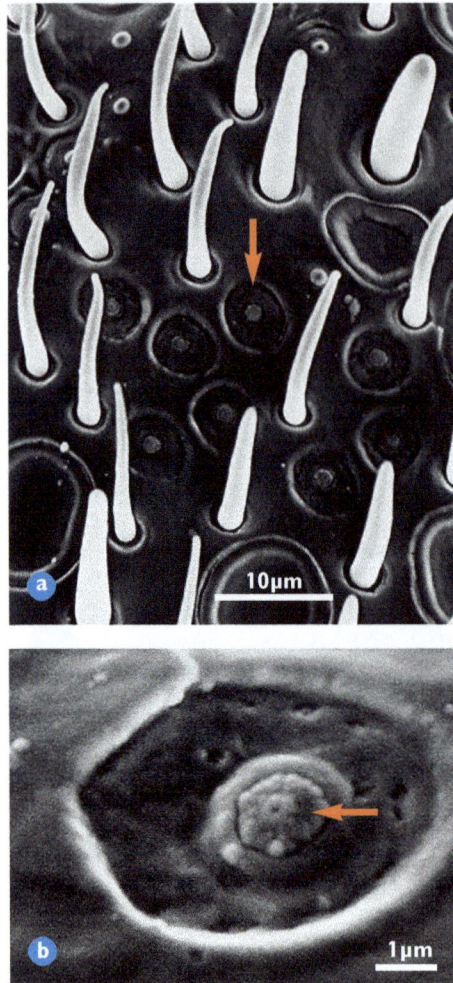

depressions in the cuticle with a central opening in which lies a small peg with a mushroom-shaped head. The peg lies just below the surface of the cuticle and has no external apertures[20] (fig. 1.15).

Each of the sensillae have at times been proposed as receptors sensitive to changes in temperature and humidity. Recent electrophysiological recordings made from the coelocapitular sensillae show the presence of three receptor cells[20]: one responds, by an increase in firing rate of its neuron, to exposure to increasing relative humidity; one responds to a decrease in relative humidity; and the third responds to increases in temperature. These receptor cells have been called, respectively, the moist receptor, the dry receptor and the thermal receptor. Sensillae with a similar structure to the coelocapitular sensillae have been shown to respond to humidity changes and increases in temperature in the cockroach and the cricket[20]. As yet, we have very little idea of the mechanism by which these receptors detect changes in humidity and temperature and transform the stimulus energy into the electrical energy used for signalling in the nervous system. In a number of insects, olfactory receptors have been reported to contain one cell that is stimulated by decreases in temperature, the 'cold' cell[12]. The bee has not so far been examined for such receptors.

The bee is very sensitive to changes in carbon dioxide levels. Most insects have carbon dioxide receptors associated with the respiratory system, but behavioural studies suggest that they are also present on the antennae of the bee. The coeloconic sensillae (fig. 1.14b) have been proposed for this role[20] and there is some electrophysiological evidence to support their sensitivity to carbon dioxide[21].

Sensitivity to temperature, humidity and carbon dioxide

On the distal edge of each of the last eight flagellar annuli of the worker there is a small cluster of coeloconic sensillae, sensillae ampullaceae and coelocapitular sensillae. The total numbers present are small, around 240 for the coeloconic sensillae and 45–60 for the coelocapitular sensillae, with the largest concentration on the outer segment[20]. The sensillae occur together in patches and are quite similar in appearance (figs 1.2, 1.14, 1.15). The coeloconic sensillae form shallow pits in the cuticle with a fluted peg, innervated by one or more sensory neurons, set in the floor. These pegs can often be seen when the sensilla is viewed externally. The sensillae ampullaceae have narrower apertures and form deeper pits in the cuticle. Their pegs cannot normally be seen externally. The coelocapitular sensillae form shallow

The tactile and positional sense of the antenna

The antennae are mobile appendages constantly used to palpate and sample the bee's surroundings, both in the hive and on

FIG. 1.16 a As well as their use of vision, hearing, smelling and tasting, humans make great use of their tactile sense in exploring and manipulating their environment. The skin contains many different types of receptor, each yielding a specific type of information. Receptors sensitive to light pressure are concentrated in areas where touch needs to be particularly sensitive, such as the finger tips, and play a large part in our ability to detect the texture of objects, such as the skin of this apple. This type of receptor adapts very rapidly and thus allows us to detect the movement of the object across the skin. Hairs are stimulated when bent and yield a similar type of information. Other receptors produce long-lasting signals while an object is touching the skin in their area. They yield information about the strength of our grip on an object, for example, the apple being passed to the child. Still other receptors adapt so rapidly they are able to detect vibrations in the skin. Without such a range of tactile information it would be difficult, if not impossible, for us to register and manipulate so delicately objects impinging upon our bodies. Free nerve endings in our skin are sensitive to chemicals released by damaged tissue and signal pain, while warm and cold receptors are also present.

b The bee needs to detect and explore objects impinging upon it in the crowded hive and to be able to manipulate objects such as wax, or palpate the queen or exchange food as in trophallaxis. However, it is covered in a hard layer of cuticle. It gains its tactile sense through the use of mechanosensitive hairs, bristles and stout pegs which require different degrees of force to bend and which show different rates of adaptation to stimulation. Some hairs can be bent easily in any direction and give information that the bee is being touched in that area. Other hairs are set asymmetrically in their sockets and can be bent in only one direction yielding directional information about the movement of objects over the bee. This range of hairs with different properties, coupled with the presence of campaniform sensillae in the cuticle, receptors that detect stress in the cuticle (see chapter 7), yield a range of information similar to that of the receptors in the human skin. The hairs are not restricted to the antennae but cover the body surface. Here two bees have aligned themselves by using their antennal tactile sense in order to exchange food by trophallaxis.

the plant. Bees scan objects within the range of their antennae with frequent, brief contacts. The presence of an olfactory stimulus, or even of water, results in co-ordinated scanning movements of both antennae towards the source[22]. When we touch an object with our fingers, the many receptors embedded in our firm but flexible skin yield information about the strength of that touch, the site being touched and the direction in which our fingers are moving (fig. 1.16a). We also know, thanks to receptors in our muscles and around our joints, where our fingers are in

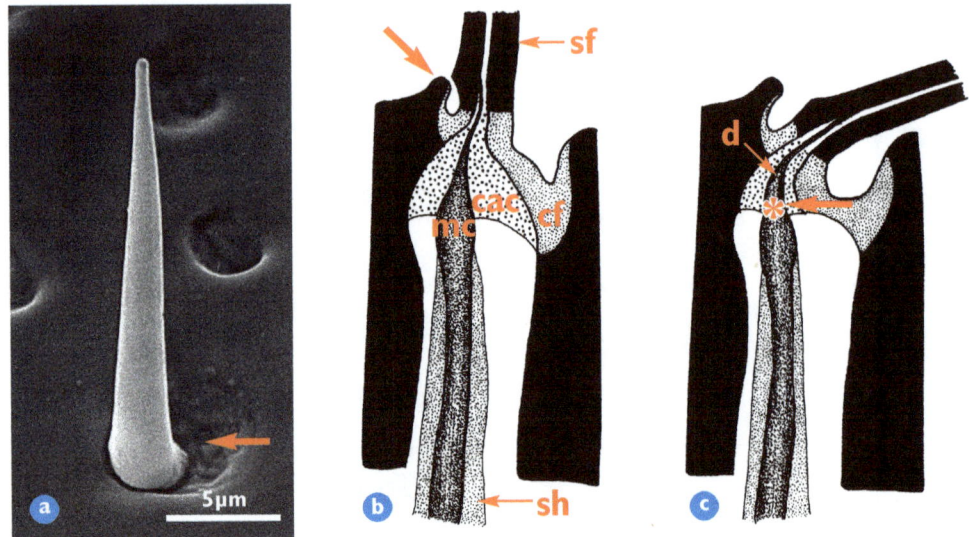

FIG. 1.17 a *A mechanoreceptive sensilla (or hair) from the hair plate at the base of the worker antenna. Note that the shaft is set asymmetrically in the socket (arrow).* b *A diagrammatic section through the base of the shaft and socket shows how the shaft (sf) is set asymmetrically within the flexible cuticle (cf) of the socket. Bending of the shaft is only possible through a restricted arc since the shaft comes up against a firm cuticular stop (arrow) in other directions. The mechanoreceptive cell (mc) enclosed in a sheath (sh) is attached to the base of the shaft. Cap cell (cac).* c *When the shaft is bent in the preferred direction, the cuticle compresses the cap cell (arrow) which, in turn, compresses the specialized end region of the sense cell (star). Deformation of this region causes ionic gates to open, giving rise to current flow and ultimately to depolarization of the dendrite (d) of the cell. This leads to the generation of nerve impulses in the axon of the cell.* After Thurm[23].

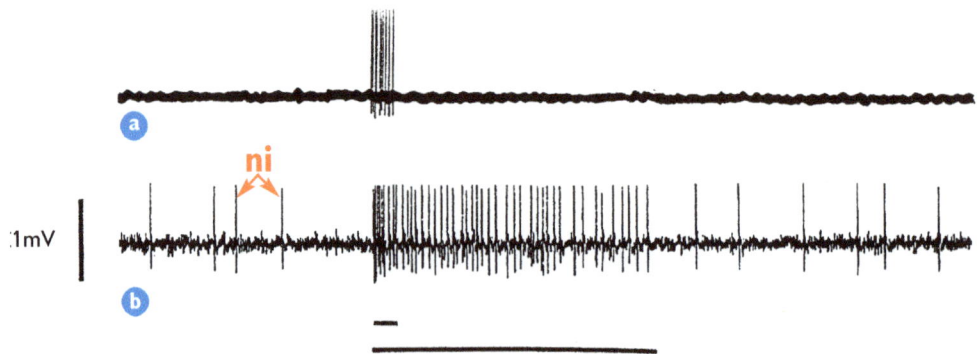

FIG. 1.18 *Adaptation rates of sensory cells in two different mechanoreceptive hairs.* a *The sensory cell of a slender, flexible hair fires while the hair is being bent but adapts very quickly, so that firing ceases when the hair is no longer being moved even though it may still be bent. This is a phasic response.* b *The sensory cell of a stiffer spine, such as those found in the hair plates, has a much slower rate of adaptation and fires all the time that the hair is bent. This is a phasotonic response. Short horizontal line indicates movement of the hairs during bending, long line indicates the period during which the hairs remain bent. Individual nerve impulses in the cell's axon (ni).*

relation to the rest of our body. The bee needs this type of information too but its body surface is covered by a hard, rigid cuticle. It gains its tactile sensitivity from the specially-adapted spines and hairs projecting from the cuticle, not only over the surface of the antennae but also over the whole body surface (fig. 1.16b). These tactile receptors, called **mechanoreceptors,** consist of hollow extrusions of the cuticle set in a narrow membranous ring, allowing the shaft of the sensilla to be bent in relation to the surface of the body. At the base of the shaft is a specialized receptor cell whose modified ending is sensitive to deformation[23,24]. Bending of the rigid shaft deforms the distal end of the cell causing ion gates to open in its membrane. In the same process that occurs in chemoreceptors, the flow of ions leads to depolarization of the sense cell which eventually leads to nerve impulses being generated in the cell's neuron. Often the shaft is set in the cuticle in such a way that bending of the shaft is only possible through a restricted arc (fig: 1.17a, b, c). If some hair shafts on the antennae can only be bent by pressure applied in a particular direction, the insect can extract information about the direction of movement of its antennae in relation to the object that it is palpating.

The human skin contains receptor-types that allow us to discriminate between a variety of tactile sensations, ranging from light touch to sustained pressure. Modifications in the hairs and in the properties of the receptor cells within them, allow the insect to make this type of discrimination. The rigidity of the cuticular extensions varies from long, fairly flexible hairs to short, stiff spines. The cells show different rates of adaptation to the stimulus. Some, usually those found in the more flexible hairs, adapt very rapidly, so that they are only responding while the hair is being bent (fig. 1.18a). The response stops when movement of the hair stops, even though the hair may still be bent. These cells are responding to the velocity of the stimulus and can signal light touch. When com-

bined with a directional insertion of the hair they can signal the direction of movement of a tactile stimulus over the surface of the cuticle. Other cells, in stouter hairs and spines, have much slower rates of adaptation and respond all the time the hair is bent; these sensillae are responding to pressure (fig. 1.18b). Spines like this are frequently found on the borders of segmented areas of the body where they are deformed if one segment is bent in relation to another. Since their neurons are firing all the time that they are deformed, the insect gains positional information about the spatial relationship of the neighbouring segments. The detail in the information can be increased by setting the

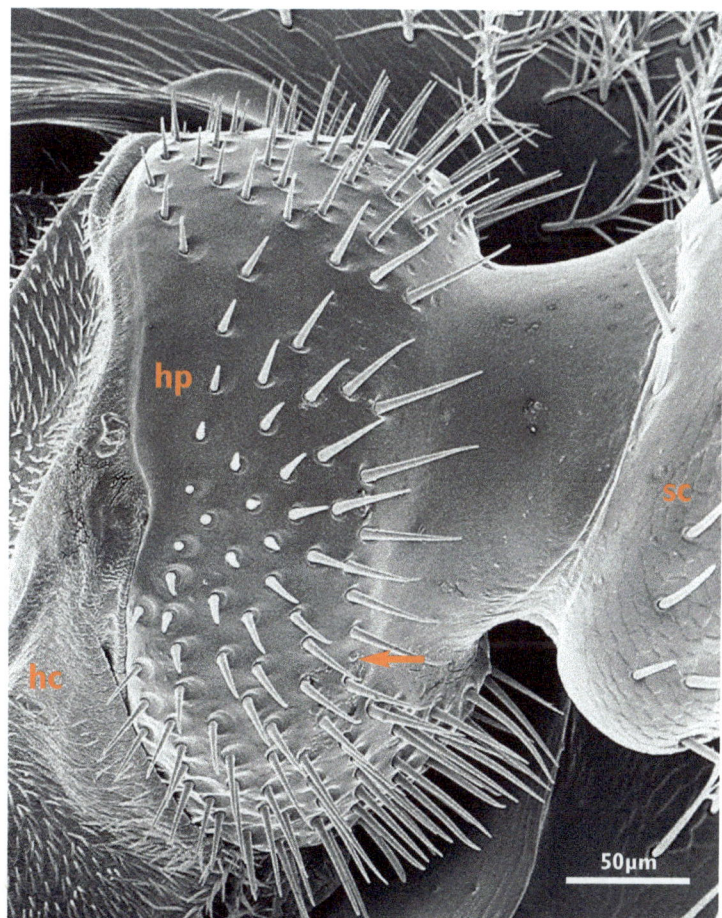

FIG. 1.19 *There is a ball and socket joint between the head capsule (hc) and the scape (sc) of the antenna allowing rotational movement of the whole antenna. The ball-like base of the scape bears a ring of stout mechanoreceptive hairs, forming a hair plate (hp). Movement of the antenna in a particular direction will cause the spines to be deflected as they come up against the head capsule. Note that the spines are set asymmetrically in their sockets and can have a preferred direction of bending. Campaniform sensilla (arrow).*

spines at different sites and at different angles around the joints. At certain sites on the body, a number of these tactile receptors are gathered together to form hair plates (figs 1.19, 1.20).

Where are the tactile sensillae on the antenna?

The cuticular extensions forming hairs and spines covering the antennae are all called trichoid sensillae and, as we have seen, it is difficult to distinguish between them externally or to assign functions to each type. Slender, flexible hairs on the flagellar annuli, especially concentrated on the ventral surface at the tip, are thought to be tactile receptors, as too are stouter, longer hairs, also concentrated at the tip which is an important area for exploration and sampling (fig. 1.9). This will be the area that first comes into contact with nestmates, with the surface of the comb and with the plant. Recently, it has been shown that surface-microsculpturing of flower petals is

taxon-specific and that bees can discriminate between the patterns on different plant species using the tips of their antennae[25]. The pattern of microsculpturing is always orientated towards the source of floral reward and it has been suggested that it may provide a textural nectar guide to the centre of the flower. The size and spacing of some of the stouter trichoid sensillae at the tip appear to be appropriate for this function.

Information on the position of the antennae is derived from the **hair plates** and the Johnston's organ. The complex antennal movements are controlled by four muscles in the head which move the scape, and two muscles within the scape which move the pedicel and flagellum together[22,26]. Groups of mechanoreceptive spines are gathered into hair plates arranged appropriately according to the type of joint. The scape is inserted in such a way as to allow rotation of the antennae and the ball-like base of the scape bears a ring of hairs, able to detect its deflection in any direction (fig. 1.19). Distributed among these hairs are campaniform sensillae, receptors sensitive to the stresses set up in the cuticle by movement, and these also contribute to signalling the position of the moving antenna. Their structure and function is discussed in greater detail in relation to the movement of the wings. There is a hinge joint between the scape and the pedicel allowing only up and down movements at this point, and here one hair plate is set on the dorsal part of the pedicel (fig. 1.20) and two separate plates are set ventrally at the median and lateral margins of the pedicel. Information from the hair plates is essential for the co-ordinated movement of the antennae and if they are destroyed such movements can no longer be made. Since the receptors in the hair-plate spines are of the slowly-adapting variety, they will respond all the time they are deformed and the insect will receive information not only about antennal movement but also about the static position of one joint relative to another.

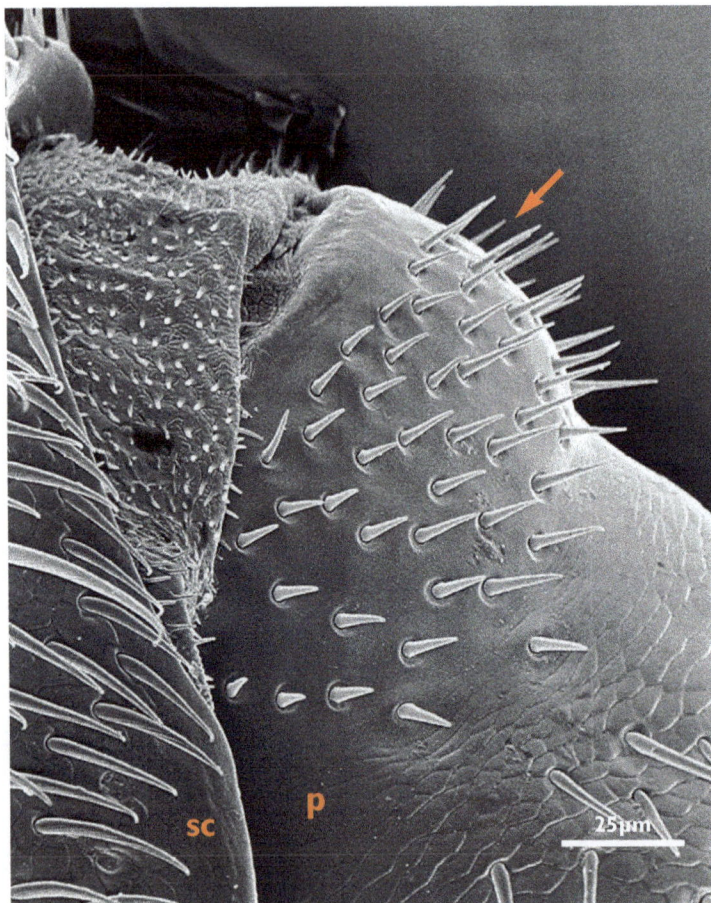

sc   p   25µm

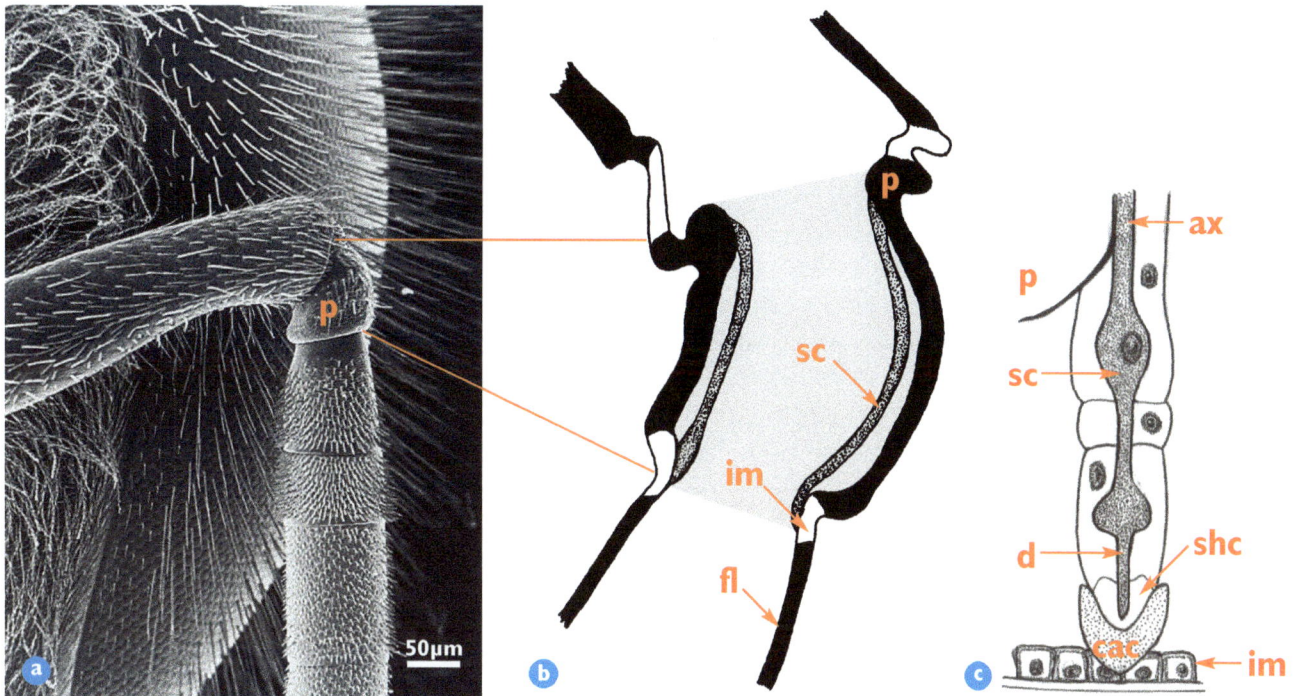

FIG. 1.21 a *Johnston's organ is situated within the pedicel (p).* **b** *The organ consists of many sensory cells, known as scolopale cells (sc) attached at one end to the proximal wall of the pedicel (p) and at the other to the intersegmental membrane (im) between the pedicel and the flagellum (fl). The sensory cells are arranged in a hollow cylinder around the inside of the pedicel and are stimulated by movement of the flagellum relative to the pedicel.* **c** *A single scolopale cell (sc) has a dendrite (d) that extends into a sheath cell (shc) which, in turn, fits into the cap cell (cac). The cap cell is inserted into the intersegmental membrane (im). Movement of the flagellum relative to the pedicel (p) is believed to cause mechanical stress in the membrane of the dendrite, leading ultimately to depolarization of the cell and the initiation of nerve impulses in the axon of the cell (ax).*

Although these receptors will be stimulated by the active movements of the antennae they will also be stimulated by the passive movements induced by air currents during flight. The flagellum and pedicel will be deflected during flight, and information received from the hair plates and Johnston's organ (described below) helps the bee to estimate and regulate its flight speed[27].

How does the antenna act as an ear?

For many years it was thought that bees were deaf to airborne vibrations. Some insects, grasshoppers and crickets, for example, can hear the chirrupings of other members of their species over some distance, but bees appeared not to have an appropriate sense organ. However, bees have a rather different type of ear from that of grasshoppers or, indeed, from that

of man. Travelling sound waves have both pressure and particle movement components. The tympanic membranes of the ears of both grasshoppers and man are sensitive to the pressure changes set up by sound waves. The motion of the tympanic membrane induced by these pressure changes ultimately excites the sensory cells of the auditory system. Recently it has been discovered that bees are able to detect the sounds made by a dancing bee using a receptor that is sensitive to the particle movement component of the sound wave[13]. Such a receptor must be a hair- or rod-like structure capable of being bent or deflected by the particle movement. The antennal flagellum is capable of being deflected in this way and the oscillations of the flagellum induced by the sound are detected by **Johnston's organ** (fig. 1.21), a large sense organ located in the pedicel and named after the scientist who first described it.

Unlike the hundreds of other sensillae on the antenna, Johnston's organ is not visible on the outer surface. It consists of many individual sensillae, attached at one end to the proximal (nearest the scape) wall of the pedicel and inserted at the other end into the intersegmental membrane of the first flagellar annulus (fig. 1.21a, b). Because of their attachment across the joint of the pedicel and the flagellum, the sensillae of Johnston's organ are able to detect vibrations of the flagellum in relation to the pedicel. The sensillae are arranged so that they form a hollow cylinder around the inside of the pedicel with their insertions in a complete circle around the base of the flagellum. This means that flagellar movements in any direction relative to the pedicel are equally effective as a stimulus. Each sensilla consists of three linearly arranged cells. Beneath the point of attachment at the flagellar intersegmental membrane is a cap cell attached to the membrane. Into this fits a small, rounded sheath cell and this, in turn, encloses the end of a sensory cell dendrite. The cell body of the sensory cell, known as a **scolopale cell**, lies near the base of the pedicel and its dendritic extension stretches most of the length of the pedicel up to the sheath and cap cells (fig. 1.21c)[13,28]. It is not known exactly how the sensory cell is stimulated by movement of the flagellum relative to the pedicel but it is assumed that this movement causes mechanical stress at the tip of the dendrite buried in the sheath cell. We have seen that such stress can initiate changes in the properties of the membrane of mechanoreceptive cells in the stout hairs on the antennae that lead eventually to generation of nerve impulses, and a similar mechanism probably operates here.

Dance communication

Returning foragers inform their nestmates of profitable food sources. The forager dances on the vertical comb, the angle that it makes with the vertical indicating the angle of the food source relative to the sun's azimuth in the field. Workers perform one of two types of dance, depending on the distance of the food source. If the forage site is under 75 metres away, the bee moves in a circle once or more clockwise and then counter-clockwise, and so on. This is the **round dance**. For greater distances the bee performs a figure of eight, wagging the abdomen as it runs in a straight line and then circling back, alternating between a left and a right return path (fig. 1.22). This is the **waggle dance**. There is also a strong correlation between the distance to the food source and the speed of dancing: the nearer the food the more dance circuits are performed per unit time[1]. The dancer is attended closely by a number of bees who follow its dance and then fly out to the forage site.

How do the follower bees perceive and interpret the dance on the crowded, dark surface of the comb?

The dancing bee emits pulses of sound during its waggle run by making tiny vibrations of its closed wings in the dorsoventral plane at a frequency of 200–300 Hz (cycles/sec) with around 15 bursts of sound every second[29]. Sounds are also made during round dances. Large pressure gradients around the edges of the vibrating wings cause oscillating air currents about the abdomen of the dancer and most of the follower bees are found with their antennae held in this zone of maximum velocity of the air particles. Within a few millimetres these near-field conditions have changed and beyond that air particle oscillations are 200 times less intense. The specialized type of ear on the bee's antenna thus operates effectively only over very short distances adjacent to the dancing bee. The duration of the dance sounds is highly correlated with the distance of the food source in both waggle and round dances[30]. There is a positive correlation between amplitude of the dance sounds and the profitability of the food source but the differences are small and may be insufficient to be of use to the bee. Use of both antennae to make a simultaneous comparison of the amplitude of the dance sounds could provide the follower bees with information about their orientation relative to the dancer's body and hence information about the direction of the food source. This is supported by the fact that recruitment in bees is significantly reduced if one antenna is removed[31].

Ingenious experiments with a model bee that can perform the waggle dance and vibrate its wings appropriately, have shown the importance of the different dance parameters. The model is able to recruit bees to a source of forage. If the vibration of the wings is stilled and no sound is emitted, the model fails to recruit bees. If the model dances and produces sound pulses by wing vibrations, but does not produce the low frequency (13–15 Hz) tail wagging movements of the dance, then it also fails to recruit. Both sound and tail wagging seem to be used to indicate distance to followers[30].

Training experiments have demonstrated that the bees can actually perceive the sounds emitted by dancing bees and ablation experiments have shown that the receptor is the Johnston's organ in the pedicel[31,32]. Analysis of the movements of the flagellum in response to simulated near-field sound, equivalent to that emitted by the bee, shows that it is particularly well suited to transmit such sound. It is deflected like a stiff rod, ensuring maximal stimulation of the scolopale sensillae attached at the base of the flagellum; it is equally sensitive in all directions and its amplitude response is linear in the range of amplitudes that occur in the natural signals. Additionally, it does not resonate and its temporal resolution will allow it to resolve the temporal structure of the dance sounds. The low frequency 'waggles' of the abdomen generate mainly a lateral air particle movement and, since the flagellum is equally sensitive to movement in all directions, it can perceive these movements as well as the dorsoventral movements of air particles caused by the vibrating wings. It remains to be seen

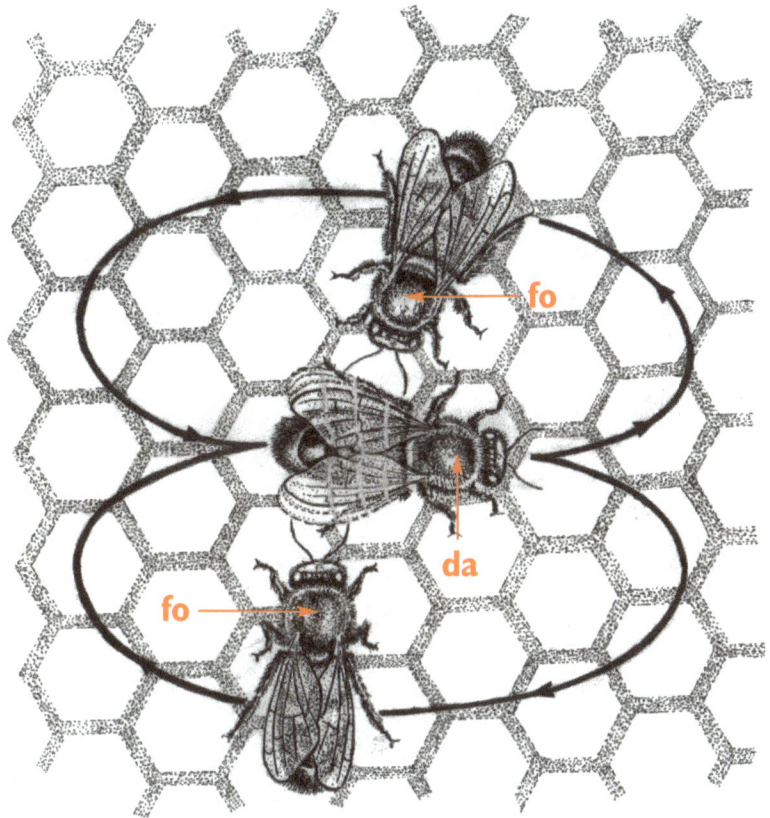

FIG. 1.22 *The dancing bee (da) emits pulses of sound by making tiny vibrations of its closed wings (exaggerated in this diagram). Follower bees (fo) hold their antennae near to the abdomen of the dancer. In this area they encounter maximum velocity of air particles resulting from the vibrations. The minute deflections of the flagellum induced by the air particle oscillations are detected by the sensory cells of the Johnston's organ. Thick black lines indicate the path of the waggle dance of this returning forager.*

exactly how the bee integrates both types of information[31].

Bees attending the dancer also communicate with it by means of vibrations, but these vibrations are transmitted through the comb and picked up by vibration-sensitive organs in its legs. Bees press their thorax against the comb and vibrate their wing muscles thereby producing minute (1 μm) displacements of the substratum at a frequency of 350 Hz. When the dancer perceives these signals, it may stop dancing and deliver small samples of its collected food to the dance attenders[33].

2. Vision in the bee:

the compound eye

The bee must collect its food at some distance from the nest or hive, relying largely on visual cues to navigate its way to the forage site. At the site it must move amongst the plants, select the appropriate flowers and find the food source within the flower, again using some visual cues. It then has to find its way back to the hive and select its own hive among others. The drone has to locate and chase the queen, a small black spot against a bright sky, during mating. Vision is thus a very important sense for the bee. The eyes of the worker occupy a large part of the head and those of the drone dominate it (figs 2.1, 2.2, 2.8, 2.9).

The compound eye

When we look closely at the eye with a hand lens or under a microscope, we see that it is composed of numerous small lenses, around 5000–6000 in the worker, 3500 in the queen and 10 000 in the drone[1,2] (fig. 2.3). Obviously the bee's eye is very different from the human eye. Humans have a camera-type eye with a concave retina (layer of photoreceptor cells) and a single lens (the corneal surface plus the underlying lens) that focuses light onto the retina (fig. 2.4a). The retinal layer is large enough to contain many thousands of photoreceptor cells (rods and cones), with which to analyse the image. All mammals have eyes like this: they are large and can accommodate two relatively large retinas in the head; also most of them can move their heads to look around. Insects, on the other hand, are mostly very small animals, usually unable to turn their heads very far. If a very small animal has to house an eye that contains enough photoreceptor cells in its retina to give it reasonable

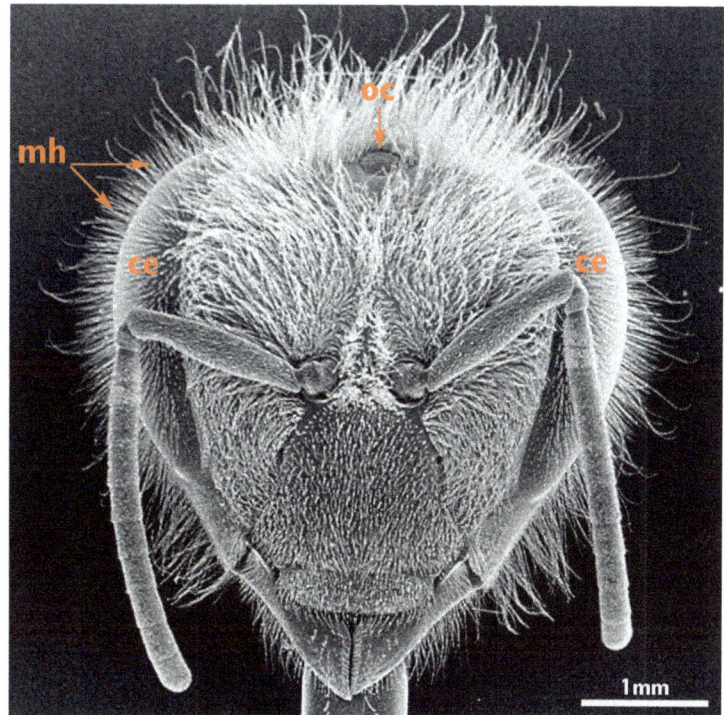

FIG. 2.2 The compound eyes of the worker bee, although smaller than those of the drone, occupy a large part of the head. Worker eyes also have mechanosensory hairs situated at intervals between the facets of the eye. Compound eye (ce); mechanoreceptive hairs (mh); ocellus (oc).

FIG. 2.3 The compound eye is composed of many small lenses overlying groups of photoreceptor cells. They are packed closely together in the bee, and the lenses are hexagonal. Part of one of the mechanosensory hairs (mh) is shown. These hairs are set in sockets and innervated at the base of the shaft. The neuron is stimulated by movement so that the bee is aware of anything touching the surface of the eye.

FIG. 2.1 (opposite) The compound eyes of the drone are very large, occupying most of the head capsule. They meet at the top of the head, displacing the three ocelli forward. The eyes are covered with fine hairs, located between the facets, many of them innervated and functioning as mechanoreceptors. Because of the presence of so many hairs it is impossible to see the individual facets, or ommatidial lenses. Note also the short proboscis of the drone.

FIG. 2.4 a *Human heads are large enough to accommodate two large retinas. The camera-type eye focuses light onto a concave retina, shown in red.*
b *Insects are small and unable to turn their heads very far, yet they need to have a panoramic view to detect predators. This has favoured the development of a convex retina (red) spread over the surface of the head.*

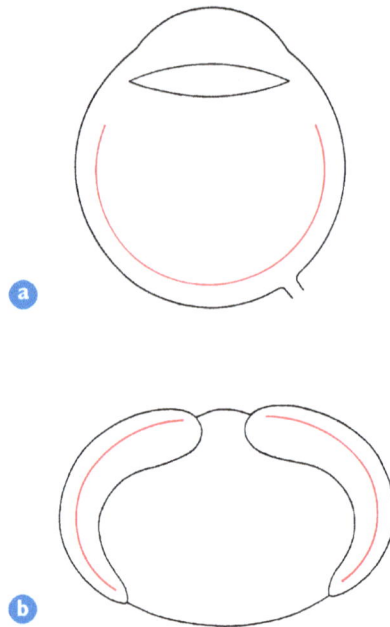

acuity over a very wide field of view, one way it can accomplish this is to spread its **retina** all over the surface of its head (fig. 2.4b). Thus, small size and the need for a panoramic view to detect the approach of predators seems to have favoured the development of a convex retina in insects. However, this poses problems with the lens. It is not possible to construct a good optical single lens for a convex retina. The lens system must be split up into many small lenses overlying the retina and this, in turn, means that the underlying photo-receptor cells must be split up into small groups beneath the individual lenses[3]. This results in the so-called 'compound' or 'faceted' eye (figs 2.3, 2.5).

How does the compound eye form an image?

To answer that question we have to examine the internal structure briefly. Beneath each small **lens**, formed from cuticle similar to that covering the rest of the insect body, lies a second transparent, cone-shaped body, the **crystalline cone**. This assists the lens in collecting light and focusing it onto the tip of the circle of photoreceptor cells lying below. The crystalline cone and the photoreceptors are surrounded by a sleeve of cells which contain black pigment gran-

ules. These pigment cells screen the receptors from light entering from anywhere except through their own individual lens system, and a basal pigment layer screens the back of the retina from stray light.

Each small lens, together with its associated receptor and pigment cells, form a single unit of the compound eye, the **ommatidium** (fig. 2.6). Each ommatidium in the bee contains nine photoreceptors, the ninth cell being shorter than the others[4]. The inner borders of the photoreceptors are modified to carry the visual pigment: the cell membranes are infolded to form many layers of small tubules or **microvilli** set at right angles to the path of the incoming light (figs 2.6a, d, e, f). The visual pigment is set in these membranes and thus presented at best advantage to the light. This arrangement is reminiscent of that in the human eye, where the lipid membranes of the rods and cones form plates bearing the visual pigment. These plates are also set at right angles to the incoming light.

The modified inner border of the insect retinula cell is known as a **rhabdomere**, and the rhabdomeres of the nine cells are fused together to form one central **rhabdom** running the length of the ommatidium. The lens system focuses light onto the tip of the rhabdom which acts as a light guide, funnelling the light down its whole length. This occurs because the difference in refractive index between the rhabdom and the rest of the photoreceptor cell traps the light within the rhabdom[5]. Thus the light entering one ommatidium is directed through all the layers of membrane carrying the visual pigment, giving the maximum opportunity for the pigment to react with light. When light reacts with a visual pigment molecule in the membrane, it undergoes a reversible conformational change in its structure which indirectly opens ionic gates in the cell membrane. The resulting movement of ions causes a brief depolarization of the cell. This depolarization spreads down the short axon of the retinal cell to its terminal where it causes the release of a neurotransmitter that stimulates the next cell in the visual pathway.

The rhabdom only measures the intensity of the light from the small area of the visual field viewed by its lens system. The nine cells forming the rhabdom do not resolve the tiny inverted image formed by that lens, rather the cells of one ommatidium act together as one receptor unit. An erect image of the outside world is built up of a mosaic of light and dark spots formed by these units. Compared to the human eye this is a very coarse-grained image with a spatial resolution about a hundred times worse than in humans.

Why is spatial resolution so poor in the compound eye?

The ability of any eye to resolve detail depends upon two factors: the fineness of the mosaic of receptor units that sample the image and the optical quality of the image that those elements receive. If the eye is to resolve fine detail, the angular separation between the receptor units must be small. In the human eye, the receptor elements are the individual rods and cones. In the region of the eye with the most acute vision, the fovea, the elements subtend an angle of 1 minute of arc, i.e. they look out at an area of the visual field 1/60 of 1° wide. In the bee eye, a single

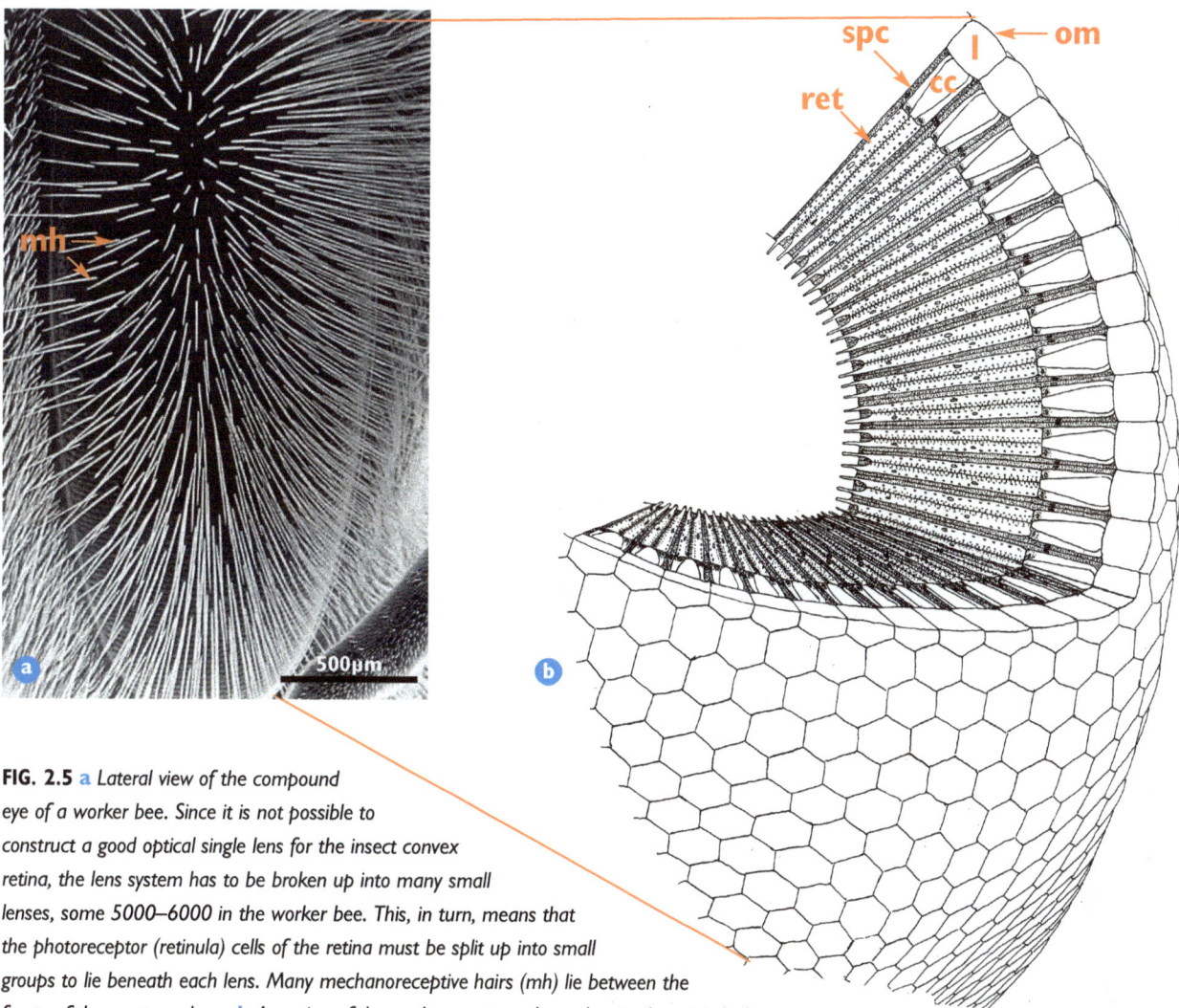

FIG. 2.5 a Lateral view of the compound eye of a worker bee. Since it is not possible to construct a good optical single lens for the insect convex retina, the lens system has to be broken up into many small lenses, some 5000–6000 in the worker bee. This, in turn, means that the photoreceptor (retinula) cells of the retina must be split up into small groups to lie beneath each lens. Many mechanoreceptive hairs (mh) lie between the facets of the compound eye. **b** A section of the worker compound eye showing how it is built up from individual units or ommatidia (om) each with its own lens system (lens (l) and crystalline cone (cc)), photoreceptor cells (ret) radially arranged around the ommatidial axis, and screening pigment cells (spc) to prevent the entry of stray light. The fine structure of the ommatidium is shown in fig. 2.6a–f.

FIG. 2.6 a *A section through a dark-adapted ommatidium of the bee compound eye. Each ommatidium forms one unit of the multifaceted compound eye. Light is focused onto the tip of the photoreceptor cells, known as retinula cells, by the transparent cuticular lens (l) and the underlying crystalline cone (cc). These two structures form the dioptric apparatus of the ommatidium. The retinula cells (ret) are radially arranged around the long axis of the ommatidium; the inner border of each cell is modified to form many short tubules, or microvilli (mv) in which the visual pigment is located. The microvilli of all the cells in the ommatidia meet centrally, forming a roughly circular area, the rhabdom (rh). The rhabdom acts as a waveguide so that the light focused onto the rhabdom tip is funnelled down its whole length and hence through all the visual pigment-bearing microvilli. There are nine retinula cells but one of them is much shorter than the others and is only seen in the bottom third of the ommatidium (star). The cells narrow to form small axons (ax) that pass through the basement membrane (bm) bounding the eye on their way to the optic ganglia. Three optic ganglia processing visual information are found between the eye and the brain (not shown). The retinula cells are enclosed by a sleeve of screening cells that isolate the individual ommatidia. These cells, the secondary pigment cells (spc), contain black pigment granules. A pair of primary pigment cells (ppc) surround the cone. Small pigment cells, attached to the basement membrane, screen the axons as they pass out of the eye. The retinula cells also contain some pigment granules but these are rather smaller than those of the screening sleeve. They are found mostly in the upper half of the eight longer cells and are particularly numerous in the zone of contact between the cells and the crystalline cone. Changes in the fine structure of the retinula cells and pigment cells occur in many insects during the course of light-dark adaptation. Most researchers have found little change in the position of the pigment granules in the bee, with the possible exception of the small pigment granules in the retinula cells, which may move inwards around the rhabdom at the top of the cell to cut down the light entering the retinula cells in bright light. One frequently-seen mechanism in insects is the appearance of large cisternae (ci) around the rhabdom when the eye is dark-adapted. These cisternae are enlarged cavities between the sheets of parallel intracellular membranes, known as the endoplasmic reticulum, that occur in most cells. The endoplasmic reticulum is intimately connected with the cells' metabolic activities. The action of the rhabdom as a waveguide is improved because of differences in the refractive indices of the rhabdom and the cisternae, ensuring that the maximum amount of light reaches the lower regions of the rhabdom. In the light-adapted eye, the cisternae are very much reduced. Cell nuclei (n); mitochondria (mi).*

g *A cross-section through an ommatidium from the dorsal rim of the eye where the ninth cell is as long as the other eight cells and thus contributes to the rhabdom (rh) along its entire length. Each star indicates one retinula cell. The borders of individual cells are joined by desmosomes (arrows) adjacent to the rhabdom. Pigment granules (pg); mitochondria (mi).*

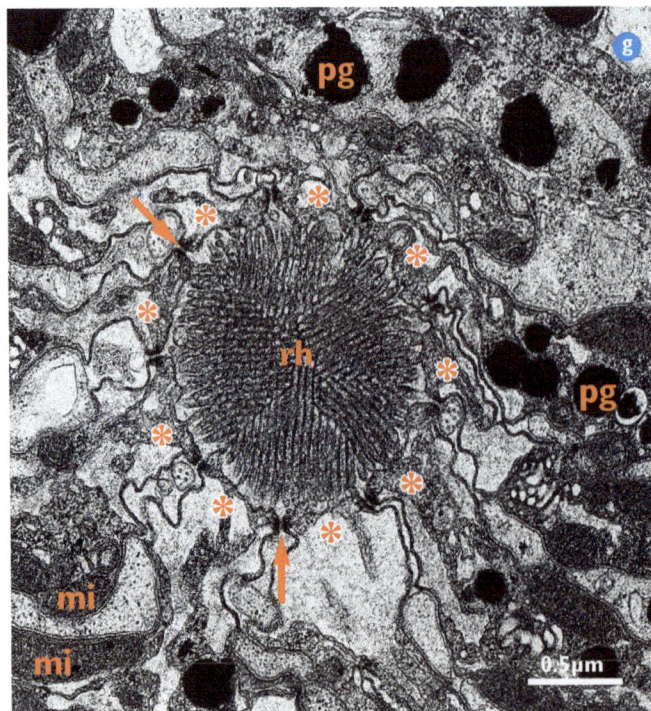

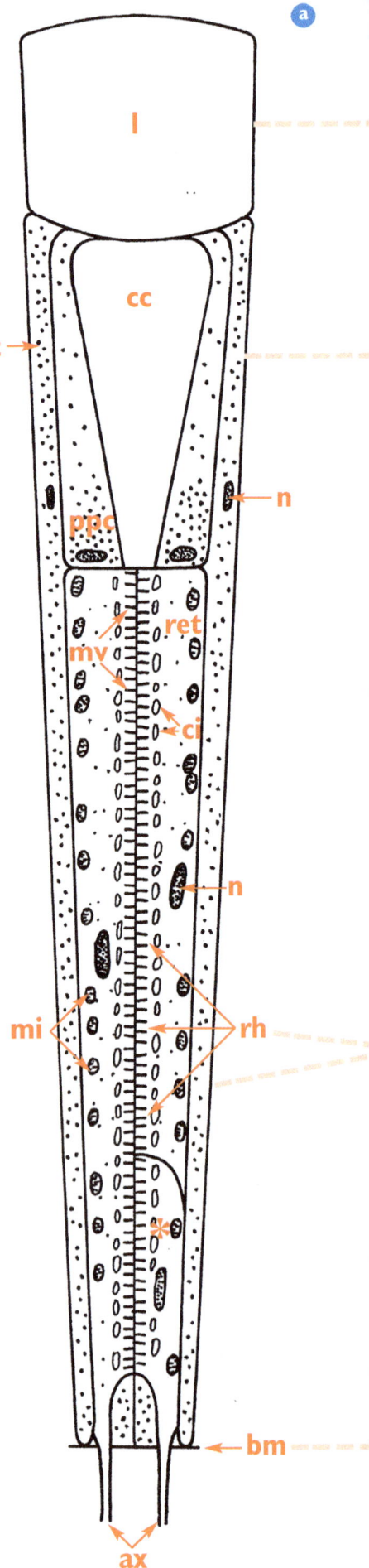

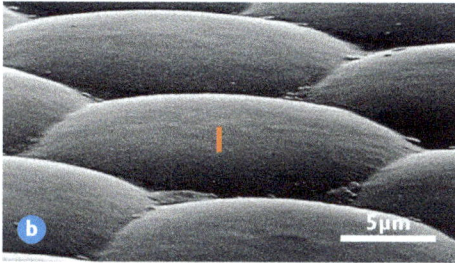

b A view of the outer surface of the cuticular lens (l) of the ommatidium. c A cross section near the periphery of the crystalline cone (cc) region of an ommatidium. The crystalline cone is composed of four cells and is surrounded by two primary pigment cells (ppc). In this region, there are few pigment granules (pg) present. The primary pigment cells are, in turn, surrounded by the sleeve of secondary pigment cells (spc) which contain many more pigment granules. d A cross section through an ommatidium, about one-third of the way inwards from the crystalline cone. In this region there are eight retinula cells (1–8), each with its modified inner border contributing to the central rhabdom (rh). (See fig. 2.11 for detail of this region.) The rhabdom is surrounded by the cisternae (ci) in the dark-adapted eye. The cells contain many mitochondria (mi) situated around their borders. A few, small pigment granules (spg) are present at this level. The cells are surrounded by secondary pigment cells (spc) containing larger pigment granules (pg). e A longitudinal section through the rhabdom of an ommatidium. The inner border of each of the two cells forms stacks of small tubules, the microvilli (mv), closely apposed to one another. The microvilli of the two cells are also closely apposed (arrow 1). The microvilli of a third retinula cell can be seen at the top of the section (arrow 2). These have been cut normal to the first two cells. The molecules of the visual pigment lie in the membrane of each microvillus. f A cross section of the nine retinula cell axons as they emerge from the ommatidium below the basement membrane.

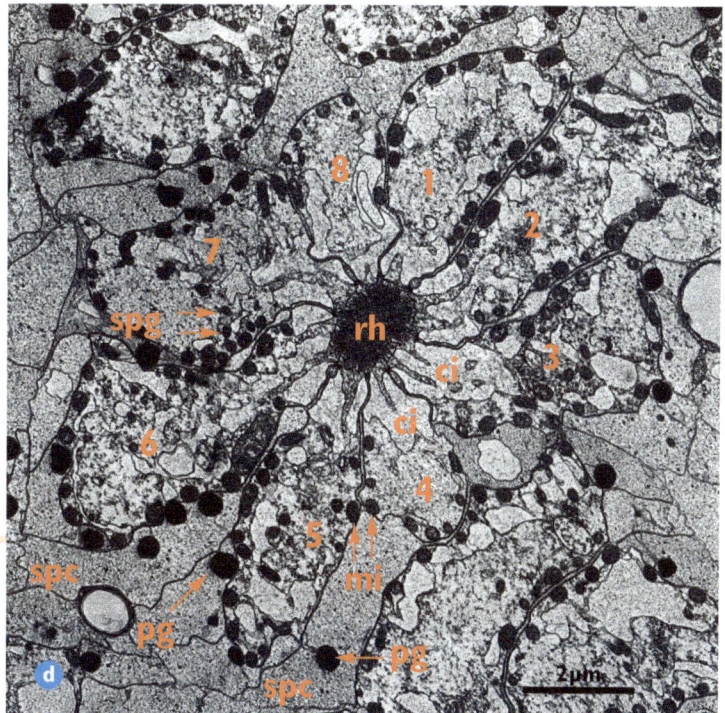

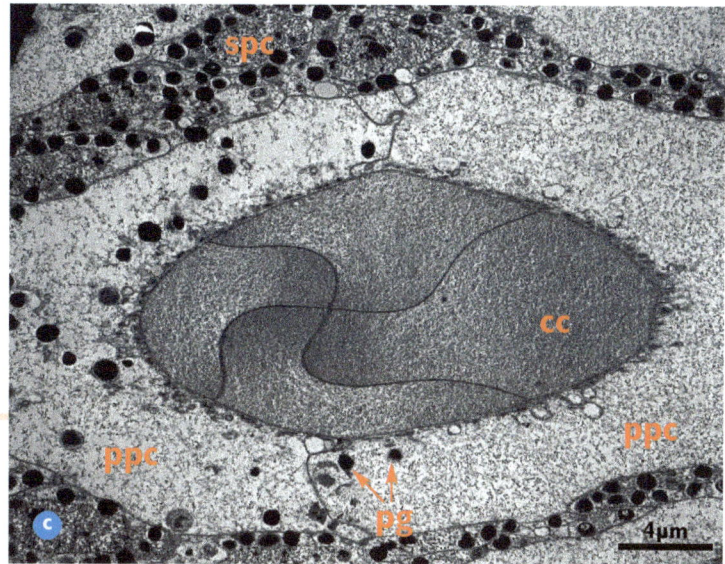

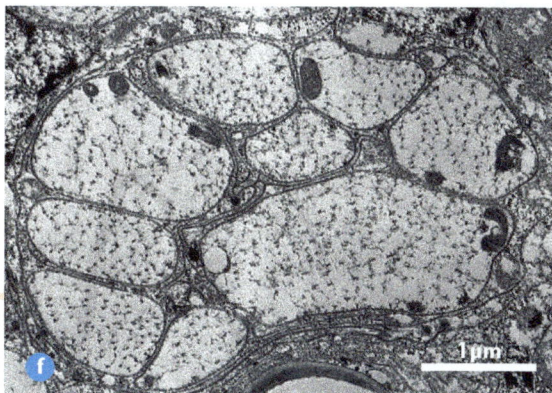

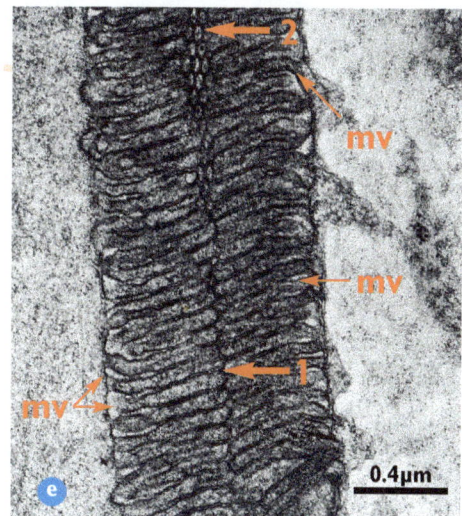

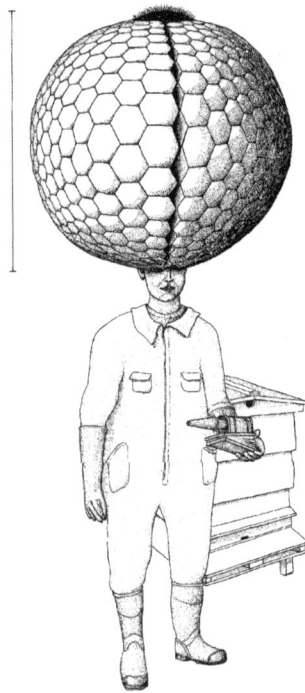

FIG. 2.7 *If a beekeeper had a compound eye like his bees, but one that had the same resolving power as the human eye, his compound eye would be impossibly large. The dependence of eye size on the square of resolution in this type of eye means that the beekeeper's compound eye would have to be around one metre in diameter to achieve the resolution of the human eye. Each of the facets shown, in fact, represents 10 000 ommatidia. Central black line equals 1 metre.* Redrawn from the original drawing by E Feinberg in Kirschfeld 1976[36] and reproduced by kind permission of the author and Springer-Verlag.

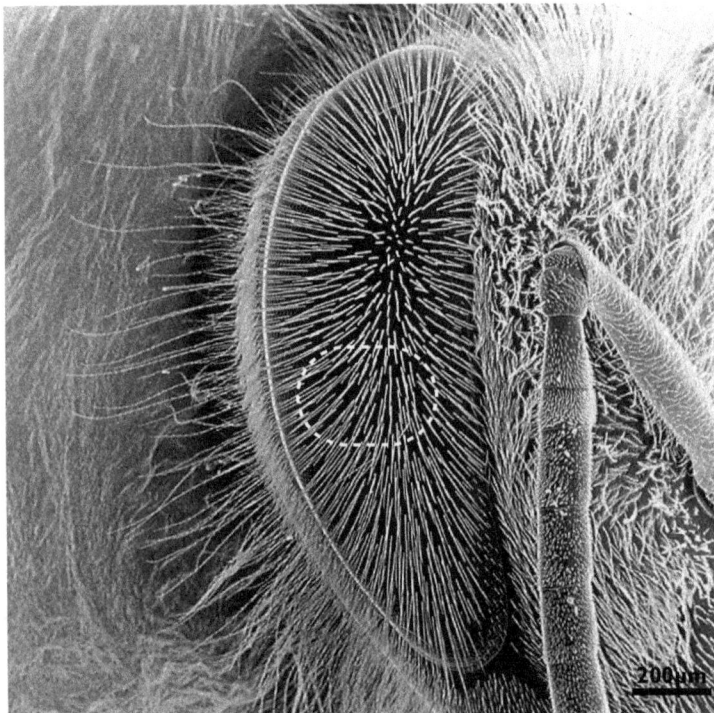

FIG. 2.8 *The eye of a worker bee, viewed laterally. The worker has a modest, frontally-directed acute zone of vision (outlined by white broken line).*

ommatidium is the operating unit and the angular separation of the receptor elements is measured by the interommatidial angle. At its minimum value, this is 1° but it is often larger, e.g. 2°–4°. The ommatidial lens diameter cannot be reduced in size without approaching the limits set by diffraction. The second requirement for acuity demands that the eye collects as much light as possible from each point in space[5]. If the ommatidial lens were to be made smaller in an effort to decrease the angular separation of receptor units, then the effects of diffraction in each ommatidium would degrade the mosaic image. The effects of diffraction would, in effect, reduce the intensity of the light focused on the tip of the rhabdom. This would reduce the degree of contrast between objects imaged and if contrast is not sufficient, then objects will not be distinguished from each other.

Diffraction sets lower limits on the size of the ommatidial lens. It would be possible to decrease the interommatidial angle, and hence reduce the angular separation of receptor units, by decreasing the curvature of the eye. However, this is not a practical solution for the insect since it would end up with a large eye too heavy for its body (cf. fig. 2.7). Many active insects that need as good a resolution as possible adopt a compromise solution and have a particular area of the eye where the curvature is less and hence the interommatidial angle smaller. The eye is flatter and the facets are larger in this region, forming an area of increased spatial resolution with the same function as the human fovea. The worker bee has a frontally-directed acute zone in the median region of the eye which is specialized in this way with interommatidial angles of 1° (fig. 2.8)[2,6].

The drone provides a classic example of an insect eye with a specialized region of acute vision (fig. 2.9). The drone uses the dorsal region of its eye to detect, fixate and approach the queen from behind and below during mating, having to locate it against the vast area of the sky. The drone's eye is divided into three distinct areas[2,7].

The dorsal two-thirds, through which it views the queen, is less curved and the facet lenses are larger resulting in inter-ommatidial angles of 1°–2° compared with 2°–4° in the ventral region. Smaller omma-tidial fields of vision would result in the entry of less light to each ommatidium and thus less contrast in the image formed, but contrast sensitivity needs to be high to dis-criminate the small silhouette of the queen. Other modifications to the eye compensate for the smaller ommatidial fields of vision. The enlarged facet dia-meter allows more light to enter, and as the photoreceptor cells are longer and wider in this part of the eye, they contain a greater volume of the visual pigment. This increase in volume produces twice the contrast sensitivity of the ventral region of the eye and twice that of the worker eye.

Although the insect compound eye has relatively coarse spatial resolution, we should not think of it as an organ provid-ing the bee with inferior information. We have only to look at the behaviour of the bee and indeed that of other flying insects, for example, the chasing behaviour of mat-ing flies and the dragonfly catching prey on the wing, to realize how well their eyes serve them and what beautiful adaptations they show to their particular lifestyles. In one attribute, temporal resolution, their eyes are superior to ours.

Temporal resolution

The eyes of good fliers, such as bees, flies and dragonflies, have much better tempo-ral resolution than human eyes. If we look at a slowly flashing light we can easily dis-tinguish the individual flashes, but if the rate of flashing is increased there will come a point at which the light will be indistin-guishable from a steady one of the same average intensity. This critical fusion fre-quency depends, among other factors, on the brightness of the light, but is around 35–40 Hz (flashes/second) in humans. This is due to the relative sluggishness of the photoreceptor response, as the response to one flash has not completely

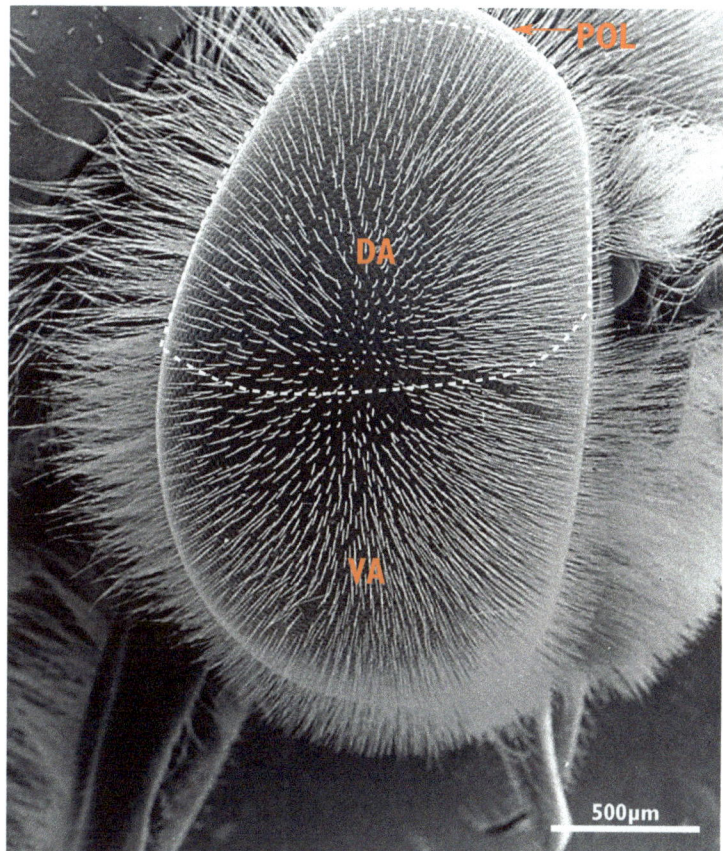

died away before the next one comes. It is for this reason that we see a continuous picture on our TV screens and at the cin-ema, and light from fluorescent light tubes appears constant and not flickering. Fast-flying insects have a much higher flicker fusion frequency. In bright light, values as high as 150 Hz and even 200 Hz have been recorded[8]. This means that a fly or a bee does see the flickering of a light bulb! This high fusion frequency is the result of inter-action between the electrical activity of the photoreceptor cells and cells in the under-lying neural layer of the eye. The advantage of a fast response like this is that it helps the flying insect to resolve objects, pre-venting 'smearing' as they pass the eye. It is a way of compensating for poor spatial res-olution, although the actual spatial acuity of the eye is not improved.

Colour vision in bees

The bee depends upon the pollen and nec-tar offered by flowers for its food. As a source of food, flowers are not very

FIG. 2.9 *As in many other male insects, the drone has an extended area of acute vision (DA) directed forwards and upwards (outlined by white broken line). This region of the eye is involved in the detection and chase of the queen, a small object against the sky. A number of specializations to improve acuity are seen in this male-specific region of the eye (see text). A small region at the dorsal rim of the eye (POL) is specialized for the detection of polarized light. The ventral region (VA) resembles that of the worker.*

reliable; their offerings vary with the season and often with the time of day. This changing food resource is scattered around in the bee's environment. The bee needs an efficient foraging strategy if it is to collect sufficient food without investigating every flowering plant in the vicinity. Clearly, insects that have a fixed preference for a food source will be at a disadvantage in a world of changing resources. On the other hand, individual insects cannot afford the time and energy to visit all plants in the neighbourhood to select the best offering each time they go out foraging. Many of the Hymenoptera adopt a strategy that permits collection from the most profitable current sources while keeping abreast of changing conditions in other flowers. Individual insects restrict their visits mainly to one type of plant, behaviour known as flower constancy, but sample a few others as well. In this way they quickly become aware if their main plant is becoming less rewarding than one of their other samples, and can switch to the more profitable source. In the honey bee, with a large hive population and scouting foragers, the colony as a whole is kept informed of fluctuating conditions over a wide area. The well-documented learning capacities of the Hymenoptera, coupled with the ability of honey bees to communicate the location of a good food source, means that they can exploit resources in their neighbourhood very efficiently.

For this kind of foraging behaviour to be successful, the bee must have sensory equipment that will allow it to distinguish one flower species from another. Several features characterize a flowering plant, including the size, shape, colour and plane of symmetry of the individual flowers; their odour; and the site and height of the plant and distribution of its blossoms. Both shape and odour are known to be used as cues by the bee, but given the bee's poor spatial resolution and the fact that odours may be mixed or dispersed at some distance from the plant, the most effective signal for flower location at a distance is

likely to be colour. Do bees see colours? Beekeepers have long thought so, marking hives with different colours to assist returning workers in locating their own hive.

Proving that bees do possess colour vision

Starting with the careful experiments of Karl von Frisch in 1914[10] and continuing over succeeding years, many behavioural studies have established that bees do see colours rather than differences in brightness. Making use of the fact that bees learn to revisit a rewarding site, von Frisch trained his bees to visit a blue square on a checkerboard of grey squares of many shades. The position of the blue square was changed at regular intervals to ensure that the bees were not being trained to a particular position. Once the bees were visiting regularly, the glass feeding dish containing sugar solution was removed, the blue square replaced with an identical, clean blue square and the checkerboard covered with a sheet of glass so that any olfactory clues were removed. On top of each square was placed an empty, clean feeding dish. The bees flew to the blue square no matter where it was placed on the checkerboard. Since attempts to train them to a particular shade of grey on an all-grey checkerboard failed, it could be concluded that they were perceiving the blue field as a colour and not as a particular shade of grey. Attempts to train the bees to other coloured squares followed, and later the method was refined to use coloured light of a carefully defined wavelength as the visual cue in training. It was found that bees trained to red squares confused them with black and dark grey ones. Bees appear to be red-blind. However, the use of coloured light of a known wavelength in training trials enabled researchers to show that the bee's visible spectrum is as large as that of man's, except that it extends further into the short wavelength end of the spectrum, i.e. into the ultraviolet region, and cuts off at the longer wavelength end (the red end). The bee's visible spectrum ranges from around 300 nm to 650 nm, while that of man extends from 370 nm to 750 nm (the wavelength of light is measured in nanometres, nm, millionths of a millimetre). Thus the bee can see in the near ultraviolet end of the spectrum. (Wavelengths from 400 nm down to 320 nm comprise the near ultraviolet or UV-A band, those from 320 nm to 280 nm comprise the UV-B band. Light in the UV-B band is so energetic that it can cause damage to protein molecules.) We know that the bee can see the near UV as a distinct colour from behavioural experiments, but we can have no idea what that colour looks like to the bee.

The mechanisms underlying colour vision

At the beginning of the 19th century Thomas Young, examining human colour vision, found that if he illuminated a small circular test field with a monochromatic (single wavelength) light, he could match this exactly by illuminating a second test field with the correct mixture of three monochromatic wavelengths of light. For example, if he illuminated his test field with light of 500 nm, he could match this by illuminating the second test field with a mixture of light at 420 nm, 560 nm and 640 nm. The German physicist, Helmholst, extended this work and proposed that, since any colour can be matched by a combination of three others, colour vision requires the presence of three colour sensitive channels within the eye[11,12]. Since we can only see colours in bright light when the cone receptors in our fovea are operating, the three channels were presumed

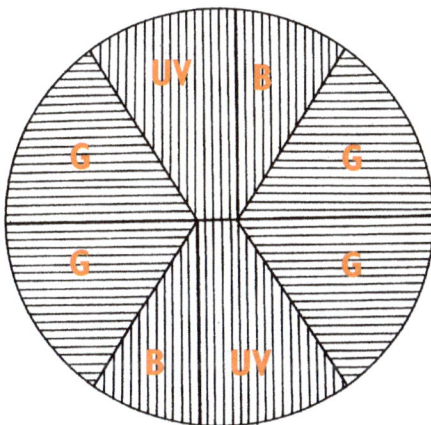

FIG. 2.10 *A transverse section through the rhabdom of a worker ommatidium in the eight-cell region. Cross hatching represents the alignment of the microvilli in each cell forming the rhabdom. The peak spectral sensitivity of each cell is indicated. There are three cells with a peak sensitivity around 340 nm, the 'ultraviolet cells' (UV). One of these is the short, ninth cell and so is not shown at this level. Two cells have a peak sensitivity around 463 nm in the blue region of the spectrum, the 'blue cells' (B); the remaining four cells have a peak sensitivity at 530 nm, the 'green cells'(G).*

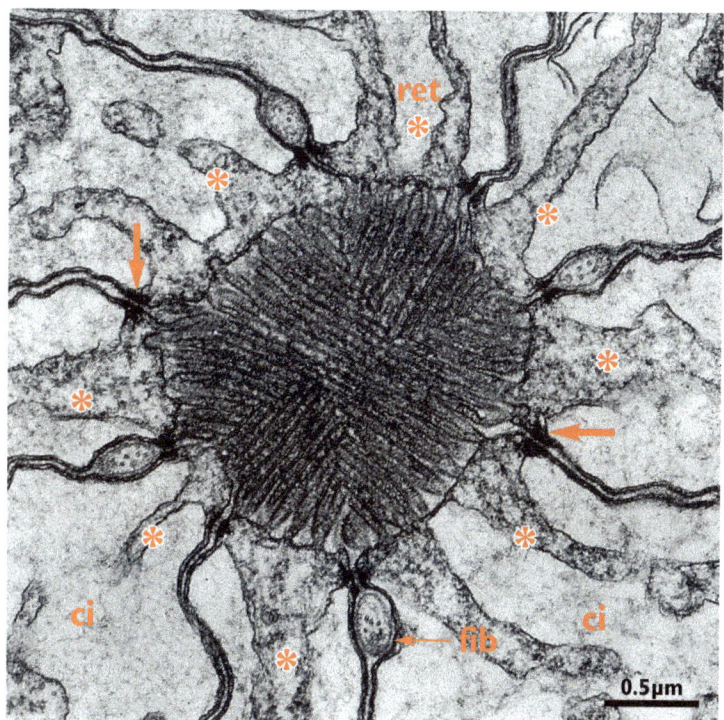

FIG. 2.11 *A transmission electron microscope section of the rhabdom at the centre of eight retinula cells of an ommatidium. The cells are joined by structures known as desmosomes (*arrows*), which makes it easy to count the number of retinula cells (ret) (*stars*) present. Lying behind alternate desmosomes are the four fibrillar processes (fib), extensions from the crystalline cone cells containing a few microtubules. These continue as far as the basement membrane and are said by some authors to tie the cone and retinula cells together[4]. The cisternae (ci), vacuole-like structures, lie adjacent to the rhabdom in this dark-adapted eye. The packing together of the microvilli of the inner border of each cell to form a rhabdom can clearly be seen; the visual pigment is borne in the membranes of these stacks of tubules.*

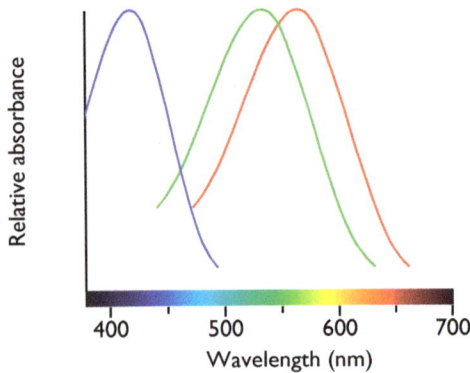

FIG. 2.12 *Measurement of the amount of light absorbed at different wavelengths across the spectrum by the visual pigment of individual human cones from the fovea shows that each cone pigment has a peak absorbency in one of three regions: blue, around 430 nm; green 530 nm; and red 560 nm. The absorption curves are quite broad, with considerable overlap of the red and green curves.* Reproduced by kind permission of J K Bowmaker[37].

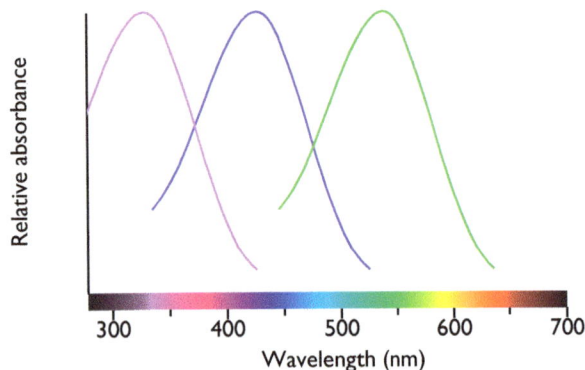

FIG. 2.13 *Measurement of the amount of light absorbed at different wavelengths across the spectrum by individual retinal cells in the ommatidia of the honey bee. Fine microelectrodes are introduced into individual photoreceptor cells within an ommatidium, the eye is stimulated by flashes of monochromatic light of equal energy across the spectrum, and the size of the cell's response measured. Like the human, the bee has three peak sensitivities; however, the peak sensitivities of the bee are shifted towards the shorter wavelengths. Each ommatidium contains three cells with maximum sensitivity in the UV at 340 nm, one of these cells is always the ninth cell. There are two cells with maximum sensitivity in the blue region of the spectrum at 463 nm; and four cells with their peak sensitivity at 530 nm in the green region.* After Menzel & Blakers[14]. *Note: the ultraviolet region of the spectrum and the peak sensitivity curve of the 'UV cells' have been represented in a lavender colour but this is not meant to imply that the bee 'sees' a lavender colour in the UV region. Behavioural studies show that the bee perceives UV as a distinct colour, but we cannot know how this colour looks to the bee.*

Measurement of the amount of light absorbed by the pigments at each wavelength shows that their absorption curves are quite broad, with considerable overlap. Light with a broad spectrum, such as sunlight, will stimulate all of the three types of cone more or less equally. We cannot distinguish between their responses and will see white light, not colour. Our perception of colour is due initially to the unequal stimulation of the three types of cone within the retina, the colour that we see depending upon the relative proportions by which the three receptors are stimulated.

Trichromacy in bees

The basic mechanism of colour vision is the same in the bee eye as in the human, although the colours perceived by the bee are different because of the shift of the bee's visible spectrum towards the shorter wavelengths. It has been possible to place a very fine recording electrode in each of the nine photoreceptor cells that are present in each ommatidium and measure the size of the response given by that cell as the bee is stimulated with flashes of monochromatic light. In this way, we can measure the absorbance spectrum of the visual pigment in the cell. There are three different peak sensitivities among the cells in each ommatidium: in the ventral part of the eye, ommatidia contain three cells with a peak sensitivity around 340 nm, the 'ultraviolet' cells, one of which is the small ninth cell. Two cells have a peak sensitivity in the blue region of the spectrum at 463 nm, the 'blue' cells, and four cells have a peak sensitivity at 530 nm, the 'green' cells[14] (figs 2.10, 2.13). Again, the absorption curves are quite broad with some overlap (fig. 2.13), and the colours

to lie within the cones. It has now been shown that the cones of the human retina do contain visual pigments with peak sensitivities in three different regions of the visual spectrum. Some cones contain a pigment maximally sensitive at around 430 nm, the 'blue' cones, others have a pigment maximally sensitive around 530 nm, the 'green' cones, and the remainder have a pigment with a maximal sensitivity around 560 nm, the 'red' cones[13,37] (fig. 2.12).

perceived by the bee will depend upon the relative stimulation of these three receptors in the first instance. Although the borders of the nine photoreceptor cells are fused together to form one rhabdom within the ommatidium, each cell has its own neuron running into the central nervous system. Light funnelled down the rhabdom will differentially stimulate the three types of cell according to its composition and these differences will be preserved in the information sent centrally by each neuron.

How do the colours seen by bees differ from those seen by humans?

The human visible spectrum is often represented as a circle in which colours that are perceptually similar are placed next to each other (fig. 2.14a). Thus, if we start with red in this figure and move anticlockwise, the colours become increasingly yellowish passing through orange to pure yellow. Moving onwards the colours become increasingly greenish until there is no trace of yellow left and we reach pure green. As we continue around the circle, blue begins to appear and then the blue turns to violet. The circle is closed by a mixture of violet and red, producing the colour we call purple, a colour not found within the spectrum. Sectors of the circle opposite one another form complementary colour pairs, e.g. red and green, blue and yellow. The bee's visible spectrum can also be represented as a colour circle (fig. 2.14b)[15], in this case extending down to 300 nm with the longer wavelengths at the red end of the spectrum missing. The bee's colour circle is closed by a mixture of yellow and ultraviolet light, forming the colour known as bee-purple. Although we cannot see ultraviolet light ourselves, we can produce mixtures of ultraviolet with light of other wavelengths and, by means of training experiments, determine whether the bee can discriminate such mixtures. In this manner, it has been shown that the bee sees bee-purple as a distinct colour, but, again, we can have no idea of how this colour appears to the bee.

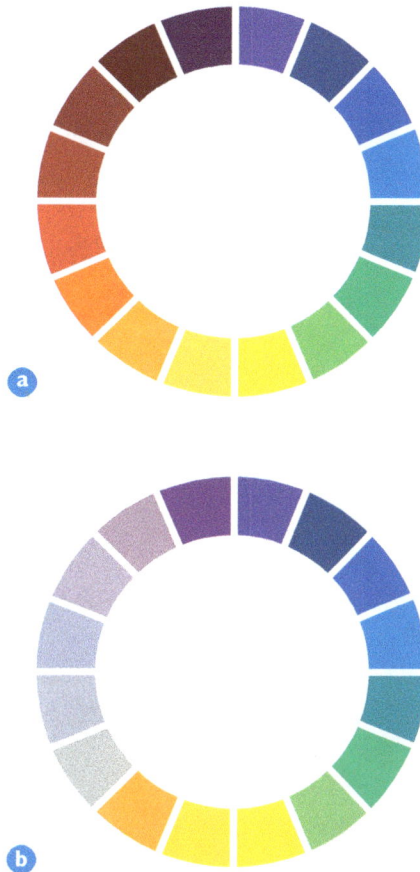

FIG. 2.14 a *The human visual spectrum represented as a circle.* **b** *The bee visual spectrum can also be represented as a colour circle.*

Flowers seldom, if ever, reflect light of only one colour and many of them reflect some ultraviolet. These flowers will present a different appearance to the bee than they do to humans. To us, the flowers of many of the Asteraceae look yellow, for example, the dandelion (*Taraxacum officinale*), *Senecio* and *Rudbeckia* species (fig. 2.15). However, these flowers reflect ultraviolet in some areas as well as yellow, and those areas will appear bee-purple to the bee. The concentric rings of small florets comprising the dandelion flower appear uniformly yellow to us, but the outer florets strongly reflect UV and thus the bee will see an outer ring of bee-purple and a central disc of yellow. The petals of *Rudbeckia* will appear divided to the bee, since the outer halves are UV reflective (see fig. 2.15 for illustration of UV reflectances in these flowers).

Bees can distinguish between bee-purple flowers, depending upon the relative amounts of ultraviolet reflected[15]. The flowers of rape (*Brassica napus*), field mustard (*B. campestris*) and wormseed mustard (*Erysimum cheiranthoides*) resemble each other in form and yellow colour to us. Rape and field mustard reflect different proportions of ultraviolet and appear to the bees as two clearly distinguishable bee-

purple hues, as demonstrated in training experiments. Wormseed mustard reflects no ultraviolet and appears as bee-yellow. Thus, three very similar flowers that grow in the same locality can be clearly distinguished by bees on the basis of colour[15].

Blue flowers can look different also, bird's eye speedwell (*Veronica chamaedrys*) and the

forget-me-not (*Myosotis sylvatica*) being two examples. The forget-me-not reflects entirely in the blue region of the spectrum and hence appears blue, while the bird's eye speedwell reflects considerable ultra-violet as well and thus appears violet to the bee. It is perhaps safer to speak of bee-blue and bee-violet since, as has been said, we cannot know anything about the subjective colour sensations of the bee. Indeed, if humans are shown a visible spectrum and asked which part of the spectrum appears to be the purest green or the purest yellow region, individuals will give different answers. Do the observers perceive colours differently or do they perceive

FIG. 2.15 *Six flowers are shown filmed first in normal daylight (*a1, b1, c1, d1, e1 *and* f1*) and then through a UV-A filter, transmitting only wavelengths between 320 nm and 400 nm (*a2, b2, c2, d2, e2 *and* f2*). The latter photographs show the areas of the plant that are UV reflective.* a1 *The pale creamy-greenish flowers of white bryony (*Bryonia dioica*) do not stand out very clearly from the surrounding foliage when viewed by the human eye.* a2 *The flower petals reflect UV light strongly and so stand out against the non-reflective foliage to an eye that is sensitive to this region of the spectrum.* b1 *The lesser celandine (*Ranunculus ficaria*), belongs to the Ranunculaceae and, like many other members of the buttercup family, has petals which humans perceive as yellow.* b2 *The peripheral part of the petal is UV reflective and so will appear bee-purple to the bee.* c1 *Creeping cinquefoil (*Potentilla reptans*) from the Rosaceae family, while appearing yellow to the human eye, appears yellow and bee-purple to the bee.* c2 *The outer rims of the petals reflect UV and hence will appear as bee-purple, while the base of the petals and centre of the flower appear yellow.* d1, e1 *and* f1 *Many members of the Asteraceae have an outer, UV-reflecting area, thus appearing as two-coloured flowers to the bee.* d1 *The coneflower (*Rudbeckia*) example shown here has long, thin petals, decorated with thin red lines to the human eye but the outer half of the petal is UV reflective.* d2 *The UV, together with the reflected yellow, will make this outer region appear as bee-purple to the bee.* e1 *The dandelion (*Taraxacum officinale*), another member of the Asteraceae, is composed of several concentric rings of small florets which appear almost uniformly yellow to the human eye.* e2 *The individual florets differ in their ability to reflect UV light according to their position: the outer yellow florets are strongly reflective, appearing bee-purple, while those nearer the centre are not.* f1 *Ragworts (*Senecio sp.*) provide another example of a flower with long, yellow petals that are UV reflective (*f2*), thus giving a conspicuous ring of bee-purple around a yellow centre.* Photographs by courtesy of J D Pye.

them in the same way but give them different names? We can never experience another person's colour sensations let alone experience the colour sensations of the bee.

Red flowers can vary in their appearance to the bee depending upon the presence or absence of reflected ultraviolet. If the flower does not reflect any ultraviolet it will appear bee-black since the bee's spectrum does not extend into the red but, if it reflects ultraviolet as well, then it will appear bee-ultraviolet[15,16]. The common poppy (Papaver rhoeas) is an example of a bee-ultraviolet flower. Some flowers that are a true red can be pollinated by butterflies whose vision does extend into the red end of the spectrum. Birds also see in the red end of the spectrum and can act as pollinators in tropical regions where bright red flowers are more prevalent.

A mixture of the bee's three primary colours, ultraviolet, blue and yellow, will yield bee-white. However, many flowers that look white or pink to the human eye do not reflect ultraviolet light and hence appear bee-blue/green to the forager since they reflect only in the blue and yellow regions of the spectrum.

The flower's function is to serve as a conspicuous visual signal to the pollinator; that means it must stand out against its background. The background colour is usually a green foliage that has a weak and fairly uniform reflectance in the ultraviolet, blue and green areas of the spectrum. The bees see this as an almost colourless grey tending to very weak bee-yellow. Most flowers will tend to stand out vividly against this background. Even flowers that to us appear to resemble the background foliage, such as the white bryony, may stand out clearly to the bee because the petals strongly reflect UV light (see fig. 2.15a2). Wind-pollinated flowers, on the other hand, often have pale greenish flowers showing little contrast with the foliage to us or to insects.

Nectar guides

Many plants provide additional visual signals to aid the pollinator in finding its reward once it has located the flower. These markings, or nectar guides, lead the bees and other pollinating insects toward the nectaries. Some of them are readily visible to us, for example, as markings on pansies (Viola spp.), or as the white-bordered purple spots on the corolla of the foxglove (Digitalis purpurea), but many of them are only visible to the insect. In creeping cinquefoil the outer edges of the petals are yellow and UV reflective, hence appearing bee-purple. The nectaries are surrounded by the pure yellow of the petal bases (see fig. 2.15c2). Usually the nectar guides contrast with the rest of the flower because they reflect little or no ultraviolet light; for example, the marsh-mallow, (Althaea officinalis), has central ultraviolet-free markings appearing bee-blue to the bee. But there are exceptions; the pheasant's-eye daffodil (Narcissus poeticus), reflects ultraviolet strongly from the red border of the yellow corolla, forming a contrasting ring around the approach to the food[10,15,16].

How well do bees discriminate colours?

Given the multi-hued array of coloured signals presented by flowers in nature, how well do bees discriminate between them? This question can be answered by experiments in which bees are trained to visit a particular colour and, after training, are then presented with this and a second colour of a different spectral wavelength, their choice being noted. The second colour presented is brought nearer and nearer to the test colour and the degree to which the bees confuse the two colours recorded. This experiment is then repeated many times using test colours ranging throughout the bee's visible spectrum. The results show that the bee's ability to discriminate between hues is comparable to that of man in its region of best discrimination. It detects differences between wavelengths only 4.5 nm apart compared to 1 nm achieved by the human eye in its best region[17]. Bees and other hymenopteran pollinators, such as bumble bees, solitary bees and wasps, are maximally sensitive to spectral differences around 400 nm and 500 nm (fig. 2.16). They all possess ultraviolet, blue and green receptors with sensitivity peaks in the spectral regions 330–350 nm, 430–450 nm and 520–540 nm. This means that they are maximally sensitive to spectral differences, not in the regions of peak sensitivity for their three receptor types, but in the regions where the spectral sensitivity curves of their receptors overlap and have steep slopes in opposite directions (fig. 2.13). In those parts of the spectrum small changes in wavelength will cause maximal changes in excitation in two of the receptor types, e.g. in the 400 nm region a shift towards shorter wavelengths will increase the response in the ultraviolet cells and reduce the response in the blue cells. These small differences are enhanced in the next stage in the colour coding process.

The bee evaluates signals from its three types of receptor cell antagonistically in a **colour opponent mechanism**. Colour opponency also underlies colour coding in the human eye, reminding us yet again that basic neural mechanisms are often similar in insects and in man. The three types of photoreceptor feed into cells that amplify their signals which in turn feed into the colour opponent cells. The colour opponent cells have a basic discharge rate at rest, i.e. when they are not being stimulated. The cells feeding into them may either stimulate the colour opponent cells, causing them to fire nerve impulses at a higher rate, or inhibit them, causing their firing rate to be reduced.

There are two types of colour opponent cell[18]. Type I cells respond antagonistically to short and medium/long wavelengths. They are either excited by ultraviolet and inhibited by blue and green wavelengths, or inhibited by ultraviolet and excited by blue and by green light (the latter condition is illustrated in fig. 2.17). Type II cells

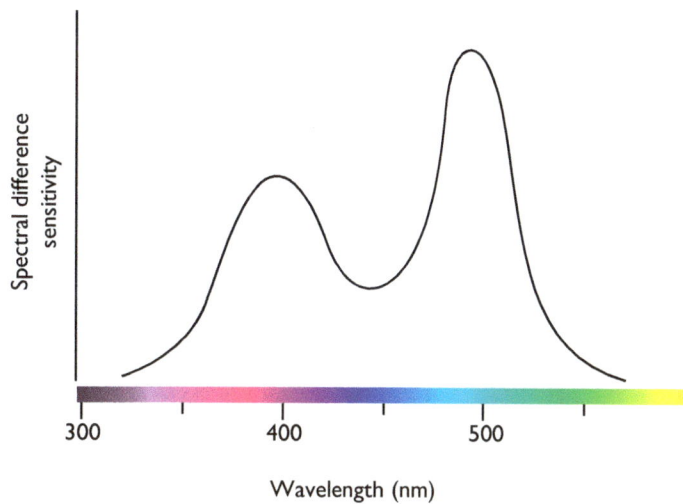

FIG. 2.16 *The spectral discrimination function of the honey bee worker demonstrates how well the bee can discriminate between two different colours[17]. Bees are maximally sensitive to differences in colours around 400 nm and 500 nm: in these regions of the spectrum they can distinguish between wavelengths only 4.5 nm apart. These peak sensitivities coincide with the regions where the spectral sensitivity curves of their three types of photoreceptor overlap and have steep slopes in opposite directions. These are the regions where small changes in the wavelength seen will result in maximal change in the response of two of the photoreceptor types.*

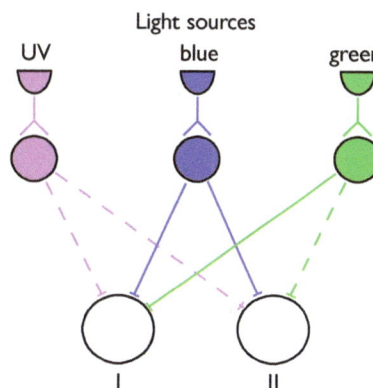

FIG. 2.17 *Colour opponency is a mechanism for enhancing the differences of signal received from the photoreceptors. The photoreceptors feed into cells that amplify their signals and these cells, in turn, may excite or inhibit the succeeding cell, the colour opponent cell. Two types of colour opponent cell have been found in the bee. Type I cells respond antagonistically to short and medium/long wavelengths, i.e. they are either excited by UV and inhibited by green and blue light (as illustrated) or vice versa. Type II cells react antagonistically to short and long wavelengths and to medium wavelengths. This arrangement maximizes the differences in the size of the signals received from the photoreceptors and so aids discrimination.*

react antagonistically to short/long and to medium wavelengths. These cells are either excited by ultraviolet and green light and inhibited by blue light or they are excited by blue light and inhibited by ultraviolet and green (illustrated in fig. 2.17). This arrangement maximizes the differences of signal received from the photoreceptors, particularly in the areas around 400 nm and 500 nm, and thus enhances the insect's ability to discriminate between colours in these regions of the spectrum.

Evolutionary tuning between flower colours and the colour vision of bees and other flower-visiting Hymenoptera

Since many flowering plants (angiosperms) rely on animals, mostly insects, to transfer their pollen to conspecifics, they compete with each other for pollen vectors. The plant provides a reward in the form of food to attract the pollinator and, ideally, needs to ensure that the pollinator will visit its species exclusively and not waste its pollen by dispersing it to other plants. The plant could restrict the pollinator type by means

of morphological adaptations but this could make it too dependent on one pollinator species and vulnerable to the loss of this pollinator. Alternatively it can advertise its reward by means of a label that is exclusive to its species. This label can potentially attract a large number of pollinator types and ensure that they find other members of the species[7]. For this strategy to be effective the plants' signal or signals must be conspicuous and readily distinguishable from those of other species. We have seen that, of the several features characterizing a flowering plant, colour is a very important cue as the insect approaches the foraging site. Tests have shown that a bee's choice of a test colour is dependent upon how near that test colour is to a colour that it has previously experienced as rewarding[19]. Thus there is a great deal of evolutionary pressure on a plant to produce a colour signal that can easily be discriminated from those of other species.

For a flower to stand out from its background and from its competitors' flowers, it must use combinations of pigments that generate sharp steps in the wavelengths of light that it reflects[9] (fig. 2.18). If moving along the spectrum a little way produces a rapid change in the amount of light of different wavelengths reflected, then there will be large changes in the relative excitation of the three types of insect photoreceptor. The insect colour vision system will be able to discriminate flower signals optimally if its spectral discrimination ability is maximal in the region where there are steep changes in the wavelengths reflected by flowers.

The spectral reflectance of the flowers of many angiosperms has been measured in Israel, where hymenopteran pollinators dominate, and the regions of maximum change in the flower's reflectance spectra determined[9]. Three regions of maximum change were found, one at 400 nm, one at 500 nm, and a third at 600 nm (fig. 2.19). As we have seen, the bee and other pollinating hymenopterans discriminate wavelengths maximally at 400 nm and 500 nm

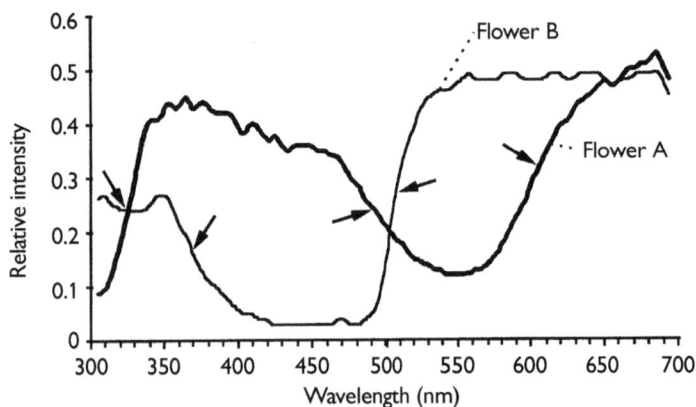

FIG. 2.18 *In order to look for the possible relationships between floral colours and hymenopteran colour vision, the spectral reflectance of the petals of a large number of flowers was measured. Since the most vivid colours are those that produce large differences between the signals of different photoreceptors, the most important characteristic of a flower reflectance spectrum is a sharp step. If these sharp steps occur where the spectral sensitivity curves of the different types of photoreceptor overlap, maximum signals will be obtained. This figure shows the spectral reflectance function across the spectrum for two flowers. A number of sharp steps occur in the case of each flower. The arrows indicate the 50% values of the steep slopes.* Redrawn from Chittka & Menzel[9].

and their colour discriminating abilities, thus, are well matched to the reflectance spectra of the flower's signals. It is not clear why there is a third peak at 600 nm. It might be that the flowers are also addressing other pollinators in the region that have tetrachromatic vision. There are insects, including some beetles, butterflies and a few hymenopterans having additional receptors with a peak sensitivity around 600 nm[20]. These insects could make use of flower signals at the red end of the spectrum. It has been suggested that another reason for the presence of the peak at 600 nm is simply that it may be easier for plants to use widespread pigments, such as anthocyans, which have both a blue and a red reflection peak, than to attempt to produce a pure blue-coloured flower[9]. As long as the blue peak is present, the presence of the additional red peak will be irrelevant to the mass of hymenopteran pollinators.

Thus, the beauty of wild flowers is not addressed to our eye, but their colours are very well tuned to the eyes of the pollinators. This evolutionary adaptation of the signal-receiver system of flowers and pollinators is of mutual benefit to plant and insect.

Navigation to and from the forage site

A bee may collect food at a considerable distance from the hive: 2000–3000 m is not uncommon, and much longer distances have been recorded. How does it

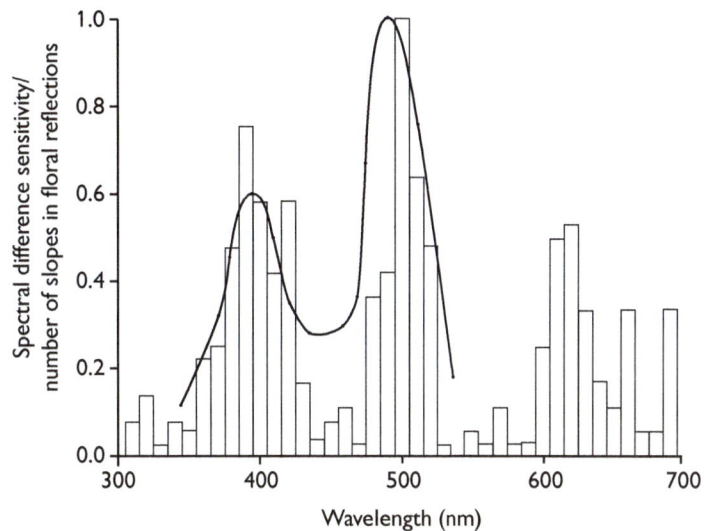

FIG. 2.19 *The histogram shows the number of steep slopes in individual flower reflectance measurements at the respective wavelengths for 180 flower species. The wavelength at the 50% value of each steep slope was used (see fig. 2.18). Superimposed is the spectral discrimination function of the bee. The bee is maximally sensitive to colour differences around 400 nm and 500 nm, and most flowers use combinations of pigments that generate sharp steps in their reflectance around the wavelengths at which hymenopteran pollinators in general are best able to discriminate colours.* Redrawn from Chittka & Menzel[9].

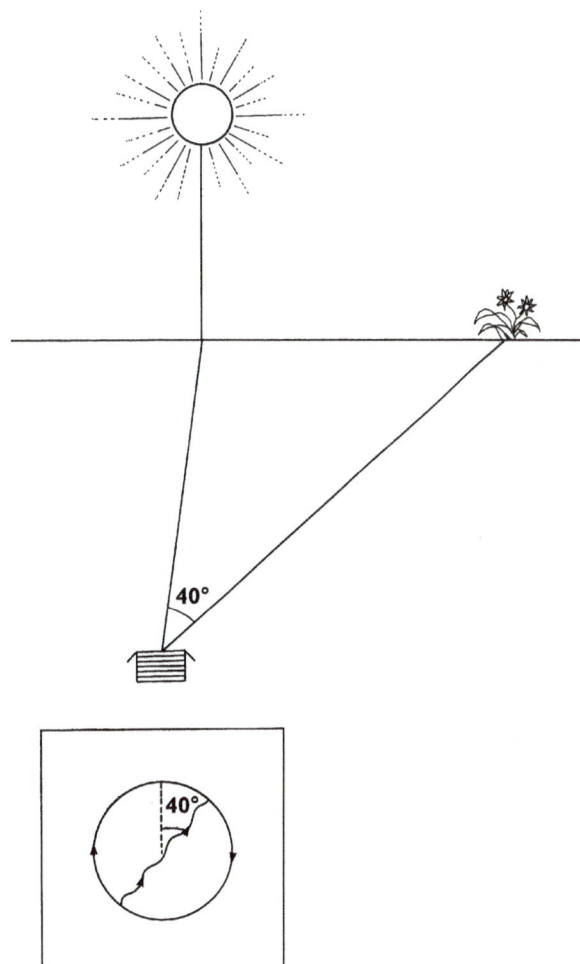

FIG. 2.20 *A bee which has found a rich nectar source some distance from the hive, dances on the vertical surface of the comb to communicate the direction and distance of its source. The angle between the direction of the foraging site and the sun's azimuth is transposed into the angle made by the bee on its straight run with respect to gravity. The direction of the sun is always represented by the vertically upward direction on the comb. In this example, the food source lies 40° right of the sun's azimuth and so the bee dances with its straight run 40° to the right of the vertical.*

find its way to and from the site? In the hive, a bee with a rich reward is stimulated to dance and communicate the direction and distance of its food source[10]. Examination of the waggle dance reveals the main cue used by the bee in its navigation. On the comb, the angle of the bee's straight run to the vertical upward direction indicates the angle that the food site makes with the sun's azimuth (the sun's azimuth is its vertical projection to the horizon). The transposition of the sun's direction to that of gravity on the comb employs the convention that the sun's direction is always represented by the vertically upward direction. A straight run, accompanied by waggles of the abdomen pointed vertically upwards means that the food source lies in the direction of the sun; a downward run, that it lies in a direction opposite to the sun. If the tail-wagging run points 40° to the right of vertical, then the food source lies 40° to the right of the sun (fig. 2.20). Experiments by many different workers have confirmed that the sun is the primary navigational tool used by the bee in long distance navigation. In fact, other insects besides the bee are able to navigate towards a goal by using the sun as a compass, maintaining a fixed angle between the goal and the sun; the ant is one example. Several celestial cues may be used to locate the position of the sun when it is not clearly visible, such as the spectral composition of the sky and the pattern of polarized light.

The sun compass

The sun subtends an angle of 0.4° and the spatial resolution of the bee's eye is poor, so poor that it is unable to resolve the sun as a distinct object. If this is so, how can it use the sun as a compass? Several cues allow the insect to locate the position of the sun[21]. The sun is normally the brightest region in the sky and, since it subtends a relatively small angle, a small very bright region in the sky is an indication of the position of the sun. The colour gradient of the sky can enable the forager to locate the position of the sun on slightly overcast days. Sunlight is scattered by small particles in the upper atmosphere; short wavelengths are scattered much more than long wavelengths, hence the blue appearance of the sky to the human eye. The shorter, ultraviolet wavelengths are scattered even more strongly, so that the sky is very rich in the ultraviolet region of the spectrum to the bee, even in conditions of light cloud cover. Indeed, the sky would appear to be predominantly of that colour to the forager. The only region of the sky containing very little ultraviolet is the region occupied by the sun. To the forager, the sun is a small, bright region of the sky containing longer wavelength light. The primary criterion for accepting a region of the sky as the sun appears to be lack of ultraviolet light. Experiments have shown that a bee experienced in navigating by the sun will accept a small, bright object reflecting green, i.e. long wavelength light, and little or no ultraviolet, as the sun[21]. The bee uses the dorsal region of its eye to detect this spectral information about the sky.

Polarized light

The bee can still navigate by the sun if a small area of blue sky is visible. Under these circumstances, it can use the polarized light pattern of the sky to assist it in finding the sun's position. Light from the sun oscillates in all possible directions perpendicular to its direction of travel; this is unpolarized light. Polarized light is light that is oscillating in one direction only, perpendicular to its direction of travel, a direction known as its e-vector. Polarization can result from scattering of light. As light enters the earth's atmosphere, it is scattered off air molecules, resulting in a net polarization which is perpendicular to the plane containing the observer, the scattering molecule and the sun. To an observer on or near the ground able to detect polarized light, as the bee can, there is a pattern of polarization across the celestial atmosphere arranged in an orderly way. Since the polarization arises from sunlight scattering, the pattern is fixed relative to the sun. The planes of polarization of light (or

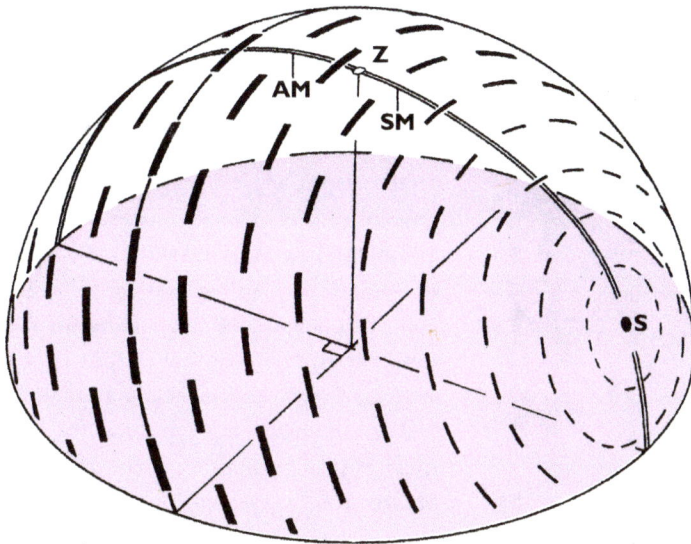

FIG. 2.21 *The celestial polarization pattern. Polarized light vibrates in only one direction normal to its line of travel, known as the e-vector direction. The e-vector directions are concentrically arranged around the sun (S) and shown here by the orientation of the black bars. The width of each bar represents the degree of polarization; maximal polarization occurs along the great circle 90° from the sun. The symmetry line of the e-vector pattern consists of the solar meridian (SM) and the anti-solar meridian (AM). Sun's zenith (Z).*

e-vector directions) seen in the sky form concentric circles around the sun (fig. 2.21), with a plane of symmetry in the pattern passing across the entire celestial hemisphere[22]. This plane of symmetry contains the solar meridian and the anti-solar meridian. The solar meridian is the arc passing from the zenith (the sun's highest point in its course each day) through the sun to the horizon and can be used as a compass reference by the bee if it can be located. The anti-solar meridian is the continuation of that arc on the other side of the celestial hemisphere.

It has been known since 1949 that bees could use the pattern of polarized light in the sky for navigational purposes, but the manner in which they did so was not fully understood[23]. Further studies showed that it was the ultraviolet region of the spectrum which was used and that it was the direction of the polarized light which was important, rather than the degree of polarization of the ultraviolet light. Recent behavioural, electrophysiological and anatomical studies have combined to reveal the mechanism underlying the bees' use of polarized light and, in so doing, has demonstrated yet another specialization of the insect compound eye. The polarization-sensitive retinal cells are located in a narrow band of ommatidia, 2–3 deep, lying around the dorsal rim of the eye[24,26]. This area is commonly known as the **POL area.** The ommatidia in the POL area are slightly larger in appearance (fig. 2.22) and their optical axes are arranged so that the majority of them view the contralateral half of the sky, looking frontally or upwards in directions close to the zenith. The corneal lenses of these ommatidia are penetrated by fine canals. Scattering or reflection of light off these canals means that light hitting the corneal lens off axis is also able to reach the rhabdom. Because of this, the ommatidia each collect light from a larger area than the ommatidia in the rest of the eye. The small dorsal rim area thus views quite a large region of the celestial hemisphere[25,39].

The dorsal rim ommatidia differ from those in the rest of the eye in other respects. The ninth retinal cell, shorter in other ommatidia, is the same length as the other cells in the POL region. Most importantly, the parallel array of microvilli on the inner border of each cell maintains its direction of orientation down the whole length of the rhabdom. The UV-sensitive cells of the POL area form the primary polarization analysers, and each UV-sensitive cell is maximally sensitive to polarized light vibrating parallel to its microvilli. Outside the POL area, the retinal cells become twisted around the longitudinal axis of the rhabdom and, as a result, the UV-sensitive cells have a greatly reduced sensitivity to polarized light. The UV-sensitive cells of

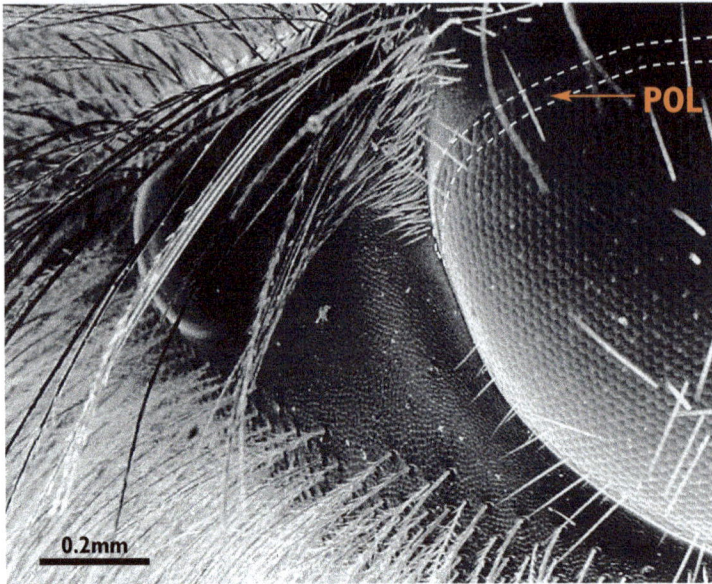

FIG. 2.22 *A small, specialized part of the bee retina, located at the dorsal rim of the compound eye, has been found necessary and sufficient for navigation by polarized light. This region is known as the POL area and is only 2–3 ommatidia deep. Part of it is shown outlined in white above, with the hairs removed from the eye and head for clarity. The ommatidia of this area are slightly larger and greyish in appearance in the living insect, due to the presence of fine, rough-walled canals in the cornea that scatter the incoming light. Because of this, the ommatidia each collect light from a larger area than ommatidia in the remainder of the eye and the small dorsal rim area in fact views quite a large region of the celestial hemisphere.*

the POL area are arranged in such a way that there is a set of cells in each ommatidium with their microvilli arranged at right angles to each other. These form the paired analysers, known as the **X- and Y-type receptors**[22,25,27].

The analysers for the detection of polarized light are thus located in the retinal cells of the 140 or so ommatidia of the dorsal rim. Based on behavioural experiments, in which the directions of polarization seen by the bee can be manipulated, and on recordings made from individual UV-sensitive retinal cells, it has been shown that the orientation of the orthogonally-arranged microvilli of the ultraviolet receptors in each POL ommatidium changes slightly as one moves from the back of the eye to the front[22,27,38]. This results in an array of polarization-sensitive analysers, whose directions of maximum sensitivity to polarized light change in a regular manner through 180° across the top of the head (fig. 2.23). The microvillar directions

of the X-type receptors in the pairs of analysers mimic the pattern of *e*-vector distributions in the sky when the sun is on the horizon. The microvilli of the other set of receptors, the Y-type receptors, are orientated at right angles. To use this array of crossed retinal analysers in celestial orientation, the bee scans the sky, turning about its vertical axis and sweeping its array of analysers across the celestial hemisphere until the summed neural output from its array of X-type analysers is maximal. This will occur when the X-type analysers make their best alignment with the pattern of polarization in the sky[25,27].

It is believed that the outputs of the two sets of polarization-sensitive cells in each ommatidium act antagonistically on a cell at the next level in the neural pathway. The UV-sensitive cells are stimulated maximally when the *e*-vector is parallel to the long axis of their microvilli. When the X-type receptors are maximally stimulated, their output excites the next cell in the pathway. This increases the rate at which nerve impulses are fired in this interneuron (fig. 2.24). The responses of the Y-type receptors are phase-shifted by 90° and, when they are maximally stimulated, their output inhibits the spontaneous rate of impulse firing in the interneuron. This arrangement has the effect of enhancing the polarization sensitivity of the analysing system, and of ensuring that the cells respond only to changes in the direction of polarization and not just to changes of light intensity[28].

In summary, the bee carries an invariant set of polarization filters in the dorsal rim of its eyes. To use them, it scans the sky until the direction of the microvilli in its X-type analysers matches the direction of the polarized light that it views in an area of blue sky (fig. 2.25). At this point in its scan, the summed output from the receptors will be maximal, and this tells the bee that it is aligned with the plane of symmetry of the pattern of polarization in the sky. The fact that it is aligned with the plane of symmetry is all the information that the bee can extract from its

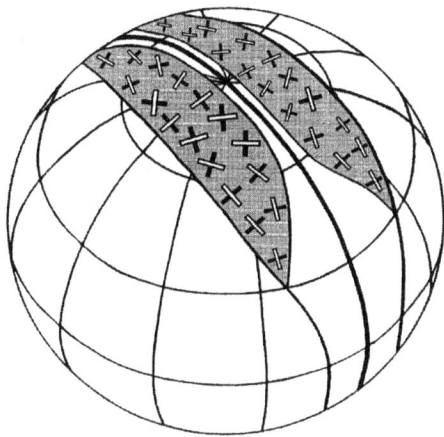

FIG. 2.23 *The orientations of the microvilli of the UV photoreceptors of the POL area (shaded) are shown projected onto the celestial sphere. The UV receptors of each ommatidium form paired analysers, X-type and Y-type receptors, with orthogonally-arranged microvillar directions. The X-type receptors are represented as open, white bars; the Y-type receptors as thin, black lines; the X-type and Y-type receptors interact antagonistically at the next neural level. Note that the orientation of the microvillar directions of the paired analysers rotates from the back to the front of the POL area, resulting in a fan-like pattern that is a close approximation to the e-vector distributions of the sky when the sun is on the horizon. This forms the bee's hard-wired e-vector map. The POL areas have contralateral fields of view. Redrawn from Rossel[38] and Sommer[39].*

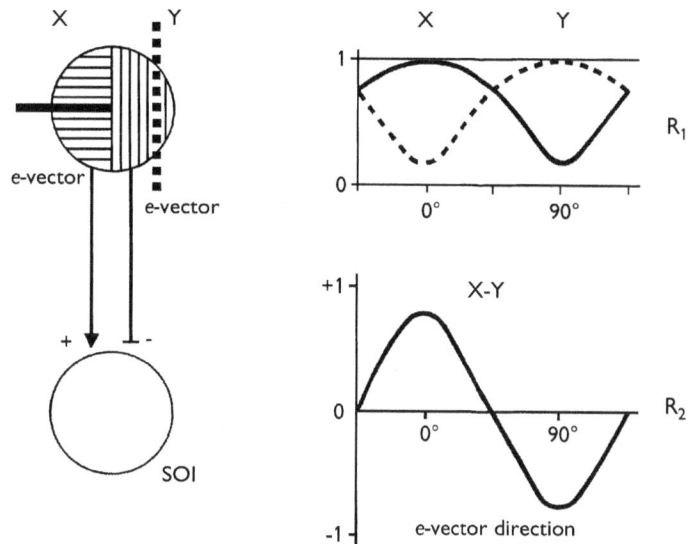

FIG. 2.24 *The long axes of the microvilli of the UV-sensitive cells in each ommatidium are arranged perpendicular to each other, forming the paired analysers, the X- and Y-type receptors. The cells are maximally stimulated when the direction of maximum polarization, the e-vector, is parallel to the long axis of the microvilli. The responses of the paired analysers to polarized light are thus phase-shifted by 90° (upper graph R_1). The X- and Y-type cells interact antagonistically at the next neural level, i.e. the second-order interneurons (SOI). The X-type receptor excites the second-order interneuron and increases its level of firing above the spontaneous rate. When the Y-type receptor is excited, its activity will inhibit the spontaneous discharge in the second-order interneuron. This arrangement enhances the polarization sensitivity of the second-order interneuron, whose output will reach a peak (lower graph R_2) when the X-type microvilli become parallel to the e-vectors in a patch of sky. It also ensures that the second-order cells respond only to variations in e-vector directions and not to intensity fluctuations. After Rossel[38] and Wehner[25].*

polarization filters. Bees have to distinguish the direction of the solar meridian from that of the anti-solar meridian to use the solar meridian as the zero point in its celestial compass. This can be done by using other cues. A gradient in the spectral composition of the light along the plane of symmetry can indicate the solar meridian: the region of the sky containing the sun will have little ultraviolet light.

The array of X-type analysers corresponds to the pattern of polarization in the sky at dawn and dusk when the sun is on the horizon. However, the direction of polarization of particular regions of the sky changes with respect to the axis of symmetry as the sun's elevation alters through the day, although there is always a mirror image symmetry of directions on either side of the axis. A complete match of *e*-vector directions with X-type receptor directions cannot be achieved except at dawn and dusk. One might assume that the bee would make errors in navigation when foraging at different times of day, but the insect does not seem unduly inconvenienced by the fact that it cannot always make a complete match, and the scanning mechanism appears to provide quite accurate compass bearings at all times of the day[25].

Compensation for the sun's movement

One drawback to using the sun as a compass is that it appears to move through the southern sky from left to right during the day in the northern temperate zone, or from right to left in the northern sky in the southern temperate zone. Furthermore, in the tropics, the sun may move either way depending on the season. A bee may spend an hour or more away from the hive foraging. On its return, it is not sufficient for the bee simply to steer a reverse course with respect to the angle of the sun, since the sun will now be in a different position in the sky. However, the bee's behaviour shows that it is able to compensate for the sun's movement, since the majority of bees do find their way back to the hive even if they have been away for some time. The food source direction, indicated by the waggle dance of a returning forager, is that appropriate for the time of return and the start of dance, not for the time of departure from the hive. Occasionally, foragers in the hive dance over a period long enough for the sun to have moved a significant distance, and in this case the bee can be seen to change the direction of its waggle run to allow for the change in angle between the sun and the food source.

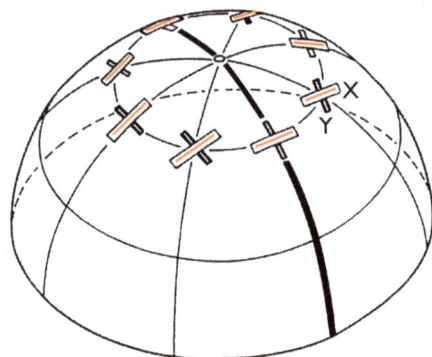

FIG. 2.25 *A ring of e-vector positions when the sun is on the horizon (orange bars), are shown projected onto the celestial hemisphere. The bee sweeps its analyser array across the sky by turning about its vertical axis until the microvillar directions of the X-type receptors align with the e-vectors in the sky. At this point, the differential response of the crossed retinal analysers passes through a peak. (See also fig. 2.24) The bee assumes that it is aligned with the symmetry plane of celestial polarization (thick black line). Having found this, other visual cues must be used to discriminate between the solar and anti-solar meridian and to set a compass course (see text). The insect's e-vector map is hard-wired for this particular position of the sun, but e-vector directions change with the sun's elevation. The insects do not appear to be unduly handicapped by the fact that there is normally not a perfect match between their e-vector map and that of the sky. X-type receptors (large open bars); Y-type receptors (small open bars). After Rossel[38].*

To compensate for the movement of the sun through the sky is not a simple matter. The path the sun takes through the sky differs with the time of year, the elevation being greater at any one point throughout the year up to the summer solstice when it declines again. It also differs with latitude. The rate of movement of the sun through the sky each day is slower near dawn and dusk than it is near midday.

Investigations indicate that the bees have to learn the direction in which the sun moves across the sky, and that this is done during their first 5–8 days' experience outside the hive[29].

Having learnt the sun's direction across the sky, how does the bee compensate for this movement? A prerequisite for sun compensation is a time sense. Bees are known to have a well-developed time sense, turning up promptly at food sites where food is available only for a restricted part of the day[29]. They have an endogenous biological clock, as do many insects, which is entrained by the diurnal light–dark cycle. There is considerable evidence to suggest that sun compensation is based on the bee's ability to learn the sun's position at intervals over the day in relation to landmarks around the hive, updating the information with the changing seasons.

When they need to determine the sun's position at some intermediate point, they might do so on the basis of linear interpolation between two learned points[30]. Recently, it has been suggested that bees may extrapolate from the rate of solar movement that they have most recently seen, in order to find the sun's position after a period in the darkness of the hive[21,30].

Thus, although we know that bees do compensate adequately for the movement of the sun, and that they do so by some simple approximation of solar movement rather than by calculations based on celestial geometry, we still do not fully understand the underlying mechanism.

Pattern recognition in bees and the use of landmarks

The celestial compass is the primary means of navigation over long distances, as the use of landmarks for navigation over these distances is limited by problems of spatial resolution of the compound eye. However, if there are prominent, well-defined landmarks present, such as the edge of a forest, a road or the edge of a lake, these features may be used in navigation by experienced foragers[23,30], but the bee is aware of the sun's position in relation to the forage site and it is this information that is communicated to recruits in the dance. When celestial cues are not visible on a completely overcast day, an experienced forager will find the way to a food source with which it is familiar using previously learned landscape features. Since it has also memorized the relationship between the sun's position and these features, it will on its return once again encode its direction of flight in relation to the sun. Bees recruited to the dance will leave the hive with only that information, although the sun is not visible. From their time-linked memory of the sun's position relative to landmarks near the nest, they are able to set their line of flight correctly[35].

Bees are strongly influenced by landmarks in the vicinity of their hive and will search for a long time in the original spot if the hive is moved only a few metres away. They are also able to learn the shapes of food-bearing plants as well as their colours and can return repeatedly to the same point in a patch of flowers. These observations have led scientists to investigate pattern recognition in the honey bee over many years[40]. Earlier researchers used a variety of training experiments to determine whether bees could discriminate between different horizontally placed patterns. For example, bees were trained to collect sugar water at one of a number of black shapes placed on a white background. The positions of shapes were changed frequently during training to make sure that the bee learnt to come to the rewarded shape and not to

a specific location. After training, the sugar water was removed and the bees' visits to all the shapes were recorded to see whether the bees could discriminate the now unrewarded shape. It was found that bees could readily distinguish between solid and open figures, such as a solid square and an open square (fig. 2.26), but not between several solid figures or several open figures[10,41], i.e. they cannot discriminate well between simple closed shapes that are similar in terms of their contour lengths but they can discriminate well between such shapes and ones that are 'dissected' and differ markedly in their contour lengths. It was found that bees showed a spontaneous preference for 'busy' patterns. These results led to the view that the decisive parameter used by bees in discriminating a pattern was its contour density, i.e. the length of contour per unit area. It was suggested that a flying bee would classify patterns by measuring the frequency of the on/off visual stimuli detected by the compound eye as it flew over the pattern. The bee eye has a high temporal resolution (see earlier), so that this would certainly be physically possible. While this view was popular for a long while, increasingly there were difficulties in explaining all of the results of pattern discrimination experiments on this basis. Bees were found to discriminate between some patterns that possessed identical

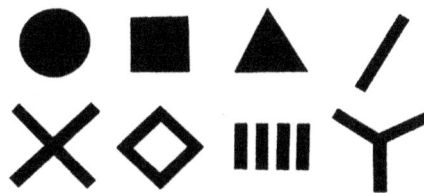

FIG. 2.26 *An example of the horizontally positioned patterns used in the first investigations of pattern recognition in bees. Bees were able to distinguish between solid patterns, such as those shown in the upper row, and broken or 'busy' patterns, shown in the lower row, but could not be trained to distinguish between the patterns in the upper row or between the patterns in the lower row.* After von Frisch[23].

FIG. 2.27 *Bees can be trained to crawl through a tube to a sugar-water reward. This arrangement can be used to test a bee's ability to discriminate between vertically-displayed patterns, placed at the end of the tube, obscuring the reward. During training to the pattern displaying the reward, the bees hover before the entrance to the tube for a few seconds. This period of fixation of the pattern is believed to allow the bee to acquire a template or neural 'snapshot' of the pattern. After Wehner[3].*

This enabled scientists to determine whether bees could distinguish between patterns that were rotated with respect to one another, something that could not be achieved with horizontally presented patterns[40]. The experiments showed that bees were good at discriminating the orientation of patterns. If trained to a disc that was half white and half black, bees could distinguish between this disc and one whose orientation differed by only 20° (fig. 2.28)[3]. It was noticed that when bees were trained to these vertically-presented patterns, they visually fixated the opening of the tube prior to landing on it and crawling inside. The bees hovered in front of the entrance for one or two seconds, adjusting their six degrees of freedom (see chapter 7) to fixed values relative to the opening of the tube which serves as their visual marker[3]. During this period of hovering in front of the pattern the bees are presumed to memorize the spatial distribution of the pattern on the compound eyes in a pictorial sense, rather like taking a neural 'snapshot' of it, i.e. they set up a stored template of the pattern. When a bee is offered a choice between the pattern to which it was trained and another pattern, it is presumed to evaluate the two patterns by comparing the extent to which each pattern matches the stored template.

A good deal of evidence has been accumulated to corroborate the template theory of pattern recognition[40,42,43,44], and support has also been obtained from studies on landmark learning. Behavioural studies, in which local landmarks were manipulated, indicated that bees memorize these landmarks as a neural 'snapshot' as viewed either from the forage site or from the nest/hive[31]. A number of landmarks, learned as a unitary array, allow the bee to fix its position. On returning to the site, the bee locates the food or nest by bringing the currently perceived image into register with the learned image, and it can do this irrespective of the direction from which it approaches the forage site[3,31]. Young bees can be seen making orientation flights on first emerging from the

contour densities, for example. Furthermore, it was not possible always to predict how bees would discriminate between patterns solely on the basis of their contour density.

Most of the pattern discrimination experiments leading to the 'flicker' hypothesis utilized horizontally-displayed patterns which the bees could approach from any direction. Presenting the bee with choices between patterns displayed in the vertical plane produced new and interesting results[40]. Bees can easily be trained to collect food by crawling through a transparent horizontal tube to an artificial feeding device. In these pattern discrimination tests, bees were trained to a patterned disc centred around the axis of the tube and positioned a short way away from its entrance (fig. 2.27). The sugar-water reward lay behind the pattern. After training, the bee was presented with a pair of such patterned discs with central tubes and had to make a choice between them[3].

FIG. 2.28 *Bees are good at discriminating the orientation of patterns. When trained to a half white, half black disc, placed at the angle represented as 0° in this diagram, bees could distinguish that disc from an identical one placed in any of the other nine positions shown, even when the test disc was turned through only 20°. Redrawn from Wehner[3].*

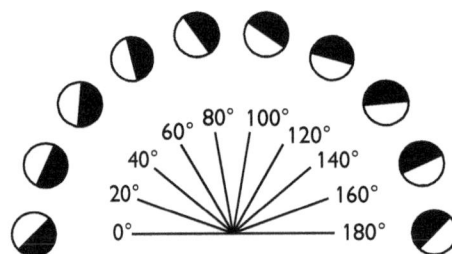

hive in which they make slow, zigzagging passes while facing the hive. During this fixation period they learn the spatial distribution of local landmarks. If a hive is displaced and worker bees leaving the hive are prevented from performing reorientation flights, these bees are not able to return to the new hive position[3]. Reorientation flights also are performed around feeding sites that are visited over a period showing that bees are able to update their snapshot images.

Although it seems that bees use neural snapshots to recognize patterns and to locate forage and hive sites using local landmarks, it is unlikely that this is the sole explanation for their abilities in this respect. The foraging bee travels from the hive to the feeding site, often over considerable distances. Once there it may have to move around among different flowers to fill its honey sac (crop) before returning to the hive. As nectar flows change on succeeding days, the bee will have to visit other sites and, if the requirements of the hive change, it may need to switch to fetching water. Scout bees in particular will be visiting new environments quite often. Thus, workers encounter new patterns that have to be learnt all the time. This makes it very unlikely that snapshots can be the only mechanism for remembering shapes, since this would require a very large memory to store all of the images. For this reason it is also unlikely that the stored images are very detailed. The problem of the great demand on storage capacity in the bee brain if pattern recognition depends only on neural snapshots led researchers to ask the question 'can bees abstract some general characteristics of a pattern that can be used to recognize it later?'. If bees could do this, then they might be able to represent patterns in a way that takes up less storage space; for example, could they abstract the orientation of a pattern and use this for recognition?[40]

This possibility was examined by training bees to distinguish between patterns of horizontal or vertical stripes. The two

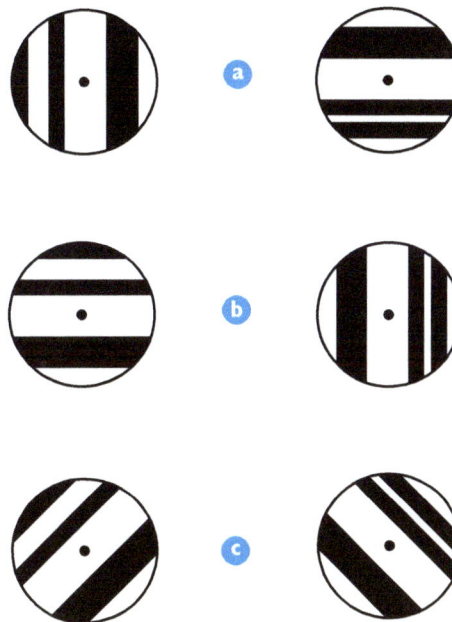

patterns were presented vertically, each one at the end of one arm of a Y-maze, with the sugar-water reward presented behind the central aperture of the training pattern[40,45]. The positions of the two patterns were interchanged during training as usual so that the bees did not learn to associate the reward with location in a specific arm of the maze. In these experiments, of course, it was necessary to prevent the bees from being able to set up a template of the rewarded stimulus. This was achieved by having the patterns at the end of long arms in the maze so that bees had to make their choice when they were still too far away from the pattern to be able to fixate it and thus create a neural snapshot. In addition, the patterns were replaced often by other patterns that, although similarly orientated, bore randomly striped patterns with different spatial structures. In this situation bees had to be able to abstract the orientation of the rewarded disc in order to learn to identify it correctly, rather than memorizing the specific pattern. The results showed that bees were able to discriminate between horizontal and vertical orientations and between two oblique directions (fig. 2.29). After training, they were even able to discriminate the orientations of other patterns never previously encountered (fig. 2.30)[40]. It

FIG. 2.29 *In order to determine whether bees could discriminate between vertically and horizontally orientated patterns when they were unable to set up a template of the training pattern, patterns were displayed at the ends of the two arms of a Y-maze. The bees had to choose which arm to take before they were near enough to fixate the training pattern. Under these conditions it was found that bees could discriminate very well between vertically and horizontally orientated random stripes whether they were rewarded on* **a** *the vertical orientation, or* **b** *the horizontal orientation.* **c** *Bees can also learn to discriminate between stripes that are orientated at +45° and −45°. Redrawn from van Hateren et al.[45]*

FIG. 2.30 *Bees that have been trained to discriminate between random stripes orientated at +45° and −45°,* **a**, *can use the orientation information they have extracted from these patterns to discriminate between other pairs of patterns,* **b**, **c** *and* **d** *which they have never previously encountered.* Redrawn from Srinivasan[40].

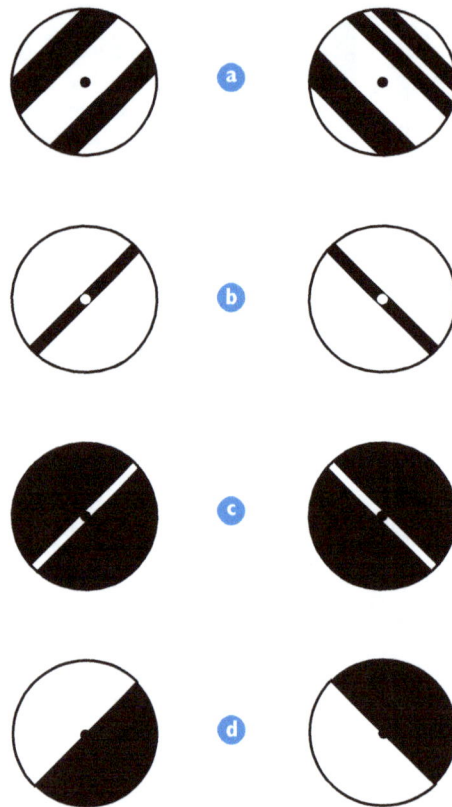

appears that bees do not need to rely on forming detailed templates for pattern recognition under all circumstances, but can abstract and make use of some general properties of shapes in order to make a rapid judgement about some of the many objects encountered in their daily life.

Is anything known about how the bee might extract information about the orientation of a pattern without forming a detailed template? It has been suggested that the flying bee might make use of directional motion cues since it possesses directionally-selective, motion-sensitive neurons (see chapter 7), but appropriately designed training tests have shown that this is not the case[40]. Of great interest is recent work which suggests that bees can analyse the orientation of an object using specific orientation-sensitive channels within its visual pathway. This is a method used by humans and other primates[46]. The photoreceptor cells of the retina of a human or an insect eye essentially signal only 'light on' or 'light off'. A great deal of processing of this basic signal is required to extract information about movement, form, etc. Both types of

eye have neural layers beneath the retina in which this processing begins and it is continued within the visual centres of the brain. Information about different aspects of a visual stimulus is processed in parallel channels within the human visual pathway, for example, motion is processed in one pathway and colour and form in another, and this is also true of the insect visual system. In the human visual cortex, cells are found that are sensitive to the orientation of a line, bar of light or contour in part of the visual field. Such cells form orientation channels; only an object with a particular orientation can stimulate them. Recent work suggests that insects have a rudimentary analysis system of orientation-sensitive channels with which to evaluate form[40]. Bees were trained to distinguish between discs with random vertical and horizontal stripes, such as those seen in figure 2.29, by rewarding at the horizontal stripes. Subsequently, when tested with the horizontally-striped disc and a plain grey disc they chose the horizontally-striped disc, but when the vertically-striped disc was paired with a grey disc they chose the grey disc. This suggests that, during training, they had not only learnt to choose the rewarded disc but had learnt to avoid the unrewarded vertically-striped disc, which implies the participation of at least two orientation channels with different preferred directions, for example, horizontal and vertical directions[40]. However, since bees can be trained to discriminate orientations at +45° and −45°, a third channel with a different preferred orientation becomes necessary to avoid ambiguity in this case. Indeed it has been shown that a minimum of three orientation channels with different preferred directions is essential to discriminate orientation unambiguously[40]. As yet there have been no electrophysiological investigations to look for the neurons that might form these orientation-sensitive channels in the bee, but neurons with the required properties have been found in the third optic ganglion (lobula) of the dragonfly[47].

Current ideas on pattern recognition in bees thus suggest that two processes may be involved. One operates at short range, requires fixation and the formation of a memorized template and allows the bee to evaluate quite precisely whether an object does correspond to a previously learned object. The second process appears to operate at long range, evaluating general features of the object, such as its colour and orientation. This processing mechanism resembles, in some ways, the processing mechanisms employed in the human visual cortex. This process enables the bee to rapidly sort out objects in its visual environment during its daily flights.

Range estimation

While knowledge of local landmarks can enable a bee to return to a particular clump of flowers to fill its honey sac, the bee will normally have to visit not only the flowers at this spot but other similar flowers in the area. It will have to move through patches of flowers and land smoothly without colliding with blossoms. A three-dimensional appreciation of the visual world is essential to the task of foraging. Vertebrates have evolved a number of mechanisms to enable them to extract the third dimension from an image, such as accommodation, or stereoscopic vision, none of which are available to the bee with its compound eye. A bee could estimate the distance of an object by its retinal size. If bees are trained to a feeding site with a conspicuous visual mark in its vicinity, and then the mark is replaced by a larger or a smaller one, the bees will search for the food either too far away or too near, showing that they have learned the distance of the mark on the basis of the visual angle that it subtended on the retina[31]. However, the bee can only use object size to estimate range if it is already familiar with the size of the object. This solution is of no use to a bee moving through unfamiliar territory.

Motion cues have been found to provide the bee's visual world with a third dimension (fig. 2.31). During locomotion, the visual field appears to move across the

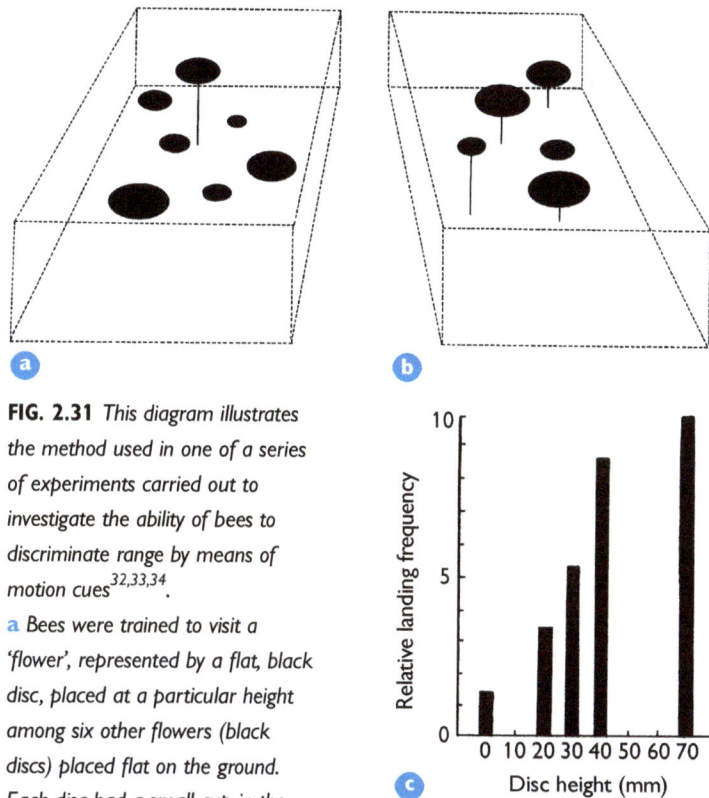

FIG. 2.31 *This diagram illustrates the method used in one of a series of experiments carried out to investigate the ability of bees to discriminate range by means of motion cues[32,33,34].*

a *Bees were trained to visit a 'flower', represented by a flat, black disc, placed at a particular height among six other flowers (black discs) placed flat on the ground. Each disc had a small cup in the centre containing sugar water as a reward in the raised disc and plain water in the remaining discs. The sizes and positions of the flowers were varied randomly between the bee's visits, but the height of the rewarded flower remained unchanged. The 'flowers' were presented in an artificial, white 'meadow', surrounded by a white wall. Under the experimental conditions, the only cue available to the bee enabling it to locate the rewarded 'flower' was the flower's height. Training was terminated after a bee reached a plateau in its learning performance, which normally took around 30 visits. **b** For subsequent discrimination tests, the bees were offered five, clean 'flowers', without reward, each of a different size, placed at a different height. Heights ranged from ground level to the height of the previously rewarded flower. The frequency of the bees' landings on each of the 'flowers' was recorded. **c** The frequency with which the bees landed on each 'flower' was shown to be strictly correlated with the flowers' height. The greatest number of bees landed on the disc placed at the original rewarded height and on the disc nearest in height. The bees were seen to fly mainly in straight lines over flowers so that their visual flow-field was rich in translatory motion components and thus in distance-related cues. The authors concluded from this and other, related experiments that bees can estimate the relative heights of the flowers, irrespective of their size or position.*
After Srinivasan et al.[32]

retina. The contours of a near object move faster and further than those of a more distant object, a phenomenon we can appreciate readily in a moving train or car. Bees have been shown to make use of these cues to discriminate the range of objects as they move about in unknown territory[32,33,34]. It has also been demonstrated that a landing response is elicited from a bee when it perceives a local increase in the speed of image movement, signalling an abrupt decrease in range as it approaches an object.

In addition to all the aspects of vision considered in this chapter, the compound eyes have a further role in extracting and processing information that is used to control the insect's flight course and speed. This function is dealt with in chapter 7.

The insect compound eye is often regarded as a very inferior type of eye, its poor acuity making it of little use to the insect in its everyday life. One has only to think of the honey bee journeying over considerable distances to specific forage sites; the dragonfly patrolling its territory and catching other insects on the wing; and the visually-mediated chasing behaviour of some flies, to realize that this view of the insect eye is very far from the truth. As we have seen, the compound eye, while of necessity constrained in size, is beautifully adapted to the needs of particular insects.

3. The dorsal ocelli:

the bee's second set of eyes

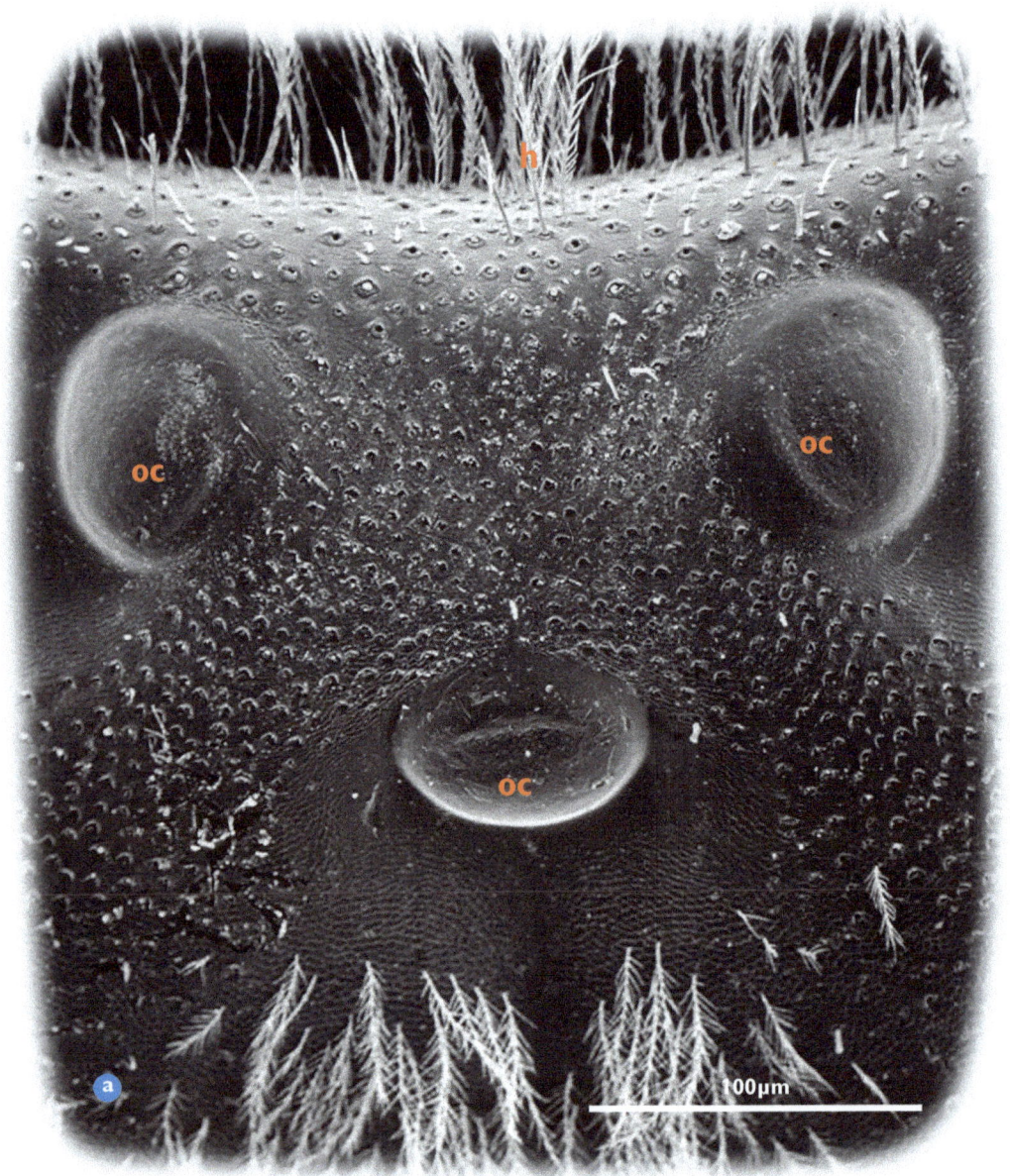

FIG. 3.1 a *The three ocelli (oc) of a worker bee are situated on top of the head capsule on the region known as the vertex. Hairs (h) covering the head have been partially removed to reveal the dome-shaped ocellar lenses.* **b** *Frontal view of the head of a worker showing the position of the three ocelli in relation to other features of the head.* Vertex, antennae, clypeus, frons, genae.

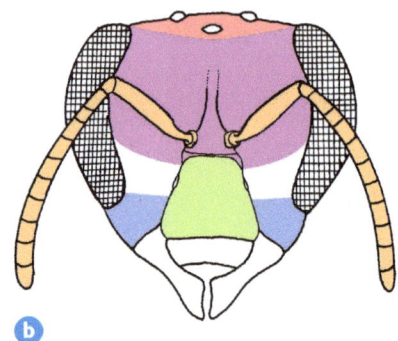

The simple eyes or dorsal ocelli

In addition to its well developed compound eyes, the bee also has three very small eyes, the **dorsal ocelli**, situated in a triangle on the top of the head in the worker and the queen, and more anteriorly in the drone (figs 3.1, 3.2). Why do bees, and other insects, need a second set of eyes as well as their well developed compound eyes? This question has puzzled scientists for many years and although we now have some idea of the functions of the dorsal ocelli, the story is by no means complete[1].

What clues do behavioural studies offer on the role of the three ocelli? The ocelli have a subtle effect on much visually mediated behaviour. Most of the bee's behaviour that requires visual input cannot be performed if the compound eyes are covered and only the ocelli are operating: for example, normal flight is impossible. Nevertheless, if the compound eyes are intact and the ocelli are painted over, many of these behaviours are affected in some way. Bees whose ocelli have been covered begin foraging later in the day and cease earlier, behaving as though they were less sensitive to light. Furthermore, orientation responses are not quite as accurate when the ocelli are occluded[2,3]. Many authors have noted that well-developed ocelli are present in strong fliers whereas they are poorly-developed or absent in weak or non-flying insects, and have suggested some role in flight behaviour for the ocelli[4].

Examination of the structure and physiology of the ocelli gives some clue as to their function. The structure of the simple eye is quite unlike that of the compound eye. Each ocellus consists of a single lens overlying a retina composed of a layer of photoreceptor cells, around 800 in the bee (fig. 3.3). When we look closely at the structure of the ocellus, it seems that it is designed for the rapid detection of changes in light intensity summed over the whole visual

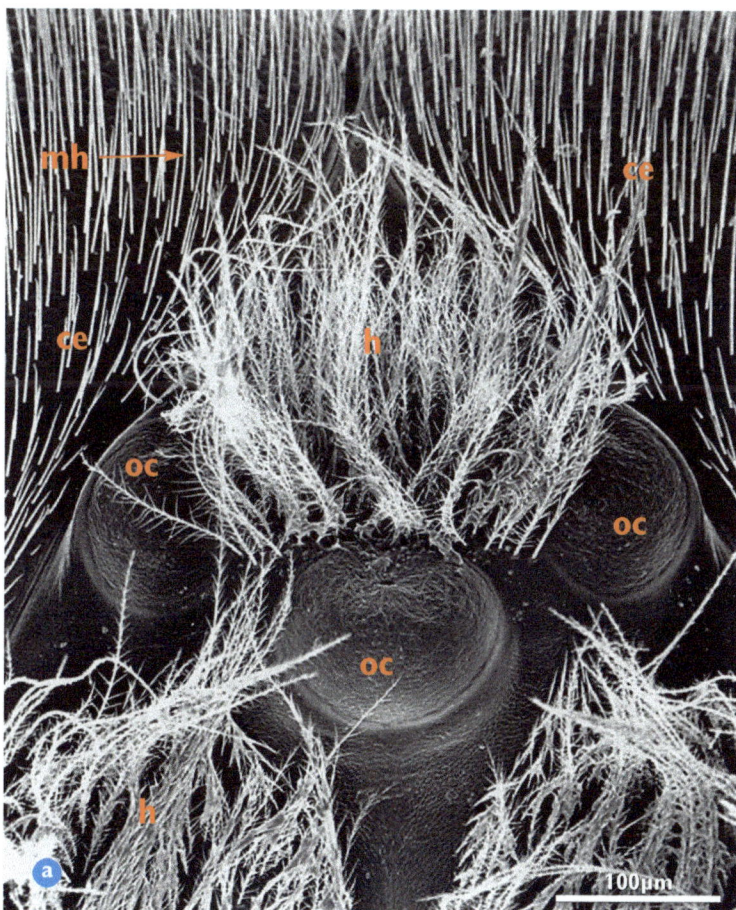

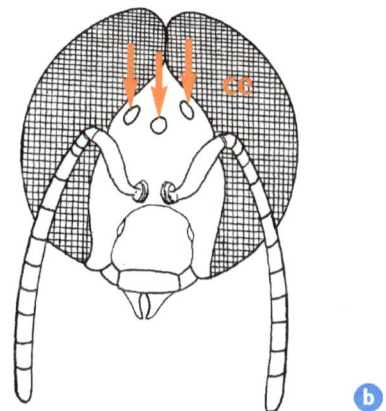

FIG. 3.2 a *The three ocelli (oc) of the drone form a triangle facing forwards with the large compound eyes (ce) meeting above them. Note the hairs (h) covering the surface of the head capsule and partially obscuring the lateral ocelli, and the mechanoreceptive hairs (mh) sensitive to touch, which cover the surface of the compound eye.*
b *Frontal view of the head of a drone showing the enormous development of the compound eyes (ce) which meet over the top of the head. The ocelli (arrows) are displaced forward.*

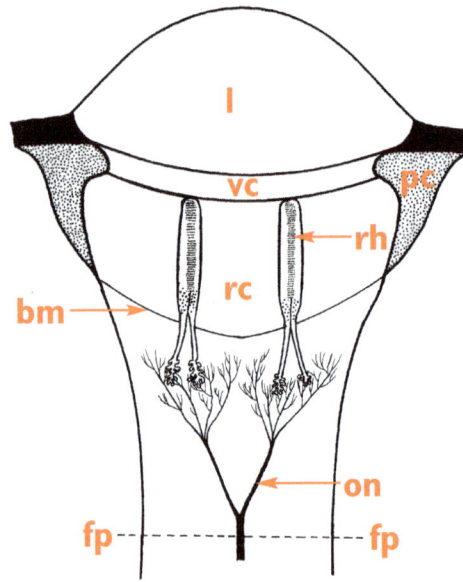

FIG. 3.3 *Longitudinal section through the median ocellus of the worker showing the lens (l) overlying a transparent layer of vitreous cells (vc) and the layer of retinal cells (rc). Two pairs of retinal cells are shown in this layer. The inner border of each cell is lined with microvilli forming the cell's rhabdomere (see fig. 3.4b, rhm). The rhabdomeres of the two cells are fused together to form the rhabdom (rh). Beneath the basement membrane (bm) the axon of each retinal cell descends to make synaptic contact with the ocellar neurons (on). These neurons run into the brain; some of them run through the brain into the thorax. Only one of the large neurons is shown here. The dotted line shows the focal plane (fp) of the lens which lies well below the layer of retinal cells. A ring of pigment cells (pc) surrounds the ocellus just below the lens forming an iris, and in bright light these cells migrate beneath the lens stopping down the aperture.*

field. This is a very different role from that of the compound eye, or indeed the vertebrate eye, which are both designed for the reception and analysis of spatial information. The ocellus has high-aperture dioptrics with the lens having a wide visual field. Investigators were surprised at first to find that the ocellar lens does not focus images on the retinal layer, since the focal plane of the lens lies beneath the retina in the ocellar nerve[5] (fig. 3.3). The ocellar lens thus appears to act as a condenser rather than an image-forming device, gathering light from a wide area and spreading it evenly over the receptor layer. The ocellus appears to be designed for sensitivity at

FIG. 3.4 a *Longitudinal section through a pair of cells from the ocellar retina (the distal region of the cell has been shortened in relation to the proximal region). The membrane of the inner border of each cell is modified to form microvilli (see fig. 3.4b). The modified border is known as the rhabdomere (rhm). The opposed rhabdomeres are fused to form the rhabdom (rh). Dark pigment granules (pg) are found below the rhabdom region helping to screen the visual cells from stray light. Mitochondria (mi), providing energy for the cell's activities, are present in large numbers; cisternae (ci), large vesicles in the cell cytoplasm; cell nucleus (n). **b** Transmission electron micrograph of the opposed borders of the two cells showing the microvilli (mv) of each cell. The visual pigment is carried in the walls of the microvilli. Together the two sets of microvilli form the rhabdom (rh, white bar). Rhabdomere (rhm).* Photo: courtesy of K C Pan.

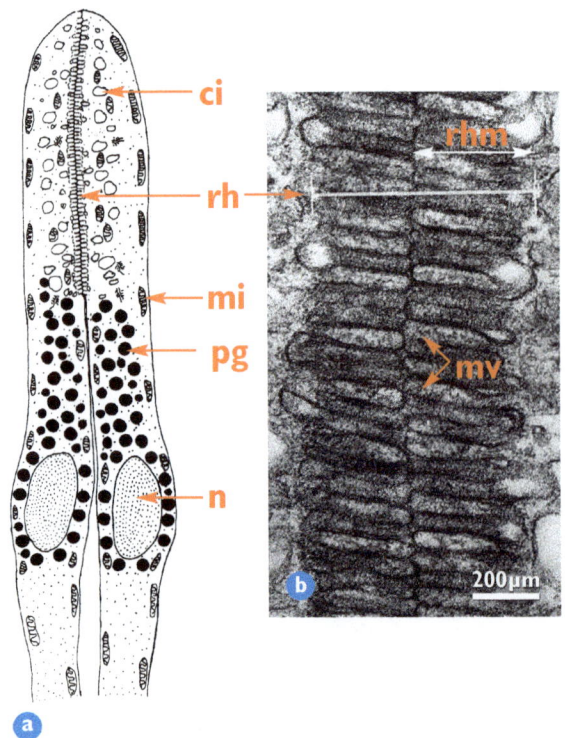

the expense of acuity. Beneath the lens, the receptor cells are not arranged in an ordered array as in the compound eye, where the rhabdomeres are presented as a mosaic of light guides for axial light. Instead, the cells are irregularly packed and their rhabdomeres form a network of baffles, a device which maximizes their ability to trap light spread diffusely through the retina, thus ensuring optimum stimulation (fig. 3.4)[6].

In some insects, particularly those that are active at night, the light-gathering ability of the ocellus is increased by a layer of granules at the back of the retina, the tapetal layer, which reflects any unabsorbed light back through the retinal cells a second time[3]. There is no tapetal layer in the bee, an insect that is active outside the hive only in daylight, but there are mechanisms for adjusting the sensitivity of the ocellus. The amount of light caught by the photo-receptor cells can be regulated in one of two ways: there is a collar of black pigment between the lens and the retina which forms an iris, stopping down the ocellar aperture under bright light conditions. In addition, the photoreceptor cells contain black pigment granules which normally lie at the inner (proximal) end of the cells. In conditions of very bright light, the pigment granules may migrate outwards along the cells, restricting the passage of light through them and hence reducing stimulation[1]. The ocellus is thus able to remain sensitive to small changes in light intensity over a very wide range. When examined for spectral sensitivity, the ocelli show peaks in the ultraviolet region, between 340 and 370 nm depending on the insect, and in the green region of the spectrum between 490 and 520 nm[6]. It appears that the ocellar retina contains two types of visual cell, UV- and green-sensitive cells.

When the neural wiring beneath the retinal cells is examined, the ocellus is found to contain two types of neuron: a small number of very large neurons, the L neurons, the largest neurons in the insect brain (fig. 3.5); and a larger number of small neurons, around 80 in the bee. The

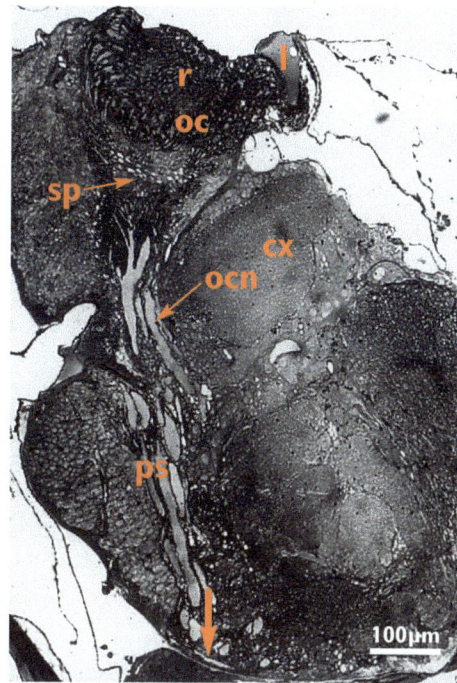

three ocelli of the bee together are innervated by just 30 large neurons, 20 of which terminate in the brain and 10 of which project down to the thoracic motor centres[7]. The input from large numbers of visual cells converge on each of the L neurons. This high degree of convergence of visual input is another device by which the sensitivity of the system is increased. Recordings from L neurons in locusts have found them to be around 5000 times as sensitive as neurons at an equivalent point in the compound eye[6]. The response characteristics of the large ocellar neurons make them particularly sensitive to rapid changes in intensity over the whole visual surround while being relatively insensitive to slow or small local changes in intensity. They are also especially sensitive to the UV part of the spectrum, possibly receiving a greater proportion of their input from the UV-sensitive visual cells[6]. Speed of transmission in insect neurons is related to the diameter of the neuron, which means that the large ocellar fibres are some of the fastest conducting neurons in the bee. Since some of them descend directly to the thoracic motor centres, it follows that sudden changes in intensity over the whole visual surround can be signalled rapidly

FIG. 3.5 *Sagittal section through the brain including one ocellus (oc) and part of the ocellar nerve (ocn) to show the conspicuously large ocellar neurons. Some of these neurons branch and make synaptic contacts with other neurons within the brain, largely in the posterior slope area (ps) where there is extensive input from the compound eyes. The remaining large ocellar neurons, while also branching in the posterior slope area, continue on through the brain and down into the thoracic motor centres (arrow). Calyx (cx), part of ocellar lens (l); retina of the ocellus (r); synaptic plexus (sp) beneath the ocellus where the visual cells synapse with both large and small ocellar neurons.* Photo: courtesy of K C Pan.

FIG. 3.6 *Photographs of the same scene taken with one camera on fine-grain release film, ASA 5–6. The sun was behind the camera;* **a** *was taken through a filter that transmitted only light in the near ultraviolet range; and* **b** *through a filter transmitting light in the blue-green range. The scene viewed by UV shows a marked contrast between earth and sky.* Photos: courtesy of J D Pye.

to the centres that control locomotory activity.

When a bee is flying with its head tilted at its normal flight angle of 14°, the median ocellus is directed towards the horizon. Insects tend to detect instability in flight by comparing current sensory input to that expected for normal flight. The horizon is the most ubiquitous horizontal reference, so any apparent motion of the horizon relative to the ocellus, such as that during downward pitch when the horizon appears to move upwards over the eye, will be readily detectable by the UV-sensitive L neurons. The lateral ocelli will be differentially stimulated if the bee rolls about its longitudinal axis or if it rotates about both its longitudinal and vertical axes as in banking manoeuvres. The sky contains a high proportion of UV light, so in the region of 370 nm, the contrast between the radiant sky and the absorbent earth will be at its maximum. The photographs of the same scene taken through UV and green filters (fig. 3.6) show the high degree of contrast between sky and earth when viewed through the filter that transmits UV light. Thus, the physiological properties of the ocelli and their associated L neurons suggest that they have a role to play in maintaining stability during flight. However, they are not solely responsible for controlling stability. The compound eyes and other sense organs have been shown to be involved in the control of pitch, roll and yaw. Neurons running from the brain to the thoracic motor centres carrying information from the compound eyes concerning apparent movement of the insect's surround due to pitch and roll, have been found to carry information from the ocelli as well[8,9]. Ocellar input reinforces that received from the compound eye if compound eye and ocellar stimulation both indicate movement in the same direction. If there is an apparent conflict in the information provided by the two systems then the signal in the neurons descending to the motor centres is reduced[10,11]. Activity in these neurons ultimately results in the correctional muscle activity necessary to restore the insect's stability. Ocellar activity appears to be particularly well developed in the bee, where there is a separate neural channel carrying ocellar information directly from the ocellus to the thorax. Thus the ocelli, with their ability to detect and swiftly signal any sudden changes in the position of the horizon, probably serve here as a rapid response system for pitching or rolling course deviation. This is followed by the more precise control exerted by the compound eyes. Yaw, since it involves no movement relative to the horizon, will not be as readily detectable by the ocelli.

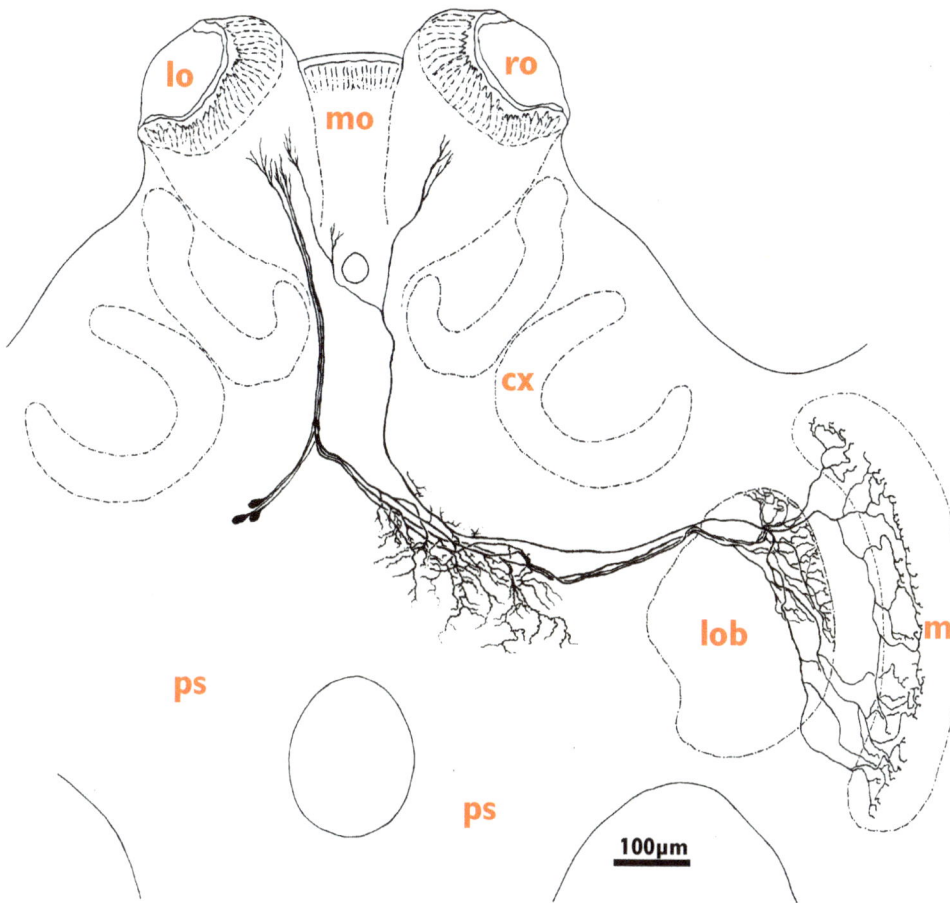

FIG. 3.7 *The small neuron system of the ocellus is very well developed in the honey bee. The pathways of the three neurons shown here are of particular interest. The neurons run between the synaptic plexus beneath each ocellus and the medulla (m) and the lobula (lob), two of the ganglia which underlie the compound eye. These ganglia process the visual information passing from the compound eye to the brain. The small ocellar neurons also branch and make synaptic contact with other neurons in the posterior slope (ps), an area at the back of the brain which receives input from the compound eyes. Thus, pathways exist that would make possible some form of interaction between ocellar and compound eye inputs to the brain. Left (lo), median (mo) and right (ro) ocelli; calyx (cx).* Photo: courtesy of K C Pan.

It is probable that the ocelli have other functions besides their involvement in controlling flight stability. The role of the many small neurons is not known, but they project to many other areas of the insect brain, including antennal centres. In the bee, the projections of these neurons to the optic ganglia underlying the compound eye are particularly well developed[1,7] (fig. 3.7), suggesting that some interaction between the two visual systems is taking place. What form this interaction might take is unknown, but it might involve ocellar illumination gating activity in certain compound eye pathways. This could explain why some behaviours mediated only by stimulation of the compound eyes are nevertheless affected by ocellar illumination.

One role proposed for the ocelli is that they determine the degree of coupling of the orientating subsystems, for example, the systems controlling flight course and flight speed. It has been suggested that coupling between the orientating systems is flexible and depends upon the context of the situation and the internal state of the bee, with the ocelli regulating coupling. For instance, foraging bees exhibit a tighter coupling between course and flight speed when they are fully sighted than when the ocelli are blinded[12].

4. The bee's response to gravity:

which way is up?

Gravity is a natural stimulus that forms an important reference system for orientation and the control of posture and equilibrium in animals (fig. 4.1). Most animals have a specialized receptor organ for detecting gravitational force consisting of a fluid-filled cavity, lined with sensory cells and containing a movable heavy body. One of the simplest examples is that of the scallop, which possesses simple sacs (statocyst organs) lined with sensory hairs, each sac containing an aggregation of calcareous material (the statolith or 'still stone'). As the scallop changes its position, the statolith shifts about and stimulates different hairs on the lining of the cavity, thereby providing information on the orientation of the animal with respect to gravity. More complex organs are found in crabs, lobsters and crayfish[1] (fig. 4.2). Here, there are at least two types of hair present, each with different response characteristics, allowing the animals to extract both static and dynamic positional information. Some of the sensory hairs present adapt very slowly when they are bent by the aggregations of sand grains that form the statolith, yielding static positional information. Other hairs appear to be unaffected by the shearing force of the statolith and instead are deflected by movement of the fluid in the statocyst. These hairs adapt rapidly after bending and signal information about the animal's rotational acceleration. An animal can control its position using only static positional information but control would be clumsy, since positional correction would be made with reference to a position that the animal had already obtained. If the statocyst contains dynamic receptors registering the rate of change of position, then it is possible to predict what the animal's position will be when the correction

becomes effective, and to apply corrections appropriate to that future time. This produces a much smoother and more accurate control of the animal's position in space.

The vertebrate organ of equilibrium, the labyrinth, employs the same principles but is more complex in structure. In humans, the labyrinth or vestibular apparatus is situated adjacent to each ear, within the bony part of the head (fig. 4.3). Part of the labyrinth is analogous to the statocysts of invertebrate animals and responds to gravitational force. This region comprises two chambers, the utriculus and the sacculus, and part of the epithelial wall of each chamber contains sensory cells bearing hair-like projections or cilia. The hair cells are embedded in a jelly-like substance which contains small particles of calcium carbonate, or otoliths, and the gelatinous

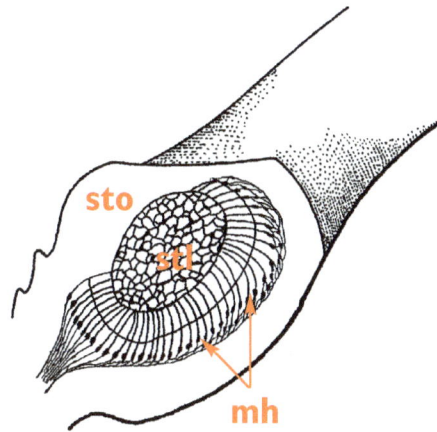

FIG. 4.2 *Each antennule of a lobster contains a statocyst organ (sto). This comprises a hollow cavity lined with sensory hairs (mh) containing a statolith (stl) formed by aggregations of sand grains.*

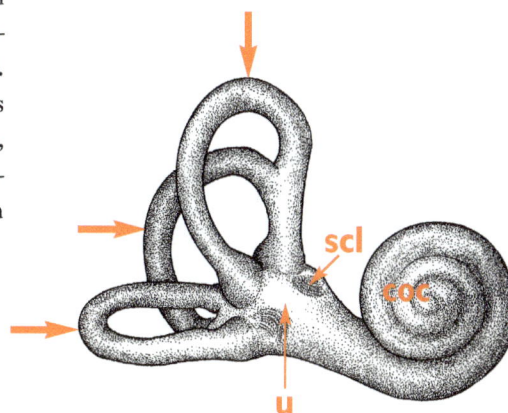

FIG. 4.1 (opposite) *Gravity is a natural stimulus that forms an important reference system for orientation and the control of posture and equilibrium in bees as in other animals. Bees are able to perform the wide range of activities involved in maintaining the colony while moving freely on the vertical combs within the dark hive.*

FIG. 4.3 *The vestibular apparatus of humans, responsible for signalling the position and movements of the head. Associated with each cochlea (coc) is a bony structure containing three semi-circular canals (arrows) and two otolith organs, the utriculus (u) and the sacculus (scl).*

mass exerts a force on the hair cells (fig. 4.4). This region is known as the macula and, with the head in the normal position, the macula of the utriculus lies approximately in the horizontal plane, while that in the sacculus lies in the vertical plane. The hair cells display directional selectivity in their response. The unstimulated cell has a constant low level of discharge of nerve impulses in its axon, known as the 'resting discharge'. The hair-like projections on the apex of each cell increase in length across the cell: bending of the hair bundle towards the longest hair leads to depolarization of the cell and an increase in the discharge rate of nerve impulses in its axon (fig. 4.4). Bending the hairs in the opposite direction leads to a decrease in

the rate of firing of the axon below the resting discharge. The cells in the macula of the utriculus do not have their longest hairs all pointing in the same direction, so that a tilt of the head in any direction will stimulate some of the cells and inhibit others, which will provide the brain with enough information to make an accurate measure of head position. These hair cells yield information on static equilibrium, i.e. the orientation of the body relative to the ground and, in conjunction with other sensory inputs such as muscle stretch receptors, they are involved in the control of posture and the positioning of the head.

In addition to these two chambers, there are three semicircular canals on each side of the head, whose receptors perceive

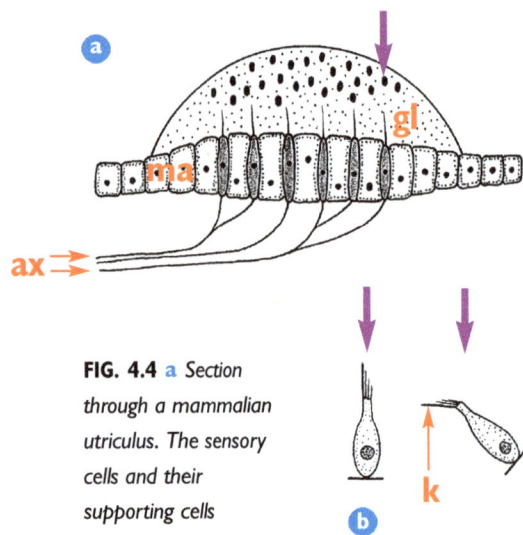

FIG. 4.4 a Section through a mammalian utriculus. The sensory cells and their supporting cells together form the macula (ma). The hair-like projections of the sensory cells are embedded in a gelatinous structure (gl), containing dense particles or otoliths (arrow); cell axons (ax). Gravity will exert more pull on the heavy otolith particles than on anything else within the organ, and the gelatinous mass in which they are embedded will be shifted in the direction of gravity, thus bending the embedded hairs. b Each cell has a number of hair-like projections or cilia at its apex, and these vary in height across the cell. If the hair bundle is bent towards the longest hair, or kinocilium (k), the cell is excited, if the hairs are bent away from it, the cell is inhibited. Arrows indicate the direction of gravitational force. The preferred directions of the sensory cells vary so that a tilt of the head in any direction will stimulate some cells and inhibit others.

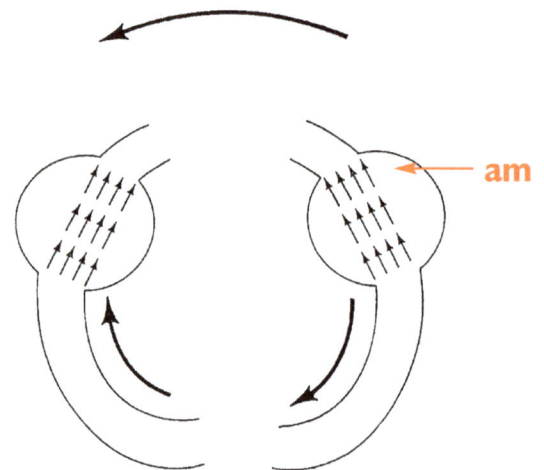

FIG. 4.5 Section through part of the left and right horizontal semicircular canals in humans. The sensory cells are located in the ampulla (am). The small arrows indicate the directional sensitivity of the individual cells. If the head is rotated to the left (top arrow), inertia will cause the fluid in the left canal to push the cupula, in which the projections of the sensory cells are embedded, to the right. Since this is their preferred direction, they will be stimulated. The cells in the right canal will be bent towards the direction in which they are least sensitive and their resting discharge will be inhibited. Lower arrows indicate direction of movement of fluid. When the head stops moving, inertia will cause the fluid in each canal to bend the cupula in the opposite direction, thus inhibiting the cells on the left side and stimulating those on the right. After Kelly[13].

rotational acceleration in the plane of each canal and function as receptors of dynamic equilibrium. Because the canals are set in three different directions, at right angles to each other, the system as a whole can detect motion in three-dimensional space. The sensory cells are located in a swelling at the end of each canal, the ampulla, and extend their sensory hairs, or cilia, into the cavity of the canal where they are covered with a gelatinous cap, the cupula, which acts as a 'swinging door'. Unlike the cells of the utriculus, these cells all have the same directional polarity. They are stimulated by the movement of the fluid in the canals when the head moves. As the head is rotated, the fluid in the canal will lag behind the turning motion because of inertia. The fluid will push against the cupula, bending it in the direction opposite to that in which the head was rotated, and in so doing, the hair cells embedded in the cupula will be stimulated. The directional sensitivity of the hair cells is such that, for example, in the horizontal canals, the hair cells in the left ampulla will be maximally stimulated when the head rotates to the left while those in the right ampulla will be inhibited (fig. 4.5). As the rotation continues and the fluid takes up the same rate of movement as the canal, the cupula returns to its original resting position and the hair cells to their rate of resting discharge. When the head ceases to move, the fluid will once again lag behind the motion of the canal and will again bend the cupula, only this time in the opposite direction, thus inhibiting the cells in the left ampulla and stimulating those in the right.

Insect gravity receptors

Unlike most animals, insects (with the exception of a few aquatic forms) do not possess statocyst organs designed specifically for the detection of gravitational force. However, perception of gravity is just as essential for insects as for other animals. In the honey bee, a precise response to gravity is critical since, in its waggle dance on the vertical comb, it

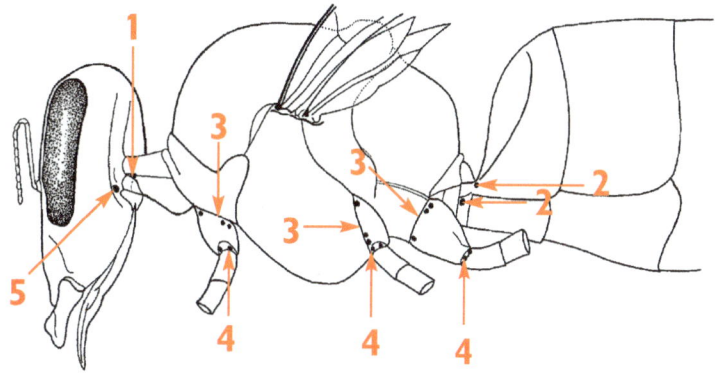

FIG. 4.6 *The hair plates (black dots) of the proprioceptive gravity receptor system of the bee include those on the episternal cone in the neck region (arrow 1), the petiole plates at the junction of the thorax and abdomen (arrows 2), and the plates at the coxal joints of the legs where they articulate with the thorax (arrows 3). The plates on the trochanter segments of the leg are also shown (arrows 4) although they are not so important in gravity reception. There is a pair of plates on the back of the head (arrow 5) whose function is not clearly understood.* After Markl[7].

transposes the angle of the food source with respect to the sun to the angle which the straight run of the dance makes with respect to gravity. Insects use cuticular mechanoreceptive hairs, gathered into **hair plates,** for gravity reception. The way in which the hairs function as mechanoreceptors has been described in chapter 1.

The hair plates concerned with gravity reception are stimulated by displacement of one part of the body relative to another. Mechanoreceptors that signal such displacement are known as **proprioceptors** (self-perception receptors). Other types of proprioceptor include campaniform sensillae, which detect stresses in the cuticle, for example, when the legs are bent, and stretch receptors within the muscles registering muscle tension. The proprioceptive gravity receptor system of insects (the PGR system) consists of hair plates on the neck, the abdomen, the legs and the antennae[3]. The relative importance of the plates in the perception of gravity differs with the insect. In the honey bee, the neck hair plates are essential, the abdominal plates and the leg plates make some contribution but the antennal plates are not involved (fig. 4.6).

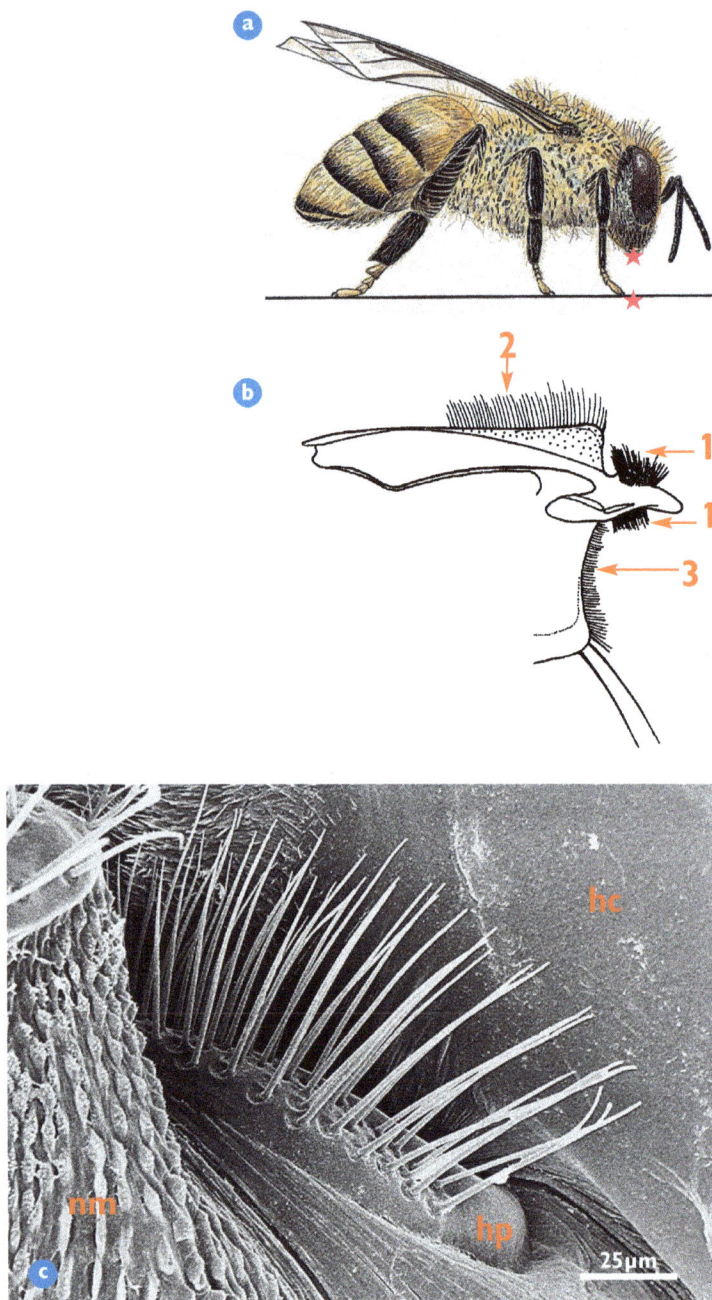

FIG. 4.7 a *When the bee is standing on a horizontal surface, the head hangs straight down (see* stars*) and the hairs around the episternal processes are in even contact with the head.* b *One of the episternal processes of the neck viewed from the inner surface. A hair plate (arrows 1) runs around the outside of the peg. There are additional hair plates on the forward upper edge of the thorax (arrow 2) and on the lateral edges of the thorax (arrow 3), but the episternal hair plates (arrows 1) are the effective organs of gravity perception.* c *Part of the hair plate (hp) on the episternal process at the right side of the neck, viewed laterally. Head capsule (hc), neck membrane (nm).*

How does the PGR system work?

Each hair plate measures only the position of a particular joint and movement about that joint. The central nervous system of the insect must be able to distinguish between the action of gravity, and other mechanical forces acting on a joint, for example, movement of a foreleg when cleaning an antenna. The presence of several hair plates means that the central nervous system can compare the input from plates at different joints. Only equidirectional change of position at several joints, such as that produced by standing on an inclined surface, would be related to gravity; a change about a single joint would be interpreted as movement independent of gravity. Experimental work on ants indicates that the central nervous system is capable of this discrimination[3,7]. Some authors believe that information from other proprioceptors, such as the campaniform sensillae and the stretch receptors of the muscles, plays some part in gravity perception, since they register the strength of stresses imposed around the joints by changes of posture[3]. Whether or not this is the case, experimental removal of the input from the hair plates shows that they play an essential role in the perception of gravity in the Hymenoptera.

The neck hair plates

The head of the bee is suspended from the thorax on two cuticular pegs (**episternal processes**) which extend forward from the thorax into a depression (occipital foramen) on the back of the head (figs 4.7, 4.8). There is a hair plate around the outside of each of these pegs, placed in such a way that when the bee is standing on a horizontal surface, the hairs of both hair plates are in even contact with the head capsule[4,5]. The head can be rotated about its transverse axis, pivoting on the cuticular pegs. The centre of gravity of the head is below the point of articulation with the thorax, so that when the bee crawls up a vertical surface the head falls downwards and the ventrally situated sensory hairs of the peg hair plates are bent more

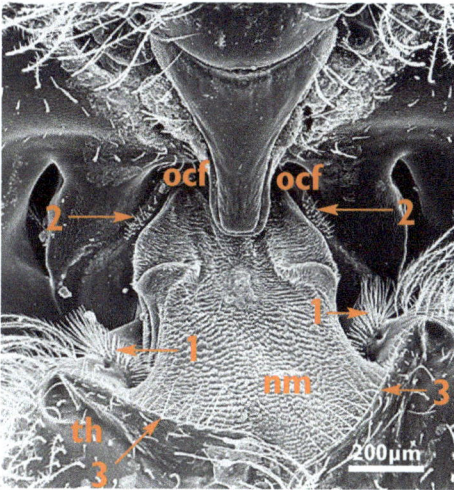

FIG. 4.8 *Ventral view of the head, unsclerotized neck membrane and thorax of a worker bee. The head has been pulled forward to show the ventral and lateral regions of the episternal hair plates (arrows 1). Hair plates on either side of the occipital foramen (ocf), where the neck joins the head, may also contribute to the PGR system of the bee (arrows 2). Mechanoreceptive hairs (arrow 3) line the ventral surface of the thorax (th) and are deflected by movements of the neck membrane (nm), but it is not known whether they contribute to the PGR system.*

the right plate will continue, gradually spreading from the lateral area to the dorsal area of the plate as the insect turns. As the insect approaches the 'vertical-down' position, the maximum shearing pressure will begin to be distributed evenly again over the dorsal region of the plates on both sides of the neck[5].

The individual hairs comprising the hair plates are set in their sockets in such a way that they are preferentially bent in one direction, giving them a directional sensitivity (figs 4.7, 4.16, see also chapter 1). The directional sensitivities of the hairs are distributed in different ways in the neck hair plates, so that certain hairs are bent by particular movements of the head. The latency of response to bending is very short, and the frequency of nerve impulses fired in response to bending increases with the degree of bending. The response of the hair never completely adapts while the hair is being bent[6]. The structural and physiological characteristics of the hairs thus enable them to rapidly register the precise position of the head relative to the thorax.

strongly (fig. 4.9a)[5]. When the bee moves downwards, the heavier, ventral part of the head hangs down, tilting the dorsal half of the head backwards against the thorax: in this position, the dorsal sensory hairs of the hair plates are stimulated (fig. 4.9b). Additional hair plates, situated just behind the pegs on the episternal region of the thorax, may also be stimulated by head movements but it is not clear whether they play any part in gravity reception[7]. If the bee is facing upward on the comb and turns to run across it at an angle, then the shearing action on the mechanoreceptive hairs on each side of the head will become asymmetrical. A clockwise turn will mean that the weight of the head will gradually press more and more on the hairs of the right hair plate and the pressure on the hairs of the left plate will be much reduced or cease as the bee approaches a transverse position across the comb. When the bee turns further towards the head-down position the shearing action on the hairs of

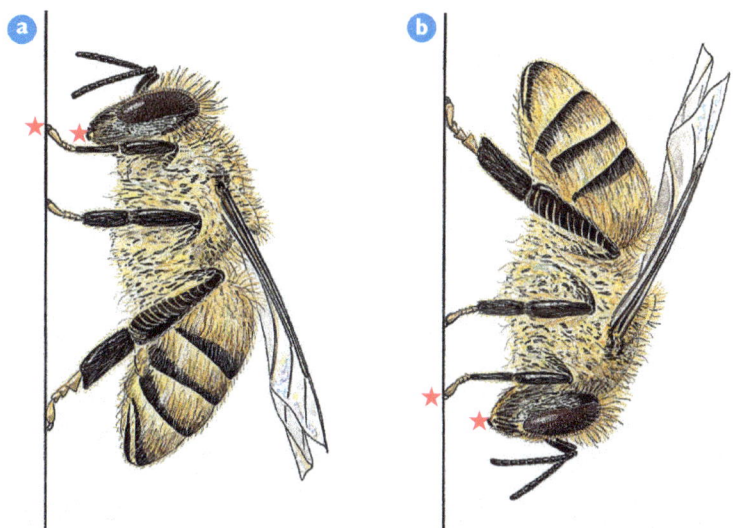

FIG. 4.9 a *The centre of gravity of the head is ventral to the episternal processes. When the bee walks upward, the head falls downwards in relation to the position it would occupy on a horizontal surface (see stars) and the ventrally situated hairs of the episternal plates are stimulated.* **b** *When the bee walks downwards, the heavier ventral part of the head hangs down (see stars), tilting the dorsal region of the head backward, and stimulating the dorsal hairs of the episternal hair plates.*

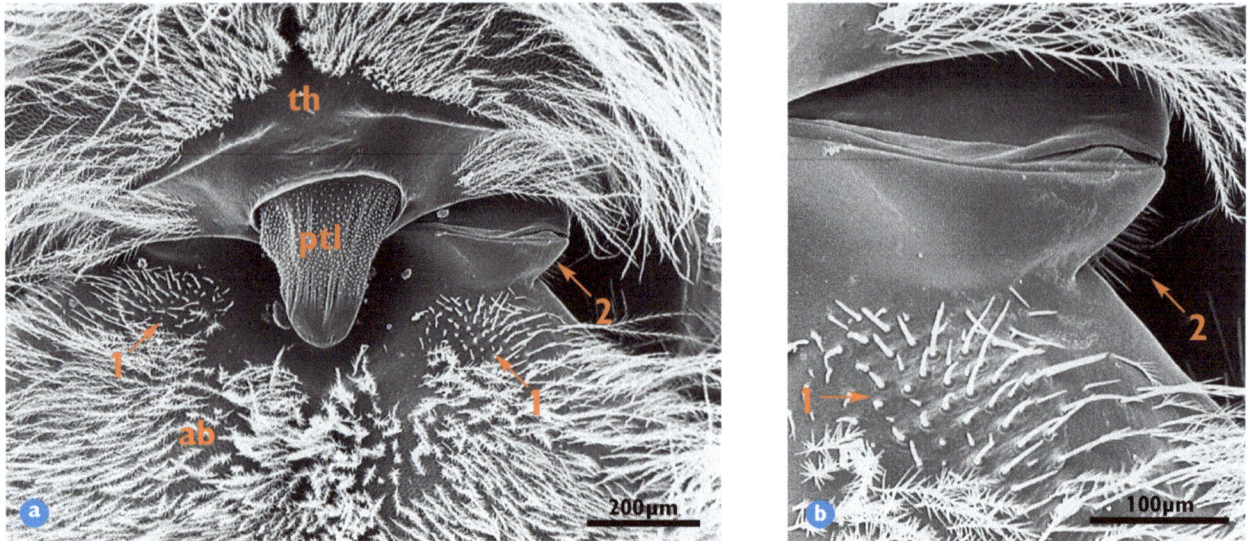

FIG. 4.10 a *The petiole hair plates. The thorax (th) and the abdomen (ab) are pulled apart here, stretching the petiole (ptl) and revealing the two dorsal hair plates (arrows 1) on the abdomen. The thorax is bent slightly to the right, revealing the site of the right lateral hair plate (arrow 2).* **b** *An enlargement of the right side shows the lateral hair plate (arrow 2) just coming into view. Dorsal hair plate (arrow 1).*

The petiole hair plates

At the point where the thorax and the abdomen join, there are two pairs of hair plates that register the movement of the abdomen relative to the thorax. One pair is situated dorsally, the other laterally (figs 4.10a, b, 4.11, 4.12). When the bee is on a horizontal surface, the abdomen hangs downwards and pushes the mechanoreceptive hairs of the lateral hair plates under a fold of cuticle in the petiole, pro-

ducing an equal shearing force on each side. If the bee moves onto a vertical surface and runs down, the weight of the abdomen presses the hairs of the dorsal hair plates against the thorax, while those of the lateral plates are left free[5]. If it turns and runs vertically upwards, the abdomen hangs down and neither the dorsal nor the lateral hair plates are stimulated. When moving at an angle across a vertical surface, the weight of the abdomen causes it

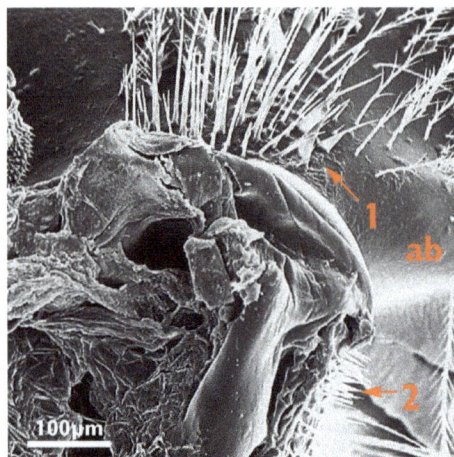

FIG. 4.11 *The left dorsal hair plate (arrow 1) and the left lateral hair plate (arrow 2) can be seen together when the thorax is removed at the petiole region and the abdomen (ab) is viewed from the front.*

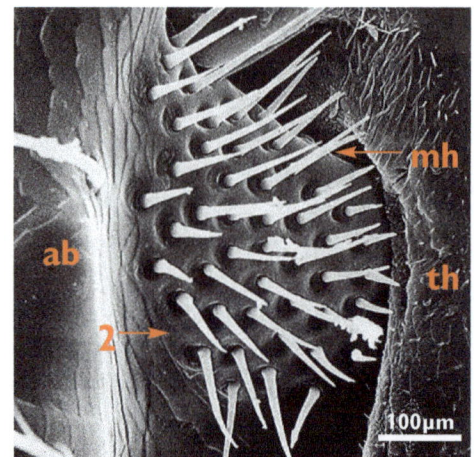

FIG. 4.12 *Right lateral hair plate (arrow 2) viewed from the side with the thorax (th) in position, showing how the mechanoreceptive hairs (mh) engage with the thorax and are bent by the relative movement of the thorax and abdomen (ab).*

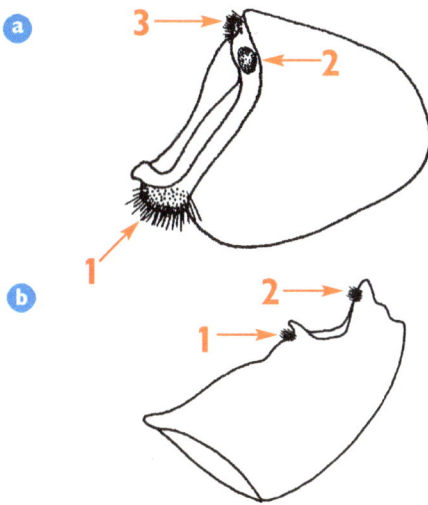

FIG. 4.13 a *Diagram showing the positions of the three hair plates (arrows 1, 2, 3) on the upper rim of the coxa of the first leg at the point where it articulates with the thorax. Movements of the coxa relative to the thorax cause deflection of the mechanoreceptive hairs.* **b** *The positions of the two hair plates (arrows 1 and 2) of the second segment of the leg, the trochanter. Relative movement of the coxa and trochanter results in stimulation of these hair plates. The second and third pairs of legs have similar coxal and trochanter hair plates.*

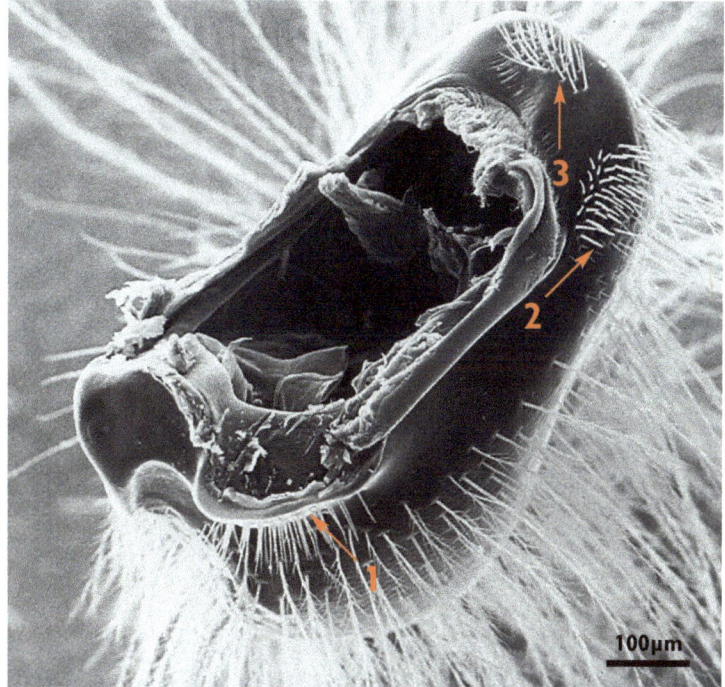

FIG. 4.14 *The coxa of the first leg has been removed from the articulation point with the thorax, revealing its upper rim. First hair plate (arrow 1), second hair plate (arrow 2), third hair plate (arrow 3).*

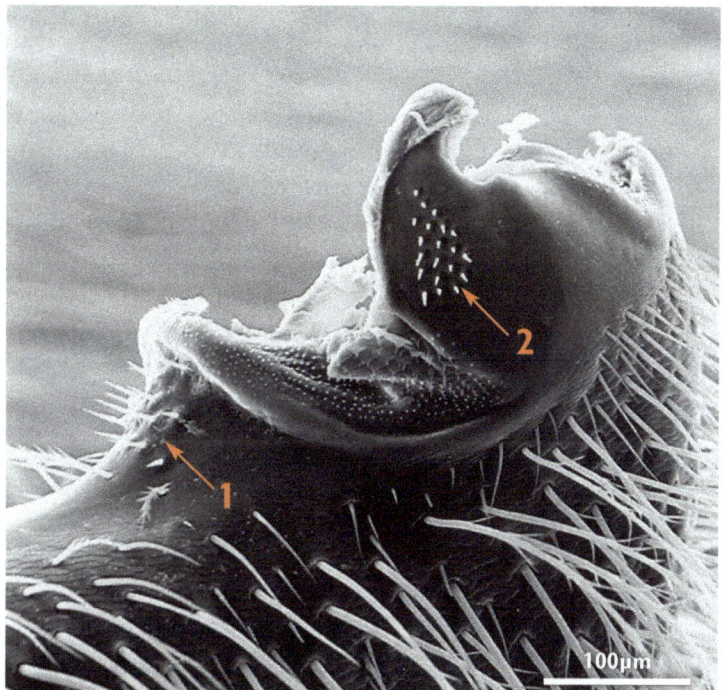

FIG. 4.15 *The trochanter of the first leg has been removed from its articulation with the coxa showing the first and second hair plates (arrows 1 and 2).*

to hang downwards and thus be bent to one side relative to the thorax, causing asymmetrical stimulation of the hair plates.

The leg hair plates

Small hair plates are also found on the legs of the bee[7]. The basal segment of the leg, the **coxa,** bears three plates on the rim where it articulates with the thorax (figs 4.13a, 4.14, 4.16). Two hair plates are also found at the joint between the coxa and the second segment of the leg, the **trochanter** (figs 4.13b, 4.15, 4.17). These hair plates are so small that they are only seen when the segments are separated from their articulation points. The coxal plates have been shown to make some contribution to gravity perception in the bee as the removal of the coxal plates from the legs on one side causes the head to be shifted from its customary position and

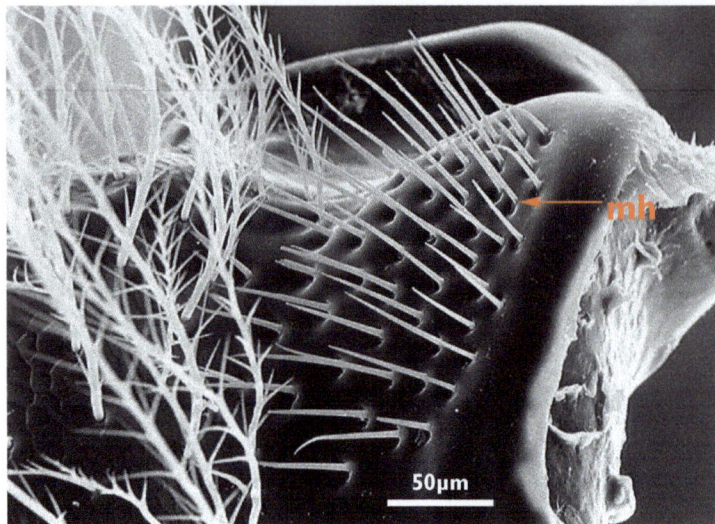

FIG. 4.16 *Hair plate 1 on the first coxa. Note that the shaft of each mechanoreceptive hair (mh) is set asymmetrically in its socket making the hairs directionally sensitive.*

disrupts the normal reflex movements of the head during walking[7].

By examining the bee's behaviour after shaving off the hairs of individual hair plates, cutting the nerves that innervate them, attaching weights to the head, thorax or abdomen, or cementing joints in abnormal positions, it has been possible to determine the relative contribution of the different plates[3,12]. The neck hair plates of the bee were found to be the most important organs for gravity perception: if the input from the neck hair plates is disconnected, there is considerable impairment in the bee's response to gravity and in its ability to orientate communication dances correctly[4,5].

Insects placed on a vertical surface in the dark commonly show an orientated response to the field of gravity, a **geotactic response.** An insect may show negative geotaxis, walking upwards on a vertical surface, or it may walk downwards, showing positive geotaxis. The angle which it makes with the vertical can vary. In honey bees, except for *Apis florea*, the angle of orientation taken by a bee on a surface is independent of the strength of gravity, i.e. of the inclination of the surface: this behaviour eliminates the effect of a variable inclination of the comb on a dancing honey

bee[3,8]. This independence is effected by the interaction of the inputs of the petiole and leg hair plates. These two systems are antagonistic in their effect. An increase in the gravitational stimulus on the hair plates of the petiole results in a decrease in the bee's angle of orientation relative to the negative geotactic direction. However, a similar increase in gravitational force on the leg hair plates results in an increase in the angle of orientation. *Apis florea* dances on a horizontal platform in the open, indicating the direction of the food source relative to the sun. Leg hair plates do not contribute to geotactic behaviour in this insect[8].

The use of gravity to signal direction in the waggle dance

The waggle dance of *Apis mellifera* involves the transposition of an angle of orientation with respect to the sun, to an angle of orientation with respect to gravity. How does the insect accomplish this? Observations of the orientation behaviour of bees and other insects have given us some clues[3,5,9]. If bees are kept on a horizontal surface in a dark room and a point source of light is switched on, many of the bees present will respond by walking towards the light. This orientated response to light is known as positive **phototaxis,** positive because the insect walks towards the light. Some bees will walk away from the light: this is negative phototactic behaviour. The sign of an insect's phototactic behaviour may change depending upon external factors, such as temperature or humidity; or internal factors, such as hormone levels or biorhythms. As well as these phototactic responses, we know that insects can show negative and positive geotactic responses when walking on a vertical surface in the dark. The sign of a geotactic response may also change in response to external and internal factors. *Apis mellifera*, for example, is more negatively geotactic in the morning while positive geotaxis predominates in the evening[10]. If bees showing positive phototaxis are placed in the dark on a vertical surface they will show a clear negative geotaxis, walking vertically up the comb. It has been suggested that the coupling of these two primitive means of orientation, a positive phototaxis with a negative geotaxis, was the starting point for the transposition of a direction relative to the sun into a direction relative to gravity[5,11].

A number of insect species are able to convert an angle walked on a horizontal surface in relation to the sun into an angle taken up with respect to gravity on a vertical surface in the dark, although many of them do this much less effectively than the bee, for example, confusing left with right[12]. Most insects can change their walking direction on an inclined surface and maintain this new orientation in relation to gravity for a long time. This type of behavioural response is known as **geomenotaxis.** The change in direction can happen spontaneously or it can be learnt. Social insects, such as bees and ants, learn a geomenotactic direction very well and can reproduce the learned direction with little error. Thus, elements of the behaviour underlying the signalling of direction in the waggle dance are present in a variety of insects. Selection pressure presumably operated on such behaviour in the honey bee so that communication of direction became unambiguous[11,12].

There are several thousand species of bee but only four are known to communicate the distance and direction of a food source by dancing. The earliest honey bees are believed to have danced in the open on a horizontal platform at the top of the comb, with the waggle run of the dance pointing in the direction of the food source, somewhat in the manner of the dwarf honey bee, *Apis florea*, of south-east Asia. The giant honey bee, *Apis dorsata*, hangs its large comb from substantial tree branches or even rocks and it has no horizontal platform on which to dance. It dances on the vertical comb, with the direction of the food represented by a transposition of visual to gravitational cues, although the bees can still glimpse the sky. *Apis cerana*, the Asian honey bee, and *Apis mellifera*, the European honey bee, live in

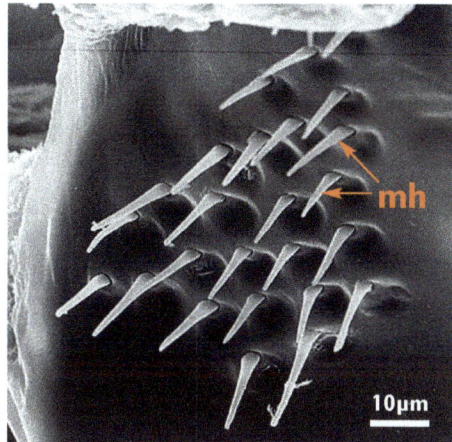

FIG. 4.17 *Hair plate 2 of the first trochanter showing the shafts of the mechanoreceptive hairs (mh) set asymmetrically in their sockets.*

enclosed spaces, where they have to dance in the dark and, lacking horizontal platforms at the top of the comb, on a vertical surface. Both these species transpose the sun angle into an angle with respect to gravity but both species, together with *Apis dorsata*, can be made to dance on a horizontal surface indicating direction by the use of visual cues. Observations of the behaviour of these four species of bees does suggest that the earliest dances of bees were on a horizontal surface in the open, and that transposition of visual cues into gravitational ones evolved later as horizontal platforms disappeared and nests were built in cavities[11]. However, recent studies on the dances of *Apis florea* and *Apis dorsata* suggest that their dances are more complex than was hitherto suspected and should not necessarily be regarded as simpler forms of dance on the way to the development of the dances of *Apis mellifera* and *Apis cerana*[12].

5. Feeding:
5.1 *Using the mouthparts*

Four resources suffice to support the honey bee colony. Nectar and pollen are collected for food, between them supplying carbohydrates and protein and some fat. Water is collected for diluting honey, in the preparation of food for the larvae and for evaporative cooling of the nest in warm weather. Resin (propolis) is used to seal up any holes or cracks in the nest or hive, to varnish the comb and to reinforce it as necessary.

It has been estimated that, on average, a colony rears 150 000 bees and that this requires 20 kg of pollen and 60 kg of honey a year. It requires about 3 million foraging trips to collect enough nectar to make 60 kg of honey and around 1.3 million trips to collect 20 kg of pollen, with foragers covering around 20 million km overall[1]. Collecting and dealing with food and feeding larvae is thus a major activity of the workers within the colony.

The bee's mouthparts

The basic insect mouthparts are designed for biting, chewing and manipulating food into the mouth. Many insects have their mouthparts modified for a specialized feeding habit, for example, the stylet-like piercing and sucking mouthparts of mosquitoes and of insects that suck sap from plants. The bee's mouthparts must cope with a variety of tasks: extracting liquid from within flowers (fig. 5.1); collecting water; transferring nectar to other workers; ripening honey; feeding the larval bees and the queen; ingesting pollen grains; manipulating wax in comb building; cleaning cells and removing debris. To this end, the mouthparts are adapted both for extension into flowers to lap and suck up liquids, and for grasping and manipulating objects.

A simple plate hinged to the front of the head capsule forms the upper lip or **labrum**

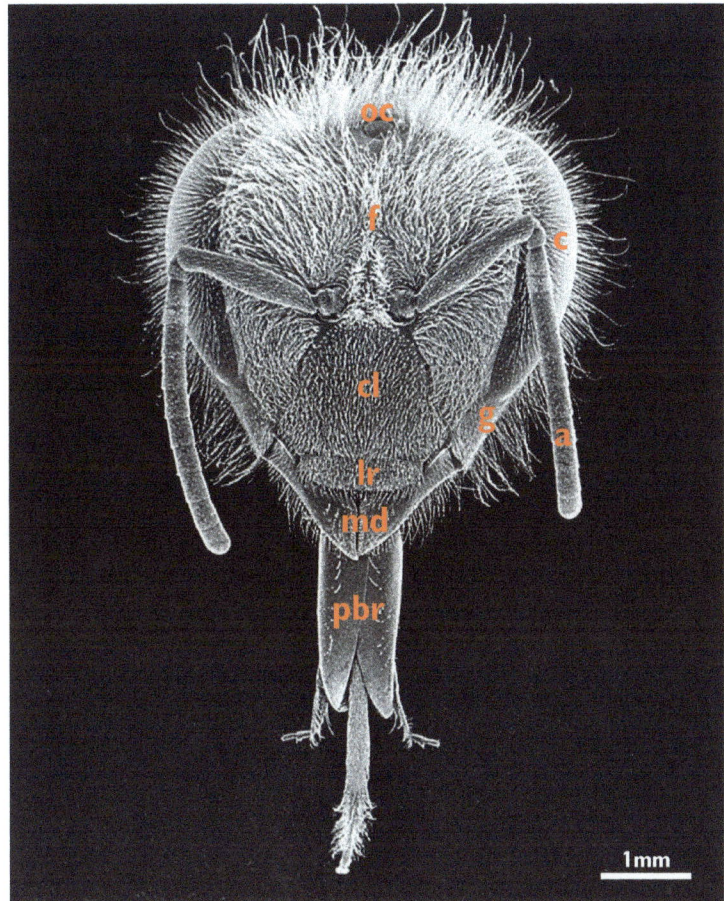

FIG. 5.2 *Frontal view of the head of a worker with the proboscis extended. The labrum (lr), or upper lip, is hinged to the clypeus (cl). The mandibles (md) articulate with the lateral 'cheeks' of the head capsule, the genae (g). The maxillae and the labium together form the proboscis (pbr). Compound eye (c); median dorsal ocellus (oc); frons region of head capsule (f); antenna (a).*

(fig. 5.2). The true mouthparts, paired segmental appendages of the head, comprise the **mandibles**, **maxillae** and the **labium**, or lower lip, where the paired appendages have fused to form one structure. The mandibles are strongly sclerotized structures that are responsible for any kind of work requiring a pair of grasping organs. The maxillae and the labium together form the **proboscis**, the labium forming an elongate, hairy 'tongue' for lapping liquids. The tongue is surrounded by a tubular food canal, formed from extended elements of the maxillae and labium, through

FIG. 5.1 (opposite) *A worker honey bee taking nectar from a floret of white clover (Trifolium repens). The proboscis is unfolded and the glossa extended to reach the nectar. Inset: the glossa ends in a spoon-shaped lobe, the flabellum or labellum.*

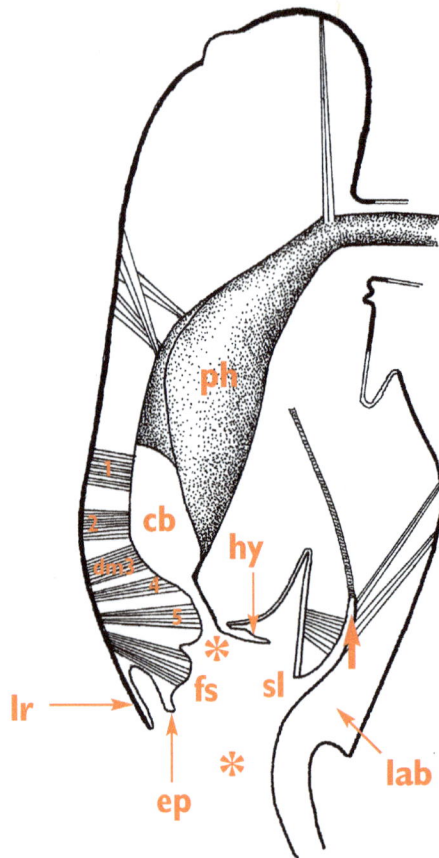

FIG. 5.3 *The food is first taken into a cavity, the preoral cavity (stars) formed from the bases of the mouthparts and the labrum. (Only the labrum (lr) and labium (lab) are visible in this longitudinal section.) The cavity is divided into frontal (fs) and posterior (sl) sacs by the central hypopharyngeal lobe (hy). The salivary glands open into the posterior sac or salivarium (arrow). The preoral cavity continues dorsally into the cibarium (cb) which, in turn, grades imperceptibly into the pharynx (ph). The dilator muscles (dm 1–5) in the cibarial wall expand this cavity drawing liquid in, while bands of transverse muscle (not shown here) compress the cavity and force the contents into the muscular pharynx. The labrum bears a ridge — the epipharynx (ep) — on its inner surface (see also fig. 5.32).*

head and thorax emerge (fig. 5.3). It is divided into a frontal sac and a smaller posterior sac, where the salivary glands discharge, by the centrally-lying hypopharynx. The opening of the preoral cavity forms the functional mouth where food is taken in and passed up into the dorsal region of the preoral cavity, the **cibarium** (fig. 5.3). The muscles in the walls of this cavity form a suction pump, helping to draw liquid up into the alimentary canal. Five pairs of large dilator muscles running between the wall of the cibarium and the front of the head capsule expand the cavity, drawing liquid into the cibarium, while bands of transverse muscle compress the cavity and force food upward into the wide, muscular **pharynx**. The true mouth lies at the top of the cibarial sac where the sac continues almost imperceptibly into the pharynx. The pharyngeal tube is also strongly contractile, helping to drive the food into the narrow **oesophagus** which leads into the greatly extensible **honey sac**[2].

The role of the individual mouthparts

The **labrum** is a transverse flap hinged to the clypeus on the front of the head capsule (figs 5.2, 5.4). It is covered with cuticular hairs of various lengths; some of these are innervated, i.e. supplied with one or more neurons, and function as sensory receptors. The longer hairs with socketed bases, particularly those overhanging the mandibles, are likely to be mechanoreceptors registering both contact with the labrum and relative movement of the labrum and the mandibles. On its inner surface the labrum bears a central ridge, the **epipharynx**, with a soft pad on either side, covered with sensory receptors of unknown function (figs 5.3, 5.32). When the proboscis is swung forward into its working position, the epipharynx fits tightly against it, thus sealing the front of the proboscis at its base so that liquid can be sucked up into the mouth without leaking out.

The strong pair of spatulate-shaped jaws, or mandibles, are hinged anteriorly to the

which the liquids can be sucked up into the mouth, the whole structure being folded back against the head when not in use. A medial structure, the **hypopharynx**, lies between the bases of the mouthparts, with the ducts of the hypopharyngeal glands opening onto its anterior surface and those of the salivary glands onto its posterior surface.

The bases of the mouthparts enclose a cavity, the **preoral cavity**, which forms a sac where the food is first ingested and into which the secretions from the glands of the head capsule at the lateral edge of the clypeus and posteriorly to the lateral plate, or gena (figs 5.2, 5.4). The insect cuticle is a very versatile structure: composed primarily of the polysaccharide **chitin**, complexed with protein, it may be thin and flexible, as in the intersegmental membranes of the abdomen, or strong and rigid, as in the head capsule and particularly in the mandibles[3]. Much of its strength comes from the bonding together of the long chains of chitin lying adjacent to each other, but additional strengthening comes from a process known as tanning or sclerotization, in which the proteins present in the cuticle become tightly cross-linked to one another. The mandibles are strongly sclerotized. Unlike human jaws, they operate transversely, being drawn apart by the large **abductor muscles** running between their outer bases and the side wall of the head capsule, and pulled together by the equally powerful **adductor muscles** attached to the inner base of the mandibles and the top and back of the head. Sensory receptors that register stresses and strains in the cuticle, the campaniform sensillae, occur at the inner base of the mandibles and monitor their movement[5].

The opposing surfaces of the mandibles are ridged and there is a slight overlap at the bottom so that one jaw slides just inside the other (fig. 5.4). Even though they are heavily sclerotized, the opposing surfaces can become quite worn in older bees (fig. 5.5). The ridges are lined with socketed hairs bending over towards the edge of the opposing jaw. These hairs are innervated by one neuron and are believed to be mechanoreceptive hairs, registering contact between the two mandibles[4]. There are other long, mechanoreceptive hairs on the front and lateral surfaces of the mandibles that will register contact between the mandibles and other surfaces. Lying alongside the medial ridges are small sensory pegs or basiconic sensillae, innervated by one neuron (fig. 5.5). A number of other fine, cuticular hairs are distributed over the surface. The inner surface of each mandible is slightly concave and spoon-

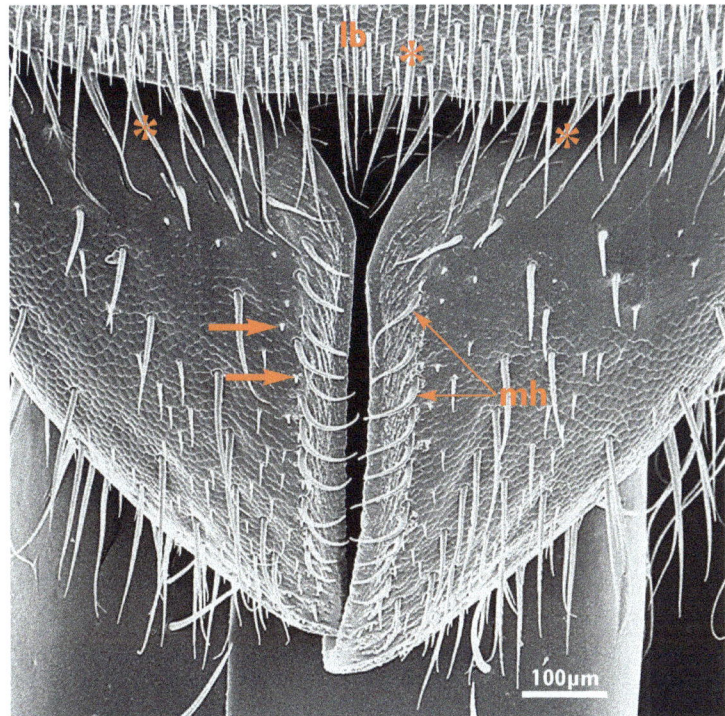

FIG. 5.4 *The cutting surface of the mandible is ridged and the opposing surfaces can just slide past each other. Mechanoreceptive hairs (mh) lining the ridges are believed to register movement of the cutting edges relative to each other. Other long hairs with socketed bases on the surface of the mandible probably have a mechanoreceptive function, as do those on the edge of the labrum (lb) overhanging the mandibles (stars). Small sensory pegs, or basiconic sensillae, lie alongside each ridge (arrows).*

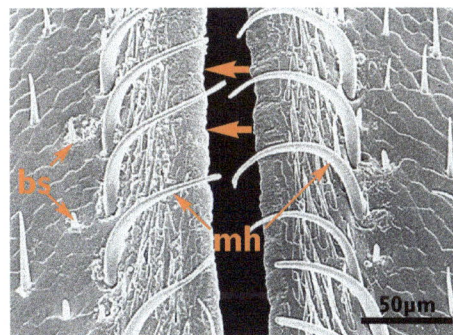

FIG. 5.5 *The cutting edge of the mandible can become quite worn in older bees (arrows). The hairs (mh) overhanging the opposing cutting surfaces are innervated by one neuron at the base and are thus presumed to have a mechanoreceptive function. Basiconic sensillae (bs).*

shaped at the outer end (fig. 5.6). A number of small domes are found on the inner surface of the cutting edge (fig. 5.7). These may be campaniform sensillae, sensory receptors that are stimulated by stresses set up in the cuticle; if so, they may be able to monitor the force exerted at the cutting edges. (For details of how campaniform sensillae function, see chapter 7, p149).

A channel fringed with hairs runs medially from the base of the mandible into the

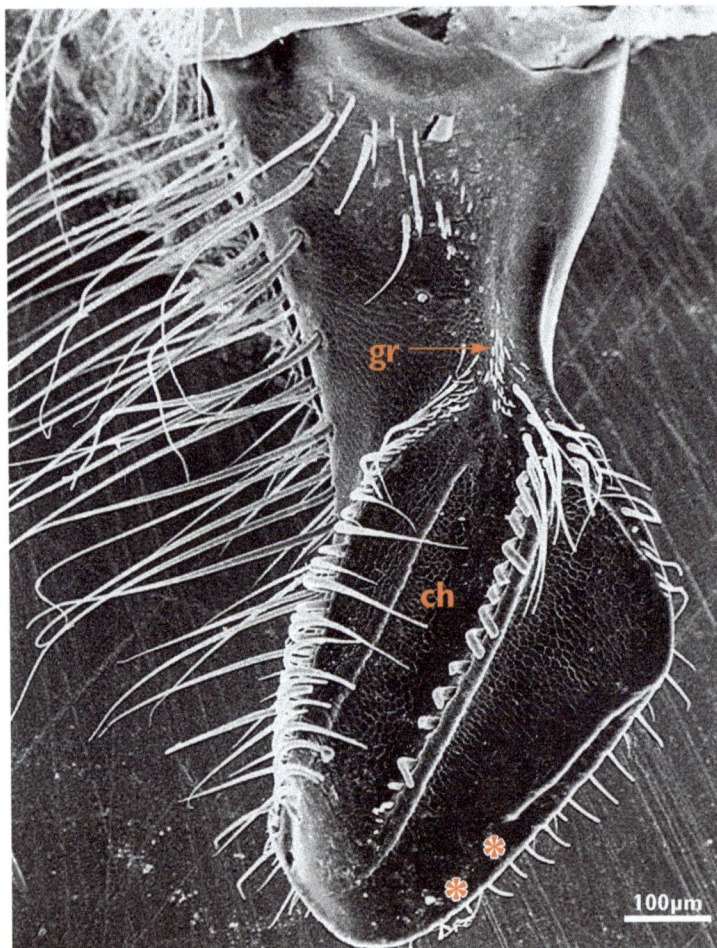

FIG. 5.6 *The inner surface of the mandible is slightly concave and spoon shaped at the lower end. The mandibular secretion runs along a shallow groove (gr) into a channel (ch) fringed with hairs, that crosses the spoon-shaped area. A number of putative chemosensory sensillae lie at the entrance to the channel (see also fig. 5.27b).*

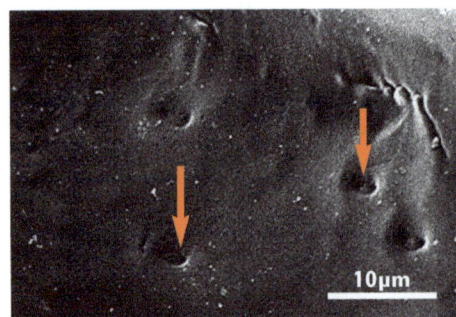

FIG. 5.7 *A number of very small domes (*arrows*) are seen in the cuticle along the inner edge of the mandible (starred in fig. 5.6). These structures may be campaniform sensillae, sensitive to stresses set up in the cuticle. Such sensillae have been reported from the mandibles of some insects, but further investigations are required to establish their presence in the honey bee.*

spoon-shaped area. The mandibular gland, lying immediately above the mandibles in the side of the head capsule, opens into the membrane at the base of the mandibles (see chapter 9, fig. 9.13), and its secretion runs through the channel onto the spoon-shaped region. In the early part of the worker's life, when it is involved with brood rearing, the mandibular glands produce brood food (see below), and are at their maximum size 5–15 days into the adult life. The bee puts its head into the larval cell, carries out an inspection for 2–20 seconds and, if feeding, parts the mandibles so that the mixed mandibular and hypopharyngeal secretion forming the brood food runs out and is deposited on the side or bottom of the cell. As workers change to guarding and foraging, the mandibular glands change over to the production of the alarm pheromone, 2-heptanone.

The mandibles are involved in the construction of the comb and the capping of cells. Wax flakes are removed from the four abdominal wax glands by the enlarged tarsal joint of the hindleg and passed forward to the forelegs and the mandibles. The wax is moistened by saliva, and the mandibles and forelegs together knead the wax until it is sufficiently malleable to manipulate into position (see chapter 8).

The plant resins, or propolis, collected to cement up holes in the nest, to strengthen the comb bases and to coat the interior walls of the hive, are handled by the mandibles. When collecting, the bee bites off pieces of sticky resin from plant buds with its mandibles. It may work the resin with its mandibles before taking the resin with the front legs and passing it back to one of the middle legs. From there it is packed into the corbicula, or pollen basket, of the hindleg on the same side[7]. On returning to the hive, the bee makes for regions at the edge of the hive where other bees are working with propolis and sits there until other bees come and pull pieces of the propolis from its corbiculae. It may take quite a while for a propolis forager to be unloaded. Propolis is not stored in great quantities but on occasion it is placed in empty cells. Taking a piece of propolis, the bee attaches it to the site where cementing is occurring. Other bees working there detach small particles of the propolis with their mandibles and collect a number of particles behind the mandibles until a small ball is formed. This is then pressed very firmly against the surface to which it is to be attached and welded to the layer of resin already there by side-to-side movements of the head and mandibles. The surface of the work is smoothed by bees nibbling off any protruding resin. Some beeswax may be mixed with the resin to make it more pliable[6].

The mandibles are also used to manipulate pollen grains within the hive, although not during their initial collection. Pollen is removed by adults from the cells in which it is stored for food, both for themselves and for feeding larvae. Although the mandibles pass pollen into the mouth, they are not normally used to break up the pollen grains, a process that must take place during digestion.

Mandibles are necessary for general housekeeping duties, when cells are cleaned and smoothed, and when debris and dead hivemates are dragged out of the hive. Mandibles may also be employed in defensive behaviour by guard bees or indeed by other workers. This behaviour includes release of the mandibular alarm pheromone to recruit other workers and repel robber bees, and biting with the mandibles as well as hair pulling in certain cases. Groups of bees may gang up and use their mandibles to grapple with other insects trying to rob the hive and forcibly remove them.

The mandibles of the queen are similar in size to those of the worker although the opposing surfaces bear a single tooth-like projection or cusp (fig. 5.8). The outer surface bears longer hairs than the worker's and they are more numerous. The inner surface lacks the hair-fringed groove running from the opening of the mandibular gland. These glands are very well devel-

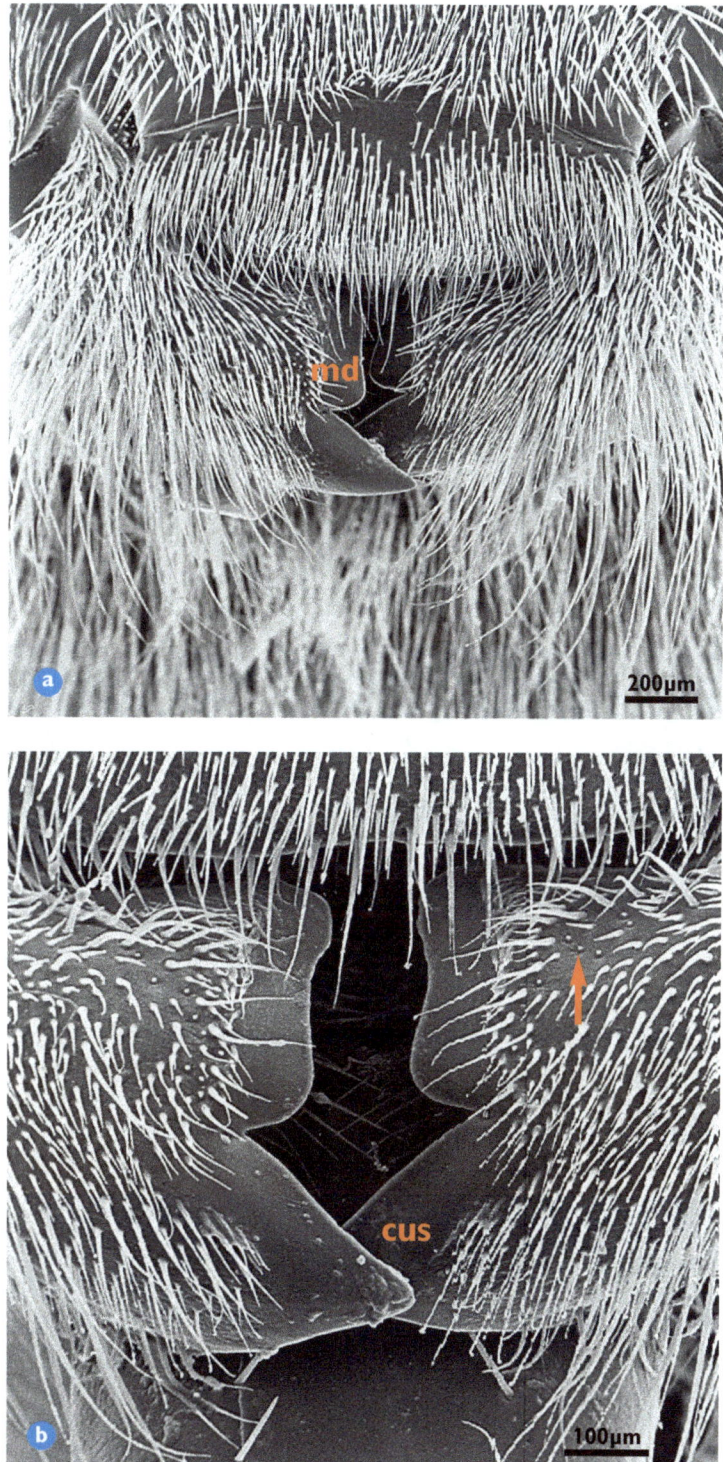

FIG. 5.8 a *The mandibles (md) of the queen are densely covered in long hairs, except at the opposing edges.* b *Each mandible has a single, large cusp (cus) on its cutting edge. There are a few very small, scattered pegs on each mandible (arrow) possibly equivalent to the basiconic sensillae of the worker mandible.*

oped in the queen and are the source of the two acids, 9-oxy-(*E*)-2-decenoic acid (9-ODA) and 9-hydroxy-(*E*)-2-decenoic acid (9-HDA), that form the major queen

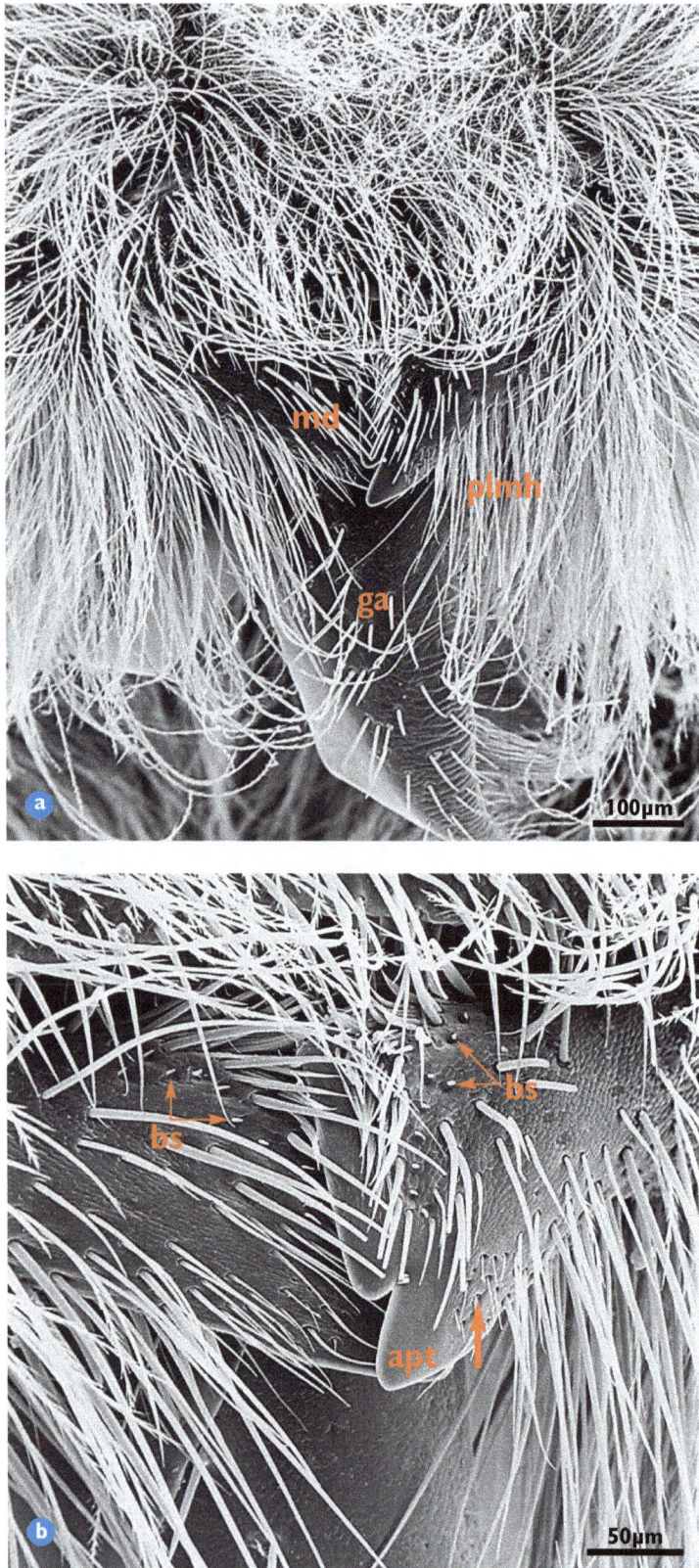

FIG. 5.9 **a** *The drone mandible (md) is smaller than the mandibles of the worker and the queen. The outer surface of the mandible bears long, plumose hairs (plmh) as does much of the head capsule. The galeae of the short proboscis are visible beneath the mandibles (ga).* **b** *The mandible bears one small apical tooth (apt). There is a row of small basiconic sensillae (bs) on each mandible and a patch of very small, socketed hairs, possibly mechanoreceptive (arrow).*

pheromones (chapter 8). 9-ODA and 9-HDA have been implicated in queen recognition; in the inhibition of queen rearing and of swarming; the prevention of development of ovaries in the workers; and the attraction of drones to the queen for mating[7]. When swarms occur, these acids are responsible for the attraction of workers to the queen, for stimulating the secretion of the Nasonov pheromone and for the stabilization of the swarm. Mandibular gland analysis has revealed the presence of small quantities of many other compounds, but it is not clear what roles, if any, these compounds play in regulating the life of the colony.

The drone mandible is smaller and narrower than those of the worker and the queen. The outer surface is covered with many long hairs (figs 5.9 and 2.1) and there appear to be fewer sensory hairs. The head as a whole is covered with long, fine hairs, some of which are plumose, i.e. having small, lateral extensions. The drone mandibular glands are small but there is some evidence that they synthesize a pheromone during the first nine days of adult life which is stored and used during mating, either to attract other drones to the congregation areas or, more probably, to attract the queens[2,7].

The medially-lying hypopharynx is situated behind the mandibles and forms the floor of the cibarium. The paired **hypopharyngeal glands**, very long, narrow structures that loop around in the sides of the head, open separately through two small pouches onto the anterior face of the hypopharynx (fig. 5.10). The hypopharyngeal secretion, together with added mandibular secretion, is fed to the developing larvae. Like the mandibular glands, the hypopharyngeal glands are well developed in the younger 'nurse' workers involved in caring for the larvae. The glands are well developed by the time the worker is three days old and nursing activity is at its peak when the worker is between 6 and 16 days old. In older workers, they become reduced in size although they continue to produce invertase, glucose oxidase, diastase and

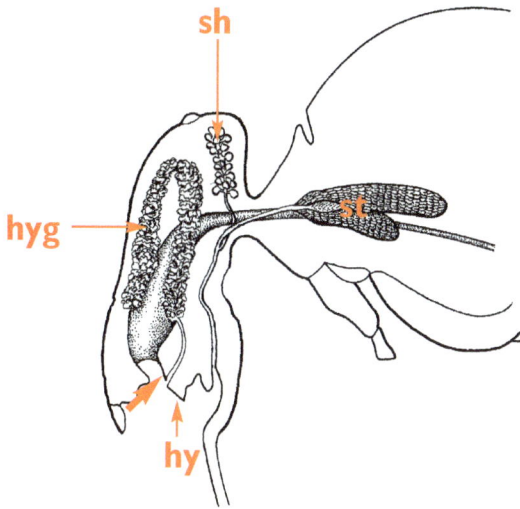

FIG. 5.10 *A longitudinal section through the head of a worker showing the position of the hypopharyngeal glands (hyg) and the head (sh) and thoracic (st) salivary glands. The latter two glands open into the salivarium, which acts as a salivary ejection pump, see text. The hypopharyngeal glands open onto the anterior face of the hypopharynx (hy). An arrow shows the position of the hypopharyngeal receptors.*

other enzymes essential for converting nectar and honeydew into honey and which have a bactericidal action protecting the stored honey[8]. Admixture of these substances to the larval food helps to protect the larvae from bacterial infection.

Just behind the laterally-situated openings of the ducts of the hypopharyngeal glands are two patches of sense organs comprising 50–60 basiconic sensillae. It is common for insects to have taste receptors on the hypopharynx as, in this situation, they are in a position to sample food taken into the mouth before it is swallowed (see fig. 5.32b, c).

Two pairs of glands open into a narrow sac or salivarium behind the posterior flange of the hypopharynx, by means of a common duct (fig. 5.10). One pair of glands, the postcerebral glands, lie against the posterior wall of the head capsule; the second pair, developed from the silk glands of the larva, extend back through the thorax to the abdomen and are known as the thoracic glands. Together they are known as the **labial** or **salivary glands**. The salivar-

ium acts as a salivary ejection pump[2]. The anterior wall of the salivarium is formed from the flexible posterior wall of the hypopharynx. A pair of dilator muscles, running from the anterior wall of the salivarium to the outer wall of the hypopharynx, expand the salivarium and suck the labial gland secretions from the salivary duct. The back wall of the salivarium is a rigid plate belonging to the prementum. Compression of the salivarium is said to be brought about by the contraction of muscles running from the corner of the pump to the proximal end of the labial prementum[2,30], forcibly expelling the saliva onto the base of the proboscis. However, other studies have suggested that compression of the salivarium is brought about by the elasticity of its thick rear wall, rather than by muscular action[29]. The outlet of the salivarium is variously described as a valve at the base of the proboscis[29] or a crescent-shaped microporous membrane[30]. The salivarium is illustrated in figures 5.14b, 5.15a and 5.16b.

Saliva is almost universally used in animals to solubilize or moisten and carry solid foods and, in many cases, for predigestive food treatment by means of enzymes added to the saliva. It may even be important in the food harvesting process itself. The dual functions of saliva are often demonstrated in the structure of salivary glands where there may be two lobes containing different cellular types or even two sets of glands, as in the bee, reflecting the diverse needs of a fluid-rich and an enzyme-rich saliva[9]. Somewhat surprisingly, in view of the great interest in honey production in bees, the role of the salivary glands is not well understood in this insect. Overall, the saliva functions to dissolve solids such as sugars or dried honeydew, provides a lubricant for any material that has to be chewed and softened and is used in grooming and cleaning the queen. The two pairs of glands produce different types of secretion. The thoracic glands secrete a colourless, watery fluid which dissolves sugars, while the postcerebral glands produce an oily secretion whose function is

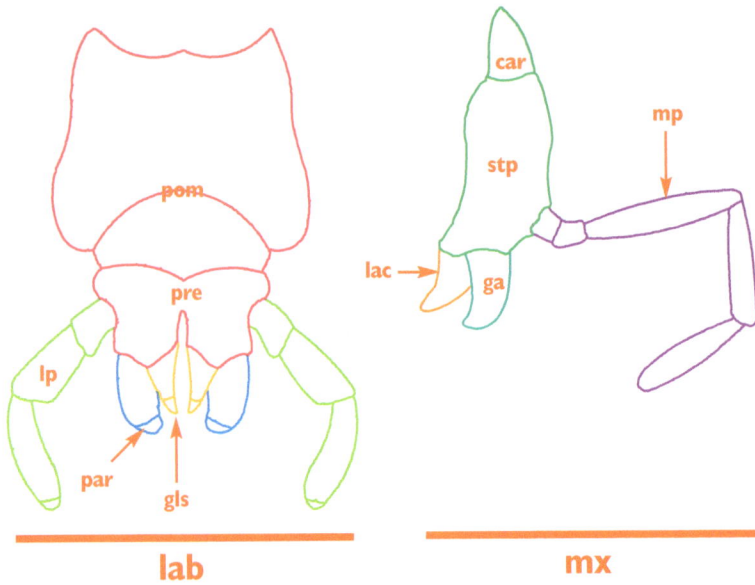

lab   **mx**

FIG. 5.11 *The labium (lab) and one maxilla (mx) from the biting and chewing mouthparts of the cockroach. The maxilla comprises a basal cardo (car) articulating with the head and a large, flat plate, the stipes (stp). Two lobes arise distally from the stipes, the outer galea (ga) and the inner lacinia (lac). These lobes assist in holding food and pushing it into the mouth. A jointed mobile palp, the maxillary palp (mp) arising from the stipes, bears sensory receptors. The labium resembles two fused maxillae. The basal region, equivalent to the cardines, is the postmentum (pom). Distal to this is the prementum (pre) equivalent to the maxillary stipes. The prementum bears four lobes — a pair of paraglossae (par) and an inner pair of glossae (gls) — and a pair of labial palps (lp).*

unknown. Most authors report that enzymes are added to the food via the hypopharyngeal glands rather than the salivary glands.

Feeding on liquids

Liquid foods are lapped up by the proboscis, formed from the pair of maxillae, situated behind the mandibles, and the posterior labium. It is easier to understand how these mouthparts function in the bee if we look at their structure in a primitive insect that bites and chews its food (fig. 5.11). In the cockroach, for example, each maxilla consists of a basal section, the cardo, hinged to the head capsule and a broad, flat plate, the stipes, bearing two lobes, the inner lacinia and the lateral galea, together with an outer segmented palp. The pointed lacinia holds the food while the mandibles cut and chew it and assist the galea and palp in pushing it into the mouth. The galeal lobe and the outer

palp bear mechanoreceptive hairs and gustatory receptors to sample the food. The labium is formed by the fusion of a pair of appendages resembling the maxillae. The basal region, equivalent to the cardo and stipes region of each maxilla, forms a large plate divided into a pre- and postmentum. The distal region bears lobes reminiscent of those of the maxillae, an inner pair of glossae and an outer pair of paraglossae, with laterally situated labial palps bearing gustatory and mechanoreceptors. The labium forms the floor of the preoral cavity.

In the bee, the glossal lobes of the labium are elongated to form the flexible, hairy **tongue** that reaches down into the pool of nectar. The **glossa** shows few signs of its origin from paired lobes, consisting of a simple hollow tube containing a long elastic rod, although its posterior wall is split by a deep groove closed by a dense fringe of small hairs. The **galeae** of the maxillae and the **labial palps** are also elongated, and these four structures are brought together around the glossa to form the tubular food canal through which liquids are drawn into the mouth (figs 5.12, 5.18). When in place for feeding, the galeae, labial palps and glossa together form the proboscis. The glossa can be extended well beyond the galeae and the labial palps to reach a pool of nectar which moves up into its covering of hairs by capillary action. The glossa is then shortened to carry the nectar into the food canal. The full action of the glossa is described later. The lacinial and paraglossal lobes of the maxillae and labium are much reduced, but in the bee still have roles to play in feeding.

Unlike many fluid-feeding insects, such as mosquitoes or aphids, bees cannot have a feeding tube permanently sealed around the mouth. The bee has to be able to use its mandibles to manipulate substances. As well as this it must be able to open its mouth so that liquids can be regurgitated for brood feeding and for feeding other adults, and to allow foraging bees to deliver regurgitated nectar into the mouths of receiving bees in the hive. The proboscis has to be moved out of the way while these activities are taking place. This is done by folding the galeae, labial palps and glossa back against the bases of the maxillae and labium (figs 5.13 and 5.14a, b) which are themselves retracted into a groove, or fossa, on the under-surface of the head. When no feeding is taking place, the folded proboscis is held in position by the mandibles clasped beneath it, with the labrum, or upper lip, brought down against the mandibles (fig. 5.13a). For the bee to reform the proboscis and commence liquid feeding, the reverse process must be carried out. The bases of the maxillae and the labium must be protracted from the fossa and the galeae, labial palps and glossa unfolded and extended forward. The front of the proboscis must be elevated to close the mouth and the galeae and labial palps brought together around the glossa. The base of the proboscis is then held firmly in place by the mandibles whose inner surfaces fit over the midribs of the galeae (fig. 5.17).

Movements of the maxillary and labial mouthparts required to form or remove the proboscis are achieved by a complex mixture of muscular activity, hydraulic activity of the haemolymph[30,31] and cuticular elasticity. Protraction and retraction of the bases of the maxillae and labium is achieved by muscular activity, protraction largely by the activity of muscles acting on the stipes and cardo of each maxilla, and retraction by muscles moving the pre- and postmentum of the labium (fig. 5.14a, b). Only the maxillae are articulated to the head capsule but the labial prementum is engaged with the maxillae by means of the **lorum**, a V-shaped cuticular structure which passes closely around the base of the postmentum while joining together the stipes of each maxilla. Because of this, the maxillae and labium move together as a single unit.

The maxillae are articulated with the head capsule by the **cardines** (plural of cardo), which form narrow rods in the bee. Each cardo articulates with a process on the lateral wall of the fossa. The cardo projects a

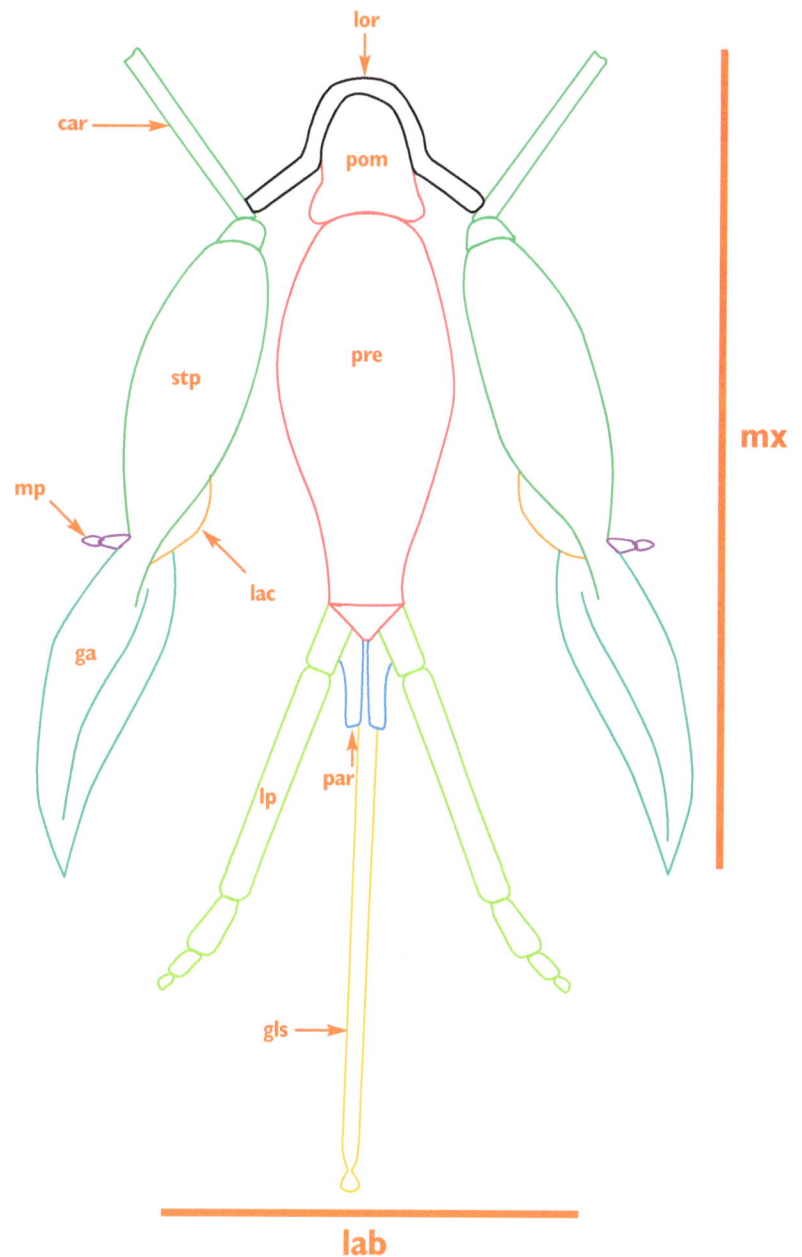

FIG. 5.12 The maxillae (mx) and labium (lab) of the worker bee, viewed from the posterior surface. Each maxillary cardo (car) is a long, rod-like structure that articulates with the head capsule. The stipes (stp) remains a broad, flat plate, but one of its lobes, the outer galea (ga) is elongated. The inner lacinia (lac) is reduced to a membranous lobe. The maxillary palp (mp) is very much reduced. The maxillae and the labium are yoked together by the lorum (lor), a V-shaped structure that runs around the base of the labium and is articulated on each side with the distal end of the cardo. The labium consists of a small, triangular-shaped postmentum (pom) and a broad, plate-like prementum (pre). The inner glossal lobes (gls) of the labium have become fused and considerably extended to form the 'tongue'. The small paraglossal lobes (par) surround the base of the tongue. The labial palps (lp) are also elongated and, together with the galeae, surround the tongue to form a food canal when the proboscis is in operation.

little way beyond the articulation point and on it is inserted a muscle running from the head wall (fig. 5.14a, 10) whose contraction swings the maxilla ventrally. Each maxilla is more strongly protracted by muscles 11, 12 and 13, inserted onto the stipes, but some assistance is given by muscle 18 inserted on the prementum (fig. 5.14b). Contraction of these muscles flattens the elbow-like angle between the stipes and the cardo and effects a protraction of the stipes on the cardo. Maxillary protraction automatically causes movement of the pre- and postmentum. The cibarial suction pump will only work to draw fluid up into the mouth if the proboscis is firmly closed up to the mouth. When muscles 12 and 13 (fig. 5.14a) protract the proboscis, they also elevate the front of the proboscis, bringing the small lacinial lobes of the maxillae tight up against the epipharynx on the under-side of the labrum (fig. 5.13c); thus, the join between the sucking tube and the mouth is made airtight.

Retraction of the prementum and the stipites (plural of stipes) into the fossa is achieved by the action of a pair of labial muscles, 17, running from the head capsule to a pair of cuticular arms, the ligular arms, at the base of the glossa and paraglossa (fig. 5.14b). This region, where the base of the glossa and the small pair of paraglossal lobes join the prementum, is known as the **ligula** and it is supported by the ligular arms on the anterior surface and a cuticular plate, the ligula plate, on the posterior surface. Contraction of muscle 17 both retracts the base of the prementum into the fossa and also pulls the ligular region into the prementum. This action also brings about retraction of the stipites and cardines because of the maxillary–labial linkage.

Flexion of the proboscis is due to the activity of muscles attached to strengthened rods and to sclerites within the galeae, labial palps and glossa, coupled with complex folding of the intersegmental membranes between the joints. Unfolding and extension of the proboscis has been attributed solely to the elastic restitution of the cuticle when the flexion muscles relax[2,11]. However, other studies suggest that it is a much more dynamic process, making use of the hydraulic action of the haemolymph,

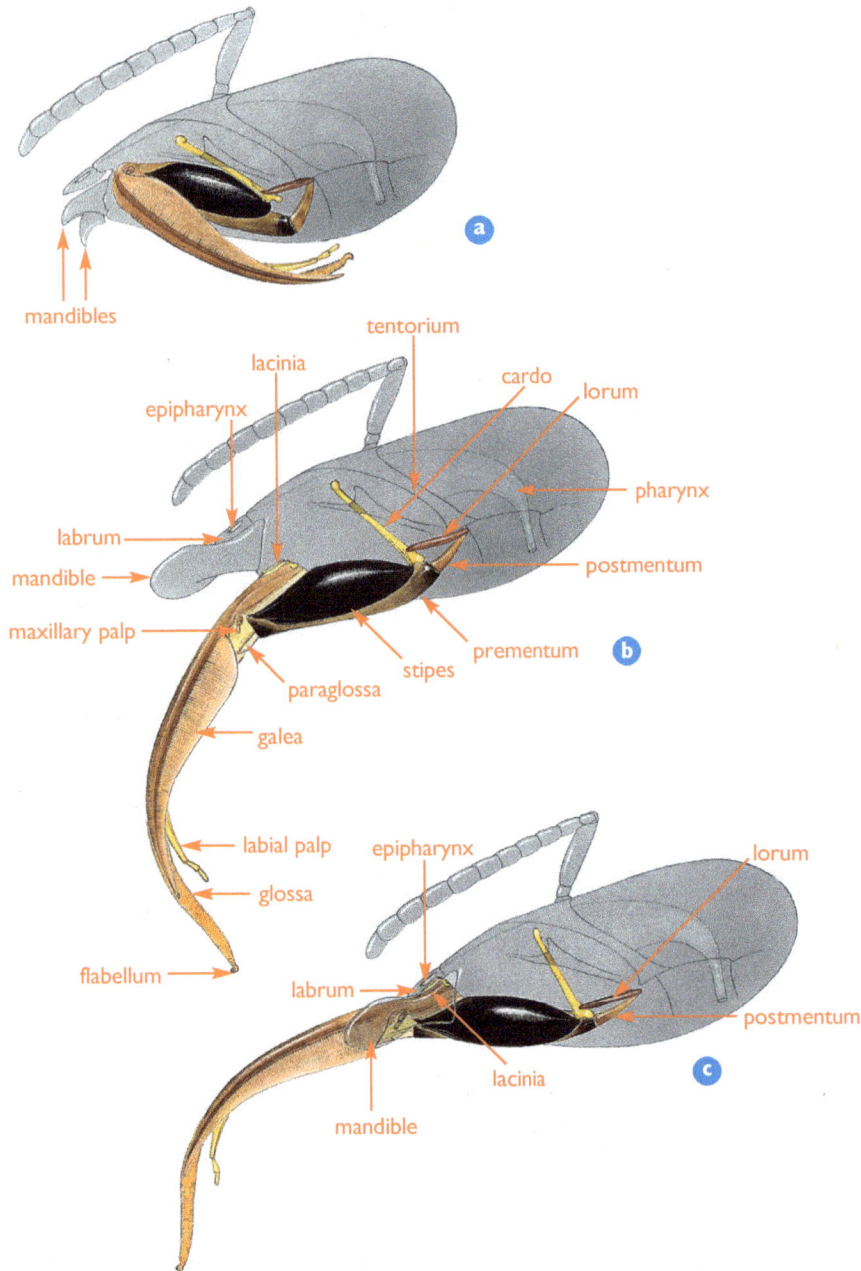

FIG. 5.13 Worker head from the side, with the proboscis **a** retracted and folded; **b** partially extended; **c** fully extended in the feeding position. When the proboscis is folded the mandibles are clasped beneath it, and when it is fully extended in the feeding position its base is held in place by the mandibles whose inner surfaces fit over the galeae. As the proboscis is fully extended it is raised so that the lacinia is pushed up against the epipharynx thus closing the mouth. The labium (consisting of the postmentum, prementum, labial palps, glossa and paraglossae) is yoked to the two maxillae by the lorum (only one maxilla visible here). Note how the movement of the cardo and stipes flattens the angle between the lorum and the postmentum of the labium, moving the labium forward as well.

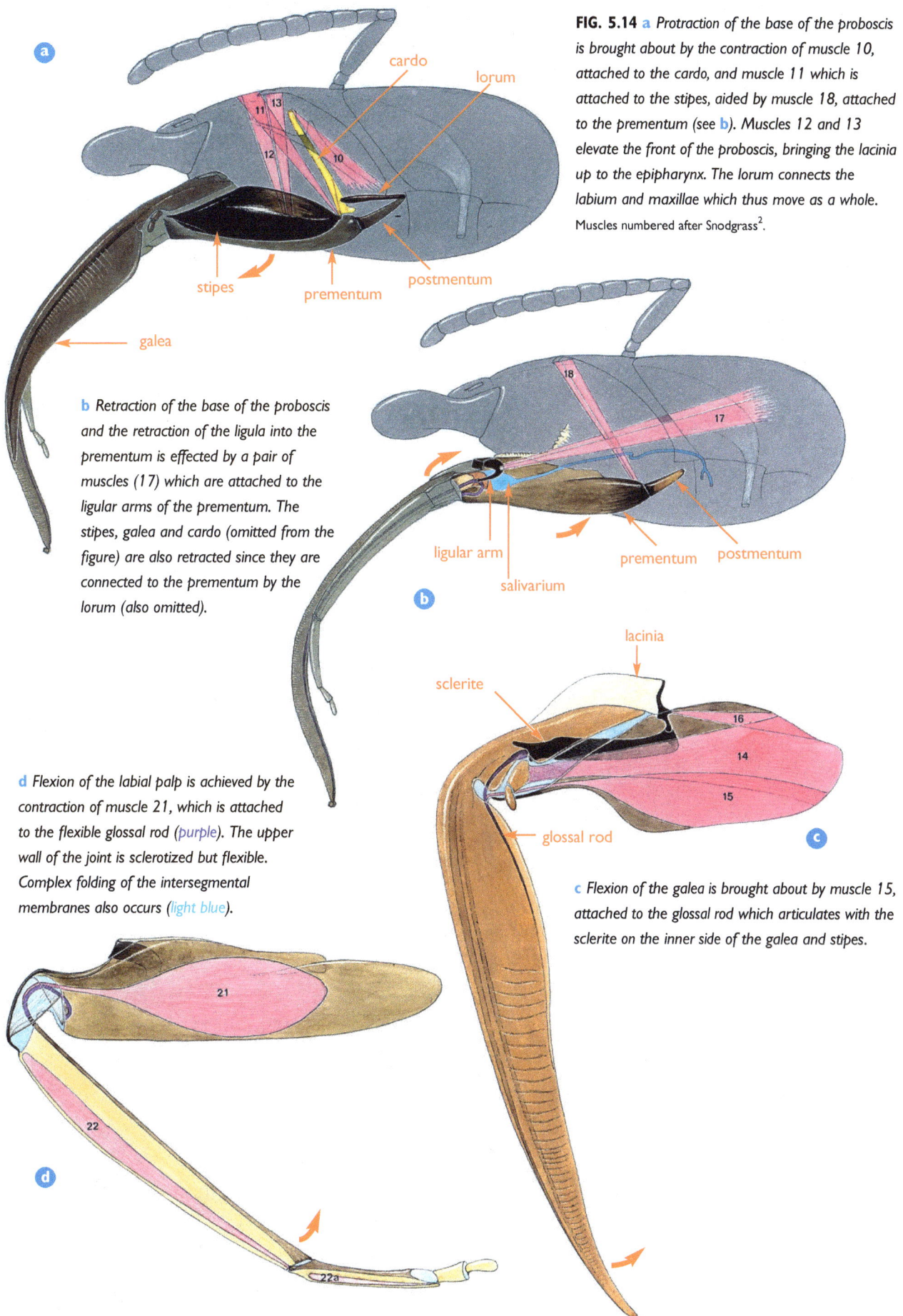

FIG. 5.14 a *Protraction of the base of the proboscis is brought about by the contraction of muscle 10, attached to the cardo, and muscle 11 which is attached to the stipes, aided by muscle 18, attached to the prementum (see b). Muscles 12 and 13 elevate the front of the proboscis, bringing the lacinia up to the epipharynx. The lorum connects the labium and maxillae which thus move as a whole. Muscles numbered after Snodgrass[2].*

b *Retraction of the base of the proboscis and the retraction of the ligula into the prementum is effected by a pair of muscles (17) which are attached to the ligular arms of the prementum. The stipes, galea and cardo (omitted from the figure) are also retracted since they are connected to the prementum by the lorum (also omitted).*

d *Flexion of the labial palp is achieved by the contraction of muscle 21, which is attached to the flexible glossal rod (purple). The upper wall of the joint is sclerotized but flexible. Complex folding of the intersegmental membranes also occurs (light blue).*

c *Flexion of the galea is brought about by muscle 15, attached to the glossal rod which articulates with the sclerite on the inner side of the galea and stipes.*

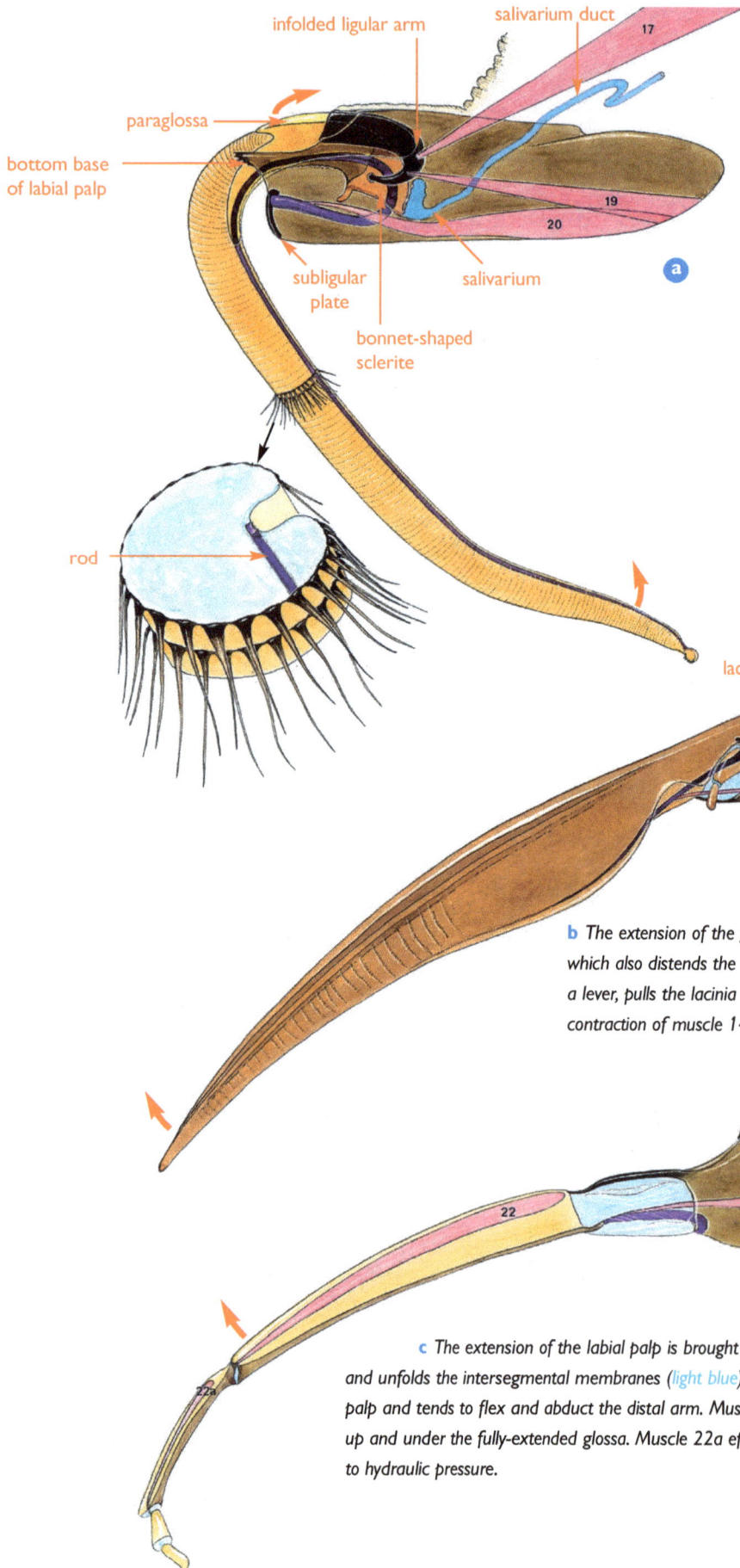

infolded ligular arm

salivarium duct

17

paraglossa

bottom base
of labial palp

subligular
plate

salivarium

bonnet-shaped
sclerite

19

20

a

rod

FIG. 5.15 a *Flexion of the glossa. Retraction of the ligula into the prementum is effected by muscles 17 and 19 acting on the ligular arms of the prementum, folding them inwards. The ligular arms are attached to the paraglossae and also extend to articulate with the bonnet-shaped sclerite at the base of the glossa. The glossal rod (purple) and the subligular plate are pulled in and the glossa shortened and recurved. Muscles 20 (paired, but only one shown) are rendered slack and ineffective, as are the salivarium muscles (not shown). The whole glossa is densely furnished with hairs. These arise from sclerotized rings which alternate with membranous spaces (orange). Only two bands of hairs are shown, and also enlarged (inset). The glossal rod (purple) lies in a groove running posteriorly along the length of the glossa. The groove additionally serves to convey saliva to the tip of the glossa. Haemolymph (light blue) may be under less pressure than shown, sucking the membrane and rod further in, or under more pressure, causing the rod and the membrane to move outside the main body of the glossa (see fig. 5.16c).*

lacinia

lever

16

14

15

b

b *The extension of the galea is effected hydraulically by haemolymph pressure, which also distends the intersegmental membrane (light blue). Muscle 16, acting on a lever, pulls the lacinia taut and causes further extension and adduction. The contraction of muscle 14 also causes some extension and abduction.*

c

21

22

22a

c *The extension of the labial palp is brought about hydraulically by haemolymph pressure which distends and unfolds the intersegmental membranes (light blue). This same pressure extends along the length of the labial palp and tends to flex and abduct the distal arm. Muscle 22 acts to extend and adduct this segment, bringing it up and under the fully-extended glossa. Muscle 22a effects movement of the distal segments, acting in opposition to hydraulic pressure.*

FIG. 5.16 a *Protraction of the ligula, is achieved hydraulically by haemolymph pressure, which pushes the paraglossae and the base of the glossa out of the prementum and unfolds the ligular arms and subligular plate. Rods extend from the ligular arms and articulate with the bonnet-shaped sclerite at the base of the glossa. This braces the latter against the pull of muscles 20. These muscles become extended and functional, as do the salivarium muscles (not shown). The glossa (normally enclosed by the galeae and labial palps; not shown) is lengthened by haemolymph pressure within its membrane (light blue). This also causes procurving of the tip.*

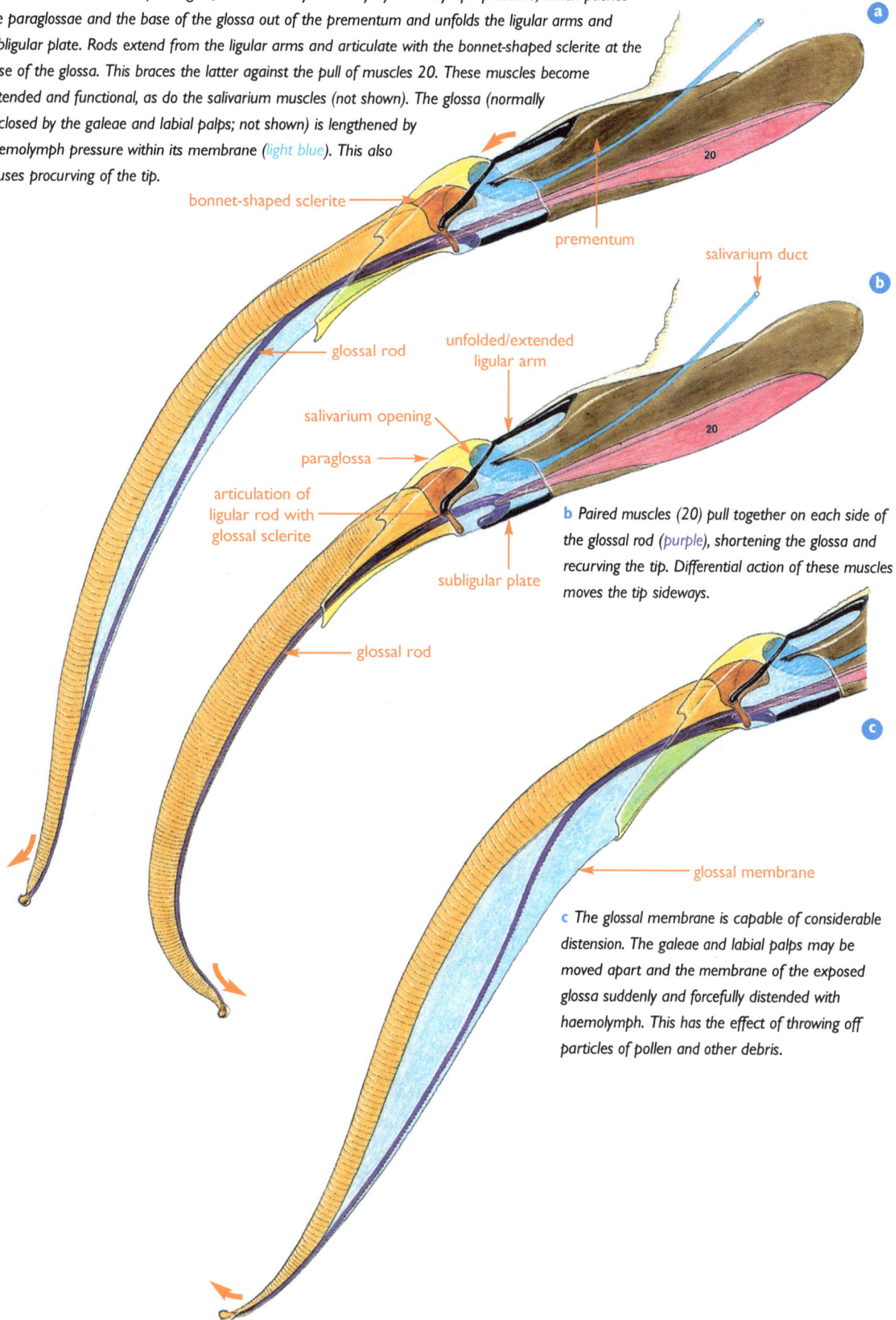

bonnet-shaped sclerite

prementum

salivarium duct

glossal rod

unfolded/extended ligular arm

salivarium opening

paraglossa

articulation of ligular rod with glossal sclerite

subligular plate

b *Paired muscles (20) pull together on each side of the glossal rod (purple), shortening the glossa and recurving the tip. Differential action of these muscles moves the tip sideways.*

glossal rod

glossal membrane

c *The glossal membrane is capable of considerable distension. The galeae and labial palps may be moved apart and the membrane of the exposed glossa suddenly and forcefully distended with haemolymph. This has the effect of throwing off particles of pollen and other debris.*

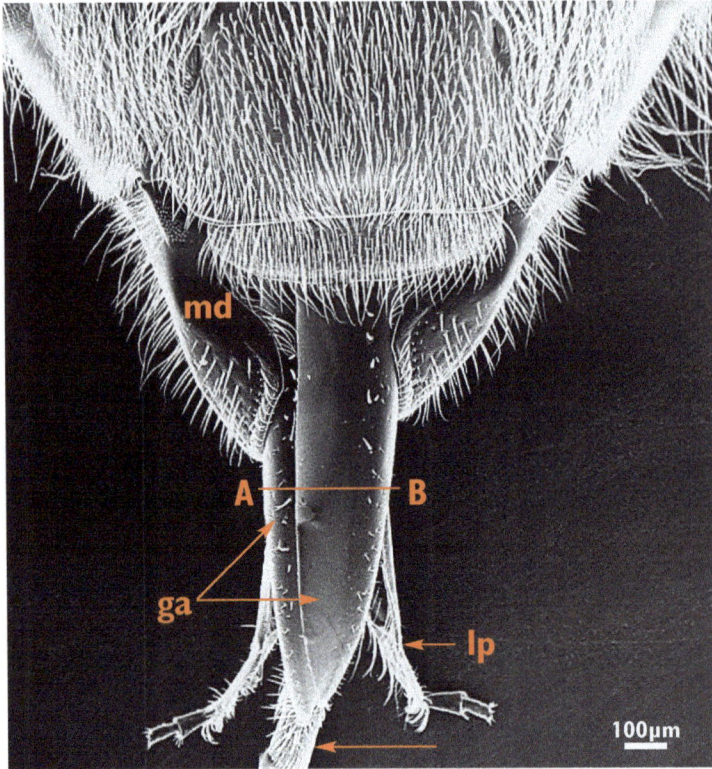

FIG. 5.17 Anterior view of the worker head showing the mandibles (md) clasping the base of the proboscis in the feeding position. The mandibles fit over the ridges on the outer surface of the galeae (ga). The labial palps (lp) can be seen in the rear together with a small section of the glossa (gls) emerging from the food canal formed by the galeae and labial palps. A—B shows the position of the section through the food canal shown diagrammatically in fig. 5.18.

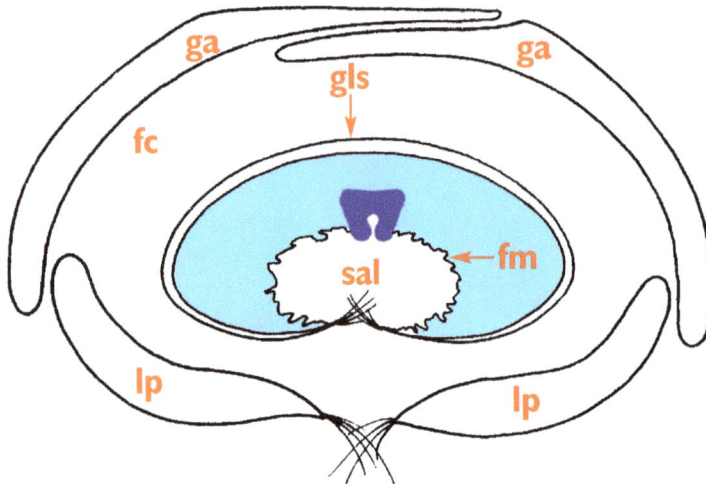

FIG. 5.18 Cross-section of the proboscis in the region A—B in fig. 5.17, showing the food canal (fc) formed around the glossa (gls) by the anterior paired galeae (ga) and the posteriorly situated labial palps (lp). Hairs on the labial palps close the food canal at the rear. The glossa, with its flexible rod (purple) lies centrally. The glossa is extended in length when haemolymph (light blue) is pumped in. Flexible glossal membrane (fm) which allows some expansion when haemolymph enters. Saliva flows down the salivary canal (sal) closed by hairs on the posterior surface. When the glossa is retracted, it draws the liquid held in its outer layer of hairs up into the food canal (hairs on glossa omitted).

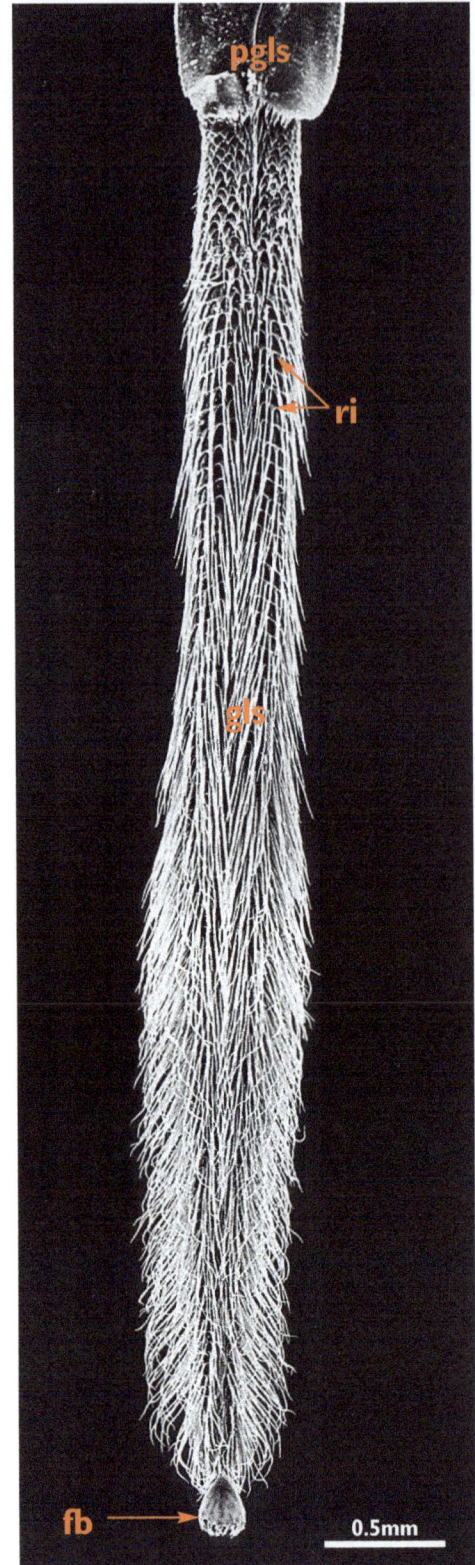

FIG. 5.19 Anterior surface of the glossa (gls). The small lobes of the paraglossa (pgls) surround the base of the glossa. The glossa is composed of rings of strengthened (sclerotized) cuticle (ri) bearing long, distally-pointing hairs, alternating with bands of membranous cuticle (see also fig. 5.20). The glossa terminates in a small, spoon-shaped flabellum (fb).

coupled with some muscular activity[30,31]. (Haemolymph is the 'blood' of the insect which is not confined to blood vessels but bathes all the tissues and organs within the body.) The stipites and the pre- and post-mentum are hollow structures containing haemolymph in communion with that of the head capsule and the remainder of the body. Haemolymph is forced into the hollow but firm galeae and labial palps bringing about their extension. Cuticular elasticity probably also helps extension. The bee is not alone in using this method to extend its mouthparts. A combination of muscular activity and haemolymph pressure extends the lepidopteran (butterflies and moths) proboscis and, together with air, plays a part in extending the fly proboscis[28,32]. The glossa is also hollow, but its posterior wall is not completely closed. However, its walls are lined with a membrane which encloses the haemolymph. Haemolymph pressure protracts the base of the glossa and paraglossa, the ligula, so that the ligula is pushed out of the prementum and the strengthening ligula arms and plate are unfolded. The glossa becomes enclosed by the galeae and the labial palps so that a further increase in haemolymph pressure results in an extension of its length so that it stretches well beyond the food canal to reach down into the nectaries. Occasionally the bee holds the galeae and the labial palps away from the glossa and, by a sudden forceful injection of haemolymph, distends the glossal membrane so that it is seen outside the glossa. This is done to shake off pollen and debris from the glossa. The complex movements of the various parts of the maxillae and the labium involved in forming the proboscis and in folding it away when not in use are best understood by reference to the diagrams (figs 5.14c, d; 5.15; 5.16) illustrating the actions of each part.

Lapping and sucking up liquids

The glossa, or tongue, is composed of a series of rings of cuticle bearing many long hairs pointing distally, interspersed with bands of more flexible cuticle. The hairs

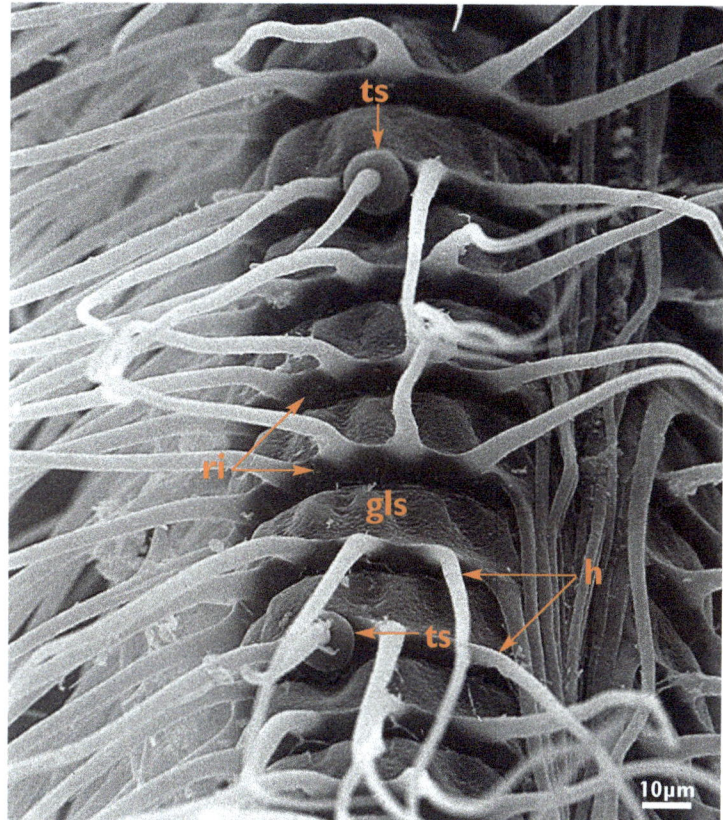

FIG. 5.20 Anterior surface of the glossa (gls) showing the rings of sclerotized cuticle (ri) alternating with membranous cuticle. This arrangement permits the lengthening, shortening and bending of the glossa. Taste receptors (ts) are interspersed between the long, distally-pointing hairs (h) borne on the rings of sclerotized cuticle.

become longer towards the tip (figs 5.19, 5.20) and form an integral part of the nectar-gathering process. At the tip, the glossa opens out into a small spoon-shaped lobe, the **flabellum** or **labellum**, edged by small branched hairs (figs 5.21, 5.22). The glossa can be extended well beyond the food canal formed by the folding of the galeae and labial palps around the glossa or it can be shortened so that it lies entirely within the canal. The exposed region of the glossa is very mobile and can be moved in all directions by the differential action of the pair of muscles, 20, acting on the base of the flexible glossal rod (fig. 5.16b). When dipped into a pool of nectar, the liquid moves onto the glossa and up between the hairs by capillary action. The hairs expand outwards as the glossa becomes laden with fluid, the radius of the glossa increasing by up to 40–50%. One study suggests that the hairs are attached to the glossal walls in such a way that they are swung outwards automatically when the glossa extends to increase its capacity to hold nectar[29]. The glossa is now retracted, carrying the liquid up into the food canal. Here it is removed from the glossa and drawn up through the mouth into the cibarium (cibariopharyngeal sac) by the action of the cibarial pump. Muscular compression of the expanded cibarium forces the nectar onwards into the oesophagus.

The repeated extension and retraction of the glossa forms the bee's **licking cycle**, comprising three phases: loading, retraction and unloading[12]. Capillary forces are important in the loading phase when the tip of the glossa is dipped into nectar and the hairs present contribute substantially to these forces by increasing the contact length at the interface between the nectar and the air. The interstices between the hairs create many fine tube-like gaps in which capillary forces arise, drawing the fluid up and onto the glossa. The forces that drive the nectar onto the glossa and influence the size of the load will be increased with increasing length of the hairs, whereas the viscous forces that resist unloading of the glossa by the cibariopharyngeal pump increase with the surface area of the hairs. The density and length of hairs found on the bee's glossa presumably represents a compromise between these two factors. The viscosity of the nectar solution influences the volume of nectar acquired in one lick, the volume decreas-

ing with increasing concentration of the nectar. The licking cycle is quite rapid; values of 4–5 cycles/second are seen in bumble bees and the honey bee's rate is likely to be of the same order.

If there is a large quantity of nectar available which the bee can reach without extending the glossa, it will dip the food canal itself into the nectar and the suction pump will operate directly to draw up the fluid into the cibarium. Losing the intermediate licking cycle makes this a more efficient means of taking up nectar[12].

The manner in which nectar is ingested is well understood in the bee but the role of the saliva in feeding is less well known and accounts of its use vary considerably. Some authors suggest that it travels to the tip of the glossa when the bee is feeding on nectar and is mixed with the nectar when that liquid is withdrawn into the food canal[11]. The saliva is said to be discharged from the anteriorly situated salivarium at the base of the paraglossal lobes. From here it is guided around the glossa by the lobes into the open end of the glossal canal. The saliva passes down the glossal canal emerging onto the posterior surface of the flabellum. While accepting that this is the route by which saliva reaches the tip of the glossa, other studies suggest that saliva is discharged at the tip only when the bee is feeding on solid substances such as sugar[2]. When feeding on sugar, the mobile tip is recurved and the anterior face of the flabellum rasps back and forth to break up

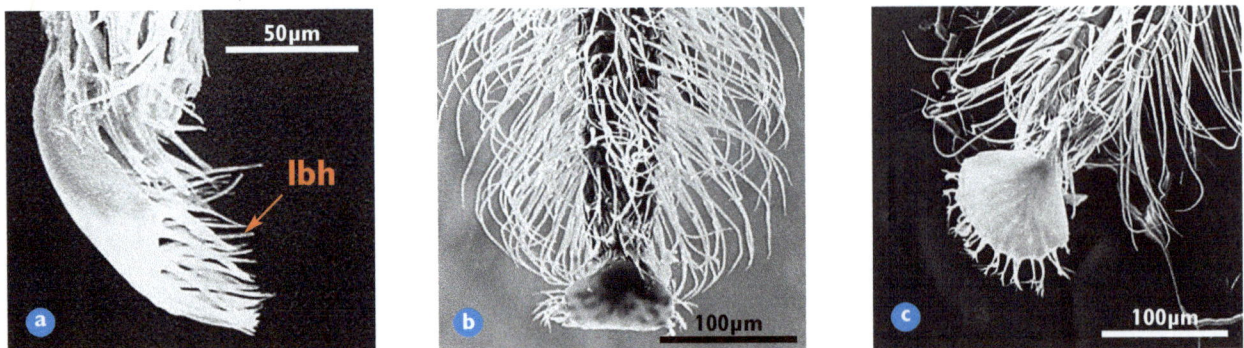

FIG. 5.21 a Lateral view of the flabellum showing its fringe of subdivided hairs (lbh). **b** The glossa viewed anteriorly showing the flabellum curved upwards to form a spoon. **c** Posterior view of the flabellum.

the sugar. The saliva moistens it and turns it into a syrup that is taken up into the hairs in the normal way. However, it has also been reported that bees feeding on sugar do so by alternately discharging saliva down the food canal, rather than the glossal canal, and then sucking up the resultant syrup through the same canal[29]. These authors also point out that saliva can be discharged from the salivarium when the ligula region of the proboscis is almost fully retracted, thus making it available to moisten materials that the mandibles are chewing.

How do bees use their mouthparts to feed themselves?

For the first three or four days after emergence, workers are fed by other bees by the method of food exchange known as **trophallaxis**. The bee begging for food palpates the antennae of another bee with its own antennae while at the same time thrusting the tip of its tongue towards the mouth of the other bee. If the other bee is going to offer food it opens its mandibles wide and, keeping its proboscis retracted beneath the head, pushes the proboscis base slightly forward and regurgitates a drop of fluid onto it. The begging bee then stretches its proboscis towards the drop and sucks it up. Antennal contact between the two is maintained throughout food exchange (fig. 1.16b), possibly for the purpose of exchanging sensory information, possibly to keep the two bees aligned with one another. These young bees are said to receive honey, a little nectar and perhaps some brood food[7]. After the first few days, the worker bees feed themselves on honey, which they lap from the cells within the comb using the tongue. Pollen is also needed for the first seven days for proper glandular development, declining at 8–10 days. The pollen is manipulated into the mouth from the pollen store by the mandibles.

Queen honey bees are fed by the workers of their retinue using the method of trophallaxis, usually by workers that are producing brood food. The adult queens

are said, by some authors[7], to receive mostly brood food, probably with some additional honey added, while others suggest that they may receive food resembling royal jelly[27]. They are able to feed themselves on sugar candy if isolated from workers, but seldom feed themselves within the hive. Drones are fed for the first few days of their life by nurse bees. The food given to them is a mixture of brood food, pollen and honey and occasionally the regurgitated contents of the honey sac. Older drones feed themselves from honey stores in the comb. Like the queen, their proboscis is shorter than that of the worker but they are able to lap from the honey stores. Larval feeding, by the discharge of glandular secretions has been described earlier (see page 74).

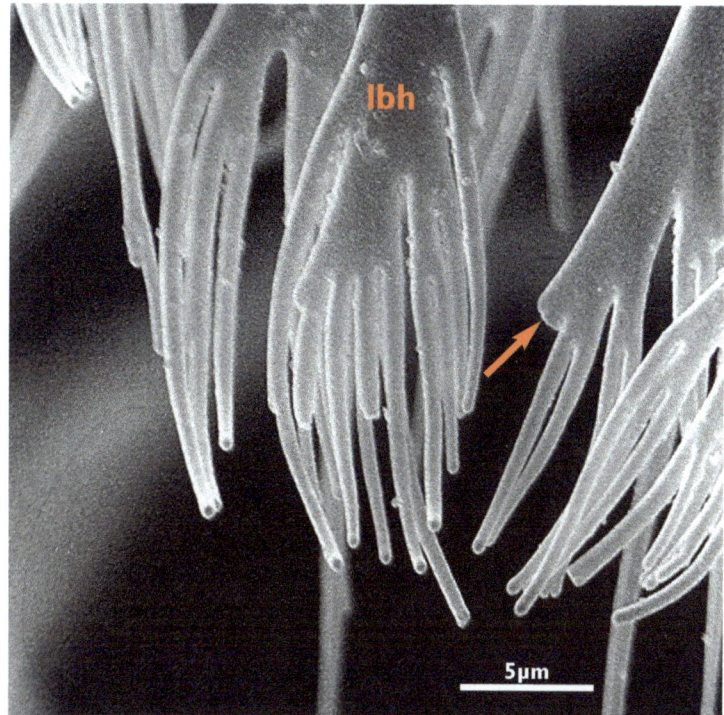

FIG. 5.22 *The lobed hairs (lbh) borne on the rim of the flabellum. In older bees these hairs may be broken or torn (arrow).*

5. Feeding:
5.2 Tasting the food

In humans, taste receptors are found in the mouth, 90% of them on the dorsal surface of the tongue (fig. 5.23a), with the remainder on the roof of the mouth and in the pharynx. They serve to sample substances that have been taken into the mouth, making the ultimate selection of potential food before it is swallowed. The receptor cells are located in little pockets, or taste buds, on papillae projecting above the surface of the tongue (fig. 5.23b). Each taste bud contains around 40–60 receptor cells (fig. 5.24) and there are some 10 000 taste buds so that food is sampled by a large array of receptors. As the food is placed in the mouth, the olfactory receptors in the nose may also be stimulated and their response will help to enhance the flavour of the food. We notice the loss of this contribution from the olfactory receptors when we have a cold! The receptor cells bear fine processes on their apical surface which carry the receptor sites for the tastant substances. Cells within the taste bud secrete mucus to bathe these processes. The tastant substances have to be dissolved in saliva in order to reach the receptor sites. If a particular tastant can react appropriately with receptor sites on a taste cell, this will lead to opening of ionic gates in the cell membrane. The subsequent movement of ions results in the cell's depolarization, leading to release of a neurotransmitter. The neurons innervating the cell are activated by the transmitter and their signals are transmitted to the brain.

We have four primary taste sensations, namely, sweet, sour, salt and bitter. If electrophysiological recordings are made from individual taste receptor cells in vertebrate taste buds, it is found that there are four types of cell giving sweet-best, sour-best, salt-best or bitter-best responses. This means, for example, that a sweet-best cell may give a small response to a salty or a bitter substance but it will give a very much larger response to a sweet substance. Although the cells within a taste bud are not exclusively of one type, there is some regional specialization on the human tongue. Sensitivity for sweetness is greatest on the front of the tongue, salt and sour on the sides, and bitter at the back of the tongue and on the soft palate. Taste buds in the centre two-thirds of the tongue seem to respond to each of the four taste qualities. There is some behavioural evidence for the existence of two other primary tastes, a metallic taste located in the middle of the tongue and a soapy taste at the very tip but this is not widely accepted. The rich variety of taste sensations that we experience is thought to be due to the differential stimulation of each of the four

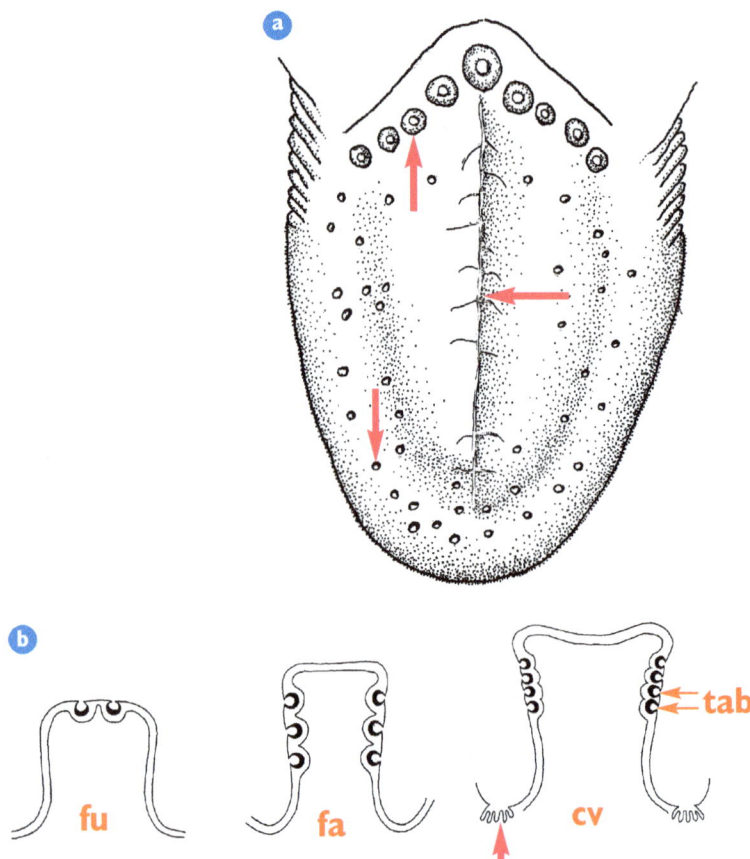

FIG. 5.23 a The human tongue is covered with papillae (arrows) which give the surface a rough appearance. There are four types of papillae: filiform, found over the entire surface; fungiform, found at the tip and sides of the tongue; foliate, forming a series of folds along the centre; and large, circumvallate papillae lying across the back of the tongue. b The fungiform (fu), foliate (fa) and circumvallate (cv) papillae bear the taste buds (tab) on their dorsal or lateral walls. The serous glands (arrow) at the base of the papillae, secrete mucus onto the tongue.

types of receptor by the material taken into the mouth. Many substances that are toxic in high concentration taste bitter, for example, alkaloids in certain plant foods, and stimulation of the bitter receptors across the back of the tongue can evoke rejection of food in the mouth. Tactile stimulation of this region can also evoke gagging, spitting and vomiting reflexes leading to the ejection of unsuitable or dangerous food. Thermal and pain (nociceptive) receptors are also present at the back of the tongue and on the soft palate, and the hot, spicy sensations associated with certain foods, such as peppers, horseradish, mustard and ginger, are due to the irritative effect of their active ingredients on the nociceptive receptors. These pungent compounds are normally rejected by animals but man can develop a 'taste' for them[13].

Insect taste receptors (contact chemoreceptors)

The receptors can occur both on the mouthparts and in the mouth; on the antennae; on the feet; on the genitalia, particularly on the ovipositor; and sometimes on the general body surface. As mentioned in chapter 1, the taste sensillae of the honey bee consist of hairs and pegs very similar in appearance to the olfactory receptors, except that here, instead of the entire surface being covered in pores, there is just a single pore at the tip (see fig. 1.14). Commonly, 4–6 neurons innervate each sensilla; their cell bodies lie at the base of the shaft and each sends an extension (a dendrite) up to the tip. The pore at the apex of the sensilla is filled with a viscous fluid, containing a mucopolysaccharide, which bathes the dendrites. The receptor sites for the stimulant chemicals lie in the membrane of the dendrites. As with olfaction, there is the problem of how these chemicals reach the receptor sites. For tasting to occur, the stimulus source has to be in contact with the taste sensillae and the stimulating chemicals in the food must come in contact with the fluid-filled pore. The molecules of the tastant dissolve

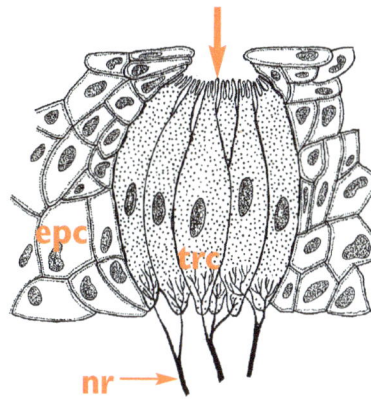

FIG. 5.24 Each human taste bud contains 40–60 receptor cells (trc). Each cell bears fine processes at its apex upon which the receptor sites for tastant molecules are situated (arrow). The cell processes are bathed in mucus secreted by cells within the taste bud. Each cell may be innervated by more than one neuron (nr) and each neuron may innervate several cells. The lifespan of each receptor cell is very short, of the order of 10 days in humans, new receptor cells being differentiated from the surrounding epithelial cells (epc).

in, or diffuse through, the fluid to reach the receptor sites on the dendrites. As in vertebrates, if the molecules are of such a nature that they can react with the specific protein acceptor sites on a particular cell, then ionic gates in the membrane will be opened, and the movement of ions into the cell will ultimately lead to its depolarization and the initiation of nerve impulses in the neuron.

Like man, insects have taste receptors sensitive to sweet, to salty, to sour or acid and to bitter compounds. Some of them, like

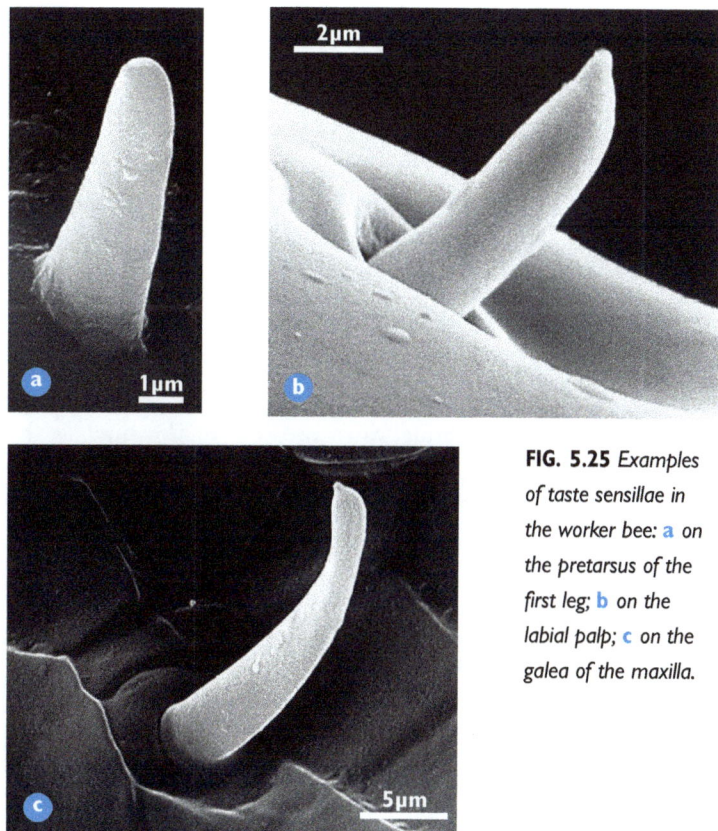

FIG. 5.25 Examples of taste sensillae in the worker bee: a on the pretarsus of the first leg; b on the labial palp; c on the galea of the maxilla.

the blowfly, also have receptors sensitive to amino acids. In addition, many insects have a cell sensitive to water or extremely dilute acids, although this is not exclusive to insects, as rats and rabbits also appear to have water receptors (osmoreceptors). Each cell present in a sensilla responds to a different range of chemicals; for example, there is usually a cell whose best response is to sugars, the sugar cell, and one whose best response is to inorganic salts, the salt cell. Depending on the insect, or on the site on the insect, there may be one whose best response is to acids, the sour cell, and/or one sensitive to behaviourally deterrent compounds, the equivalent of a bitter cell. In many insects, this deterrent cell can respond to compounds from a variety of chemical classes. In plant-feeding insects, such cells commonly respond to chemical compounds that are present in the plant to act as a protective measure, in order to reduce insect grazing. The fourth or fifth cell can be an amino acid-sensitive or a water-sensitive cell; both can be present if there are six dendrites in the shaft[13,14]. In social insects, such as the bee, where pheromones play a large part in controlling colony behaviour, some of the contact chemoreceptive sensillae may contain cells sensitive to specific pheromone molecules as mentioned in chapter 1. No individual sensilla with such specific properties has yet been identified, but the small size of the sensory pegs makes investigation very difficult.

One of the receptor cells innervating a taste receptor peg often terminates at the base of the peg. The dendrite of the cell attaches to the wall of the shaft at this point, and is stimulated not by chemical compounds but by the movement of the shaft relative to the surface of the cuticle. The sensory peg thus serves a dual function: taste receptor and mechanoreceptor. Taste receptors are normally associated with parts of the body that are concerned with exploring or manipulating food, and the mechanoreceptive cell signals contact with objects and gives positional information. Sometimes the peg is clearly sitting in a socket in the cuticle (figs 5.25 and 1.13), the base of the peg forming a ball and socket joint, suggesting the likely presence of a mechanoreceptive cell. In the absence of this clue, only investigation of the fine structure will reveal whether the receptor has a dual function.

Taste receptors in bees

Taste receptors in the bee are found on the legs, antennae and mouthparts, and in the mouth. Structurally, they consist of either smooth-walled, blunt pegs less than 10 μm long, known as basiconic sensillae, or longer, hair-like extensions of the cuticle known as sensillae chaeticae (fig. 5.25). Both types may have terminal papillae.

Receptors on the legs

Touching a drop of sugar water to the forelegs results in extension of the proboscis, and anatomical investigation has shown the presence of around 100 taste sensillae on the tarsus and pretarsus (fig. 5.26)[4]. These are mostly sensillae chaeticae, distributed evenly between the five subsegments of the tarsus. There is a rather higher concentration on the terminal, claw-bearing pretarsus. The sensillae on the pretarsus are relatively easy to see, being mainly located on the inner margins of the claws. Those on the tarsus are surrounded by the long cuticular hairs that cover the legs and by long, sharp-pointed trichoid sensillae which are innervated at the base and function as mechanoreceptors (fig. 5.26a–g). Sections of some of the sensillae chaeticae show the dendrites of four cells running to the tip of the shaft with a fifth ending at the base suggesting that there are around 400 taste receptor cells on the leg[4]. At present, it is unclear whether all of these receptors are involved in feeding behaviour or whether some of them may be involved in other behaviours, for example, in detecting the footprint pheromone of other bees (chapter 8). Detailed examination of the tarsi of the mid and hindlegs suggests that they also bear a similar number of taste sensillae[15].

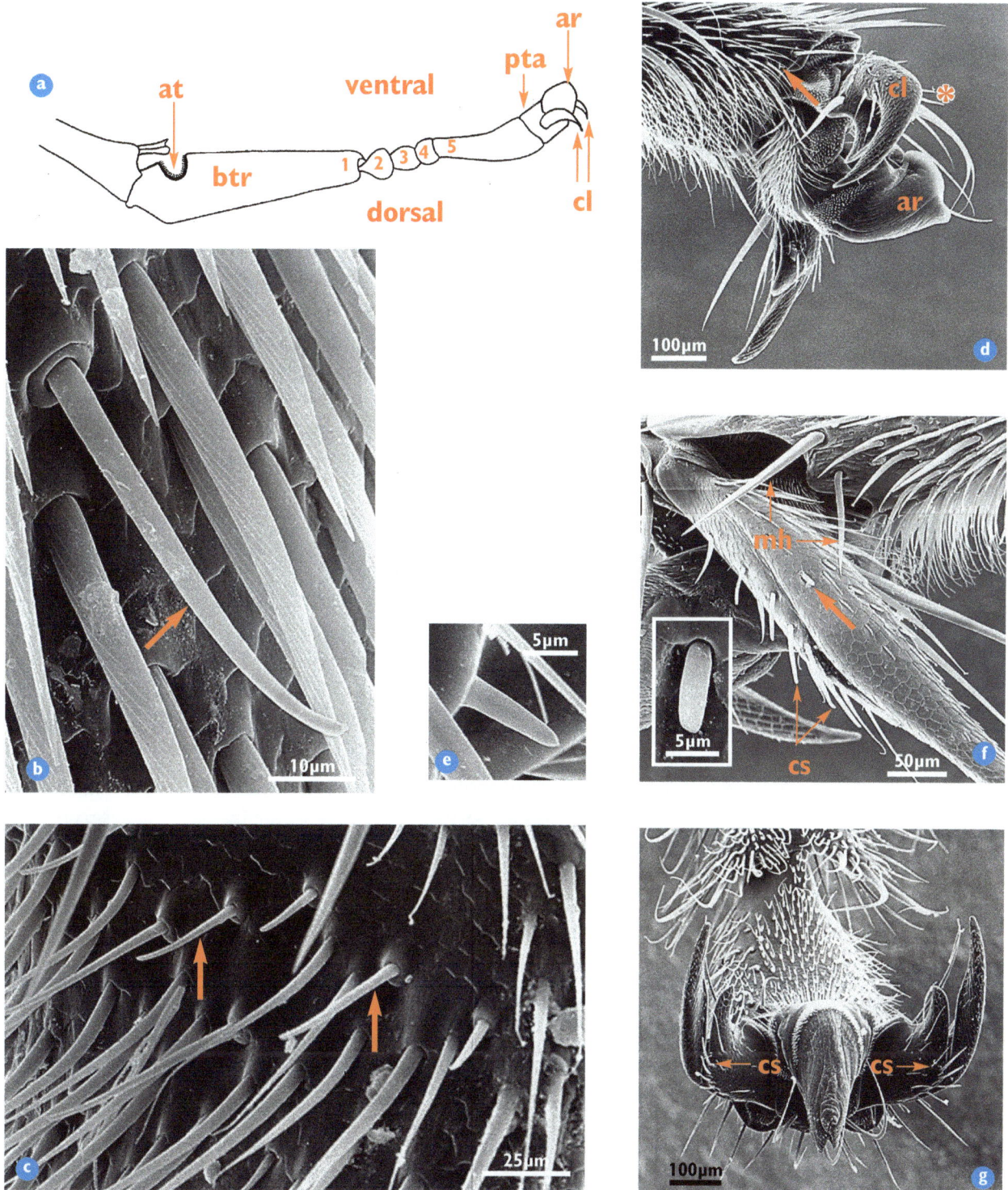

FIG. 5.26 a *The tarsus of each leg consists of five subsegments, or tarsomeres (1–5): the longer basitarsus 1 (btr) is followed by three very small segments (2,3,4) and a larger fifth. The distal pretarsus (pta) bears a pair of lateral claws (cl), or ungues on either side of a soft, apical lobe, the arolium (ar). The foreleg, shown here, bears the antennal cleaner (at).* b *Sensilla chaetica (*arrow*) on the fifth tarsomere, surrounded by cuticular hairs and sharp, pointed trichoid sensillae, probably acting as mechanoreceptors.* c *Sensillae chaeticae on the fourth tarsomere (*arrows*).* d *Side view of the distal end of the fifth tarsomere and the pretarsus showing the tarsal claw (cl) and the arolium (ar). The fifth tarsomere has small basiconic sensillae among its hairs (*arrow*) (see enlargement in* e*). The inner margins of the claws bear sensillae chaeticae (star).* e *Taste sensilla arrowed in* d. f *The claw bears a variety of sensillae. There are sensillae chaeticae (cs) along the inner rim. A small number of spatulate pegs (*arrow* and inset) have been variously described as chemoreceptors or as mechanoreceptors. Mechanoreceptive sensillae (mh) occur adjacent to the claws.* g *The underside of the apex of the pretarsus showing the sensillae chaeticae (cs) on the inner margin of the claws.*

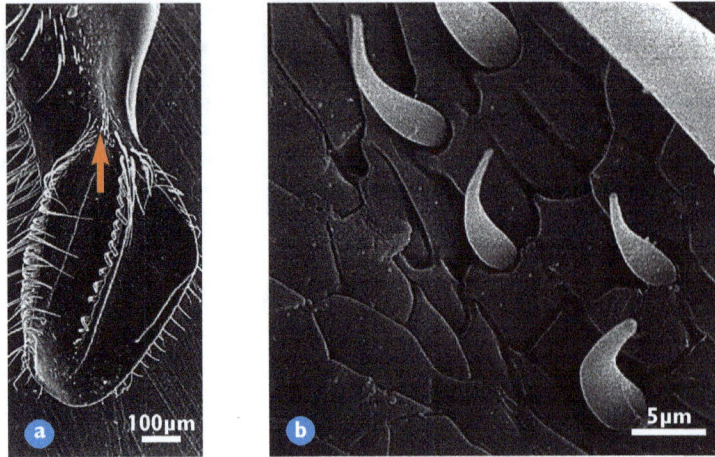

FIG. 5.27 a *The inner surface of the worker mandible showing the site of a cluster of putative chemo- receptors (arrow).* **b** *Examples of the group of pegs lying at the entrance to the groove across the spoon-shaped area of the mandible.*

Receptors on the antennae

Touching sugar water to the antennae also results in extension of the proboscis, sug- gesting that some of the taste receptors on the antennae can be involved in feeding behaviour. Around 300 sensillae chaeticae have been found distributed over the 10 annuli of the worker flagellum, with rather more present on the flattened ventral sur- face of the distal-most annulus than on the other annuli (chapter 1). These sensillae have either four or five dendrites in the shaft and a mechanoreceptive dendrite at the base (see fig. 1.13)[16,17]. There may well be many other taste receptors present on the antennae but, until the fine structure of a sensilla has been determined, it is dif- ficult to decide whether it is olfactory, gus- tatory, or mechanoreceptive in function.

Receptors on the mouthparts

Insect mandibles (fig. 5.4) do not usually bear taste receptors; however, there are a number of structures on the mandible of the worker bee that have been proposed as chemoreceptors[18]. In addition to their many mechanoreceptive hairs, the mandi- bles bear small sensory pegs, with one den- drite in the shaft[4], alongside the medial ridges on the outer surface (fig. 5.4). On the inner, spoon-shaped surface, a small group of pegs lies in the groove (fig. 5.27a, b). They would be in a position to monitor the secretions moving across the mandible. At present there is no further evidence to suggest that these structures are chemo-

receptors and their role requires further investigation.

The proboscis (fig. 5.17) is well supplied with taste receptors. The glossa (fig. 5.19) bears an average of 72 sensillae chaeticae. There are 12 long sensilla on the tip, just behind the flabellum, 10 of them lying ven- trolaterally and pointing forward (fig. 5.28). The remainder lie on the distal two- thirds of the glossa, short, peg-like struc- tures, arising from a socket, lying among the long hairs that assist in taking up nec- tar during licking (figs 5.20, 5.29)[4,19]. With the dendrites of four taste cells in each, there are around 290 taste receptors on the glossa. They are in a position to taste the nectar, both as the tip of the glossa is dipped into the pool and as it is drawn up among the glossal hairs and into the food canal. As the base of the peg is innervated, it will also function as a mechanoreceptor, being stimulated as the nectar moves among the hairs and displaces the peg rel- ative to its socket. Thus, as well as tasting, the pegs are capable of monitoring the flow of liquid onto the glossa.

There are more taste sensillae on the three distal segments of the labial palps. The much longer first segment of the palps forms part of the food canal, together with the galeae, but the three short segments protrude beyond this; they bear the chemoreceptors and are in a position to be stimulated as the bee performs its many roles within the life of the colony. The tip of the distal-most, or fourth, labial segment bears a battery of short, peg-like basiconic sensillae and longer, blunt sensillae chaeti- cae (fig. 5.30). A second group of sensillae chaeticae surround the tip of the third seg- ment, while shorter pegs are found on the outer surface. On the second segment, the sensillae chaeticae and basiconic sensillae are confined to the outer ridge, somewhat hidden among cuticular hairs. Altogether, there are some 50–65 sensillae on each labial palp.

The galeae of the maxillae have around 12 sensillae chaeticae distally along their lateral margin with basiconic sensillae interspersed between them (fig. 5.31). The lumen of the sensilla chaetica contains four dendrites and a fifth, presumably a mechanoreceptor, innervates the base. A few sensillae are found along the inner surface of the galea, with a further 40–50 basiconic sensillae near the base (figs 5.31, 5.25)[4]. The reduced maxillary palp bears no chemoreceptors.

Receptors in the mouth

Food entering the mouth will contact some 50–60 **hypopharyngeal sensillae** lying in two patches on the floor of the cibarium just anterior to the openings of the hypopharyngeal glands (fig. 5.32). Light-microscope observations suggest that they are innervated by four neurons[19]. Since their structure resembles that of cibarial contact chemoreceptors in other insects, it is assumed that this is their function in the bee. If this is so, then food is sampled by a battery of some 200–240 receptor cells before it passes on into the oesophagus. The receptors are also in a position to sample brood food and solutions regurgitated by worked bees. The **epipharyngeal pads** are reported to bear many, very small, peg-like structures (fig. 5.32a). They apparently have no apical papillae or pores, but are said, on the basis of light microscope studies, to be innervated by one neuron[19]. It is not clear whether these structures are receptors, possibly serving as mechanoreceptors detecting the closing of the mouth aperture when the proboscis is swung into the feeding position, or cuticular sculpturing.

How bees use their sense of taste

Our current understanding of taste in bees has been achieved in a number of ways. Sugars and other compounds have been presented for foraging bees to collect and their behaviour observed[20]. If the tarsi of the forelegs are touched with a suitable sugar solution, the proboscis is extended. Touching the tip of the antennae also results in proboscis extension[21]. Foraging

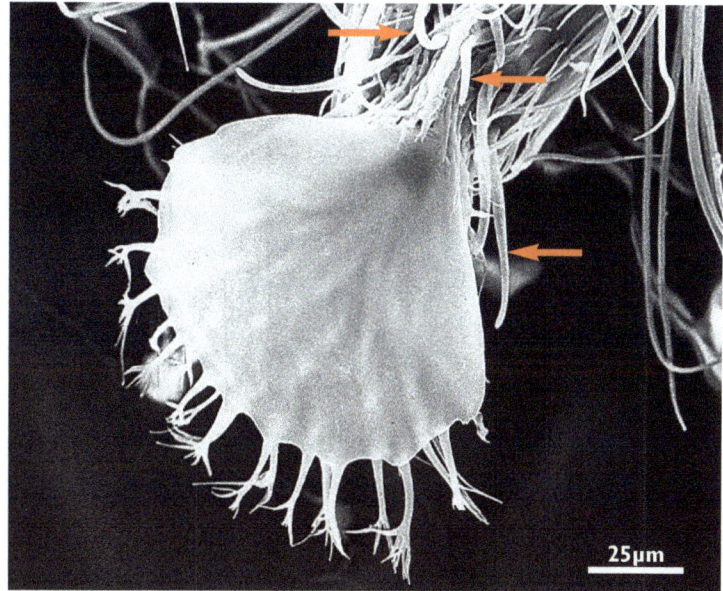

FIG. 5.28 Some of the long sensillae chaeticae (arrows) lying just behind the posterior surface of the flabellum.

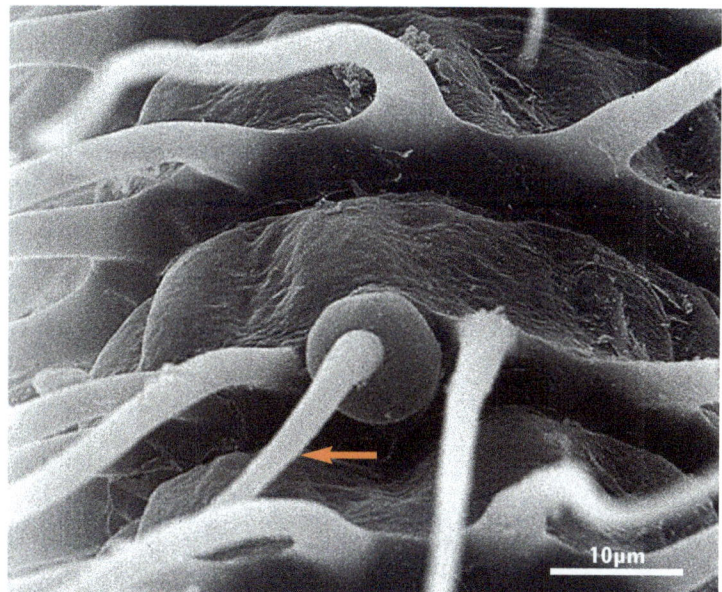

FIG. 5.29 The distal two-thirds of the glossa bears short, peg-like taste receptors (arrow) set in prominent sockets on both the anterior and posterior surfaces. Four dendrites run to the apical region of the peg and a fifth innervates the base[4,19] suggesting that the sensillae act as both contact chemoreceptors and mechanoreceptors.

behaviour and the proboscis extension reflex have been used to study the bee's ability to taste different sugars, to determine the threshold of response and to examine the response to other compounds. The stimulus in these cases has to be sufficient to trigger a behavioural

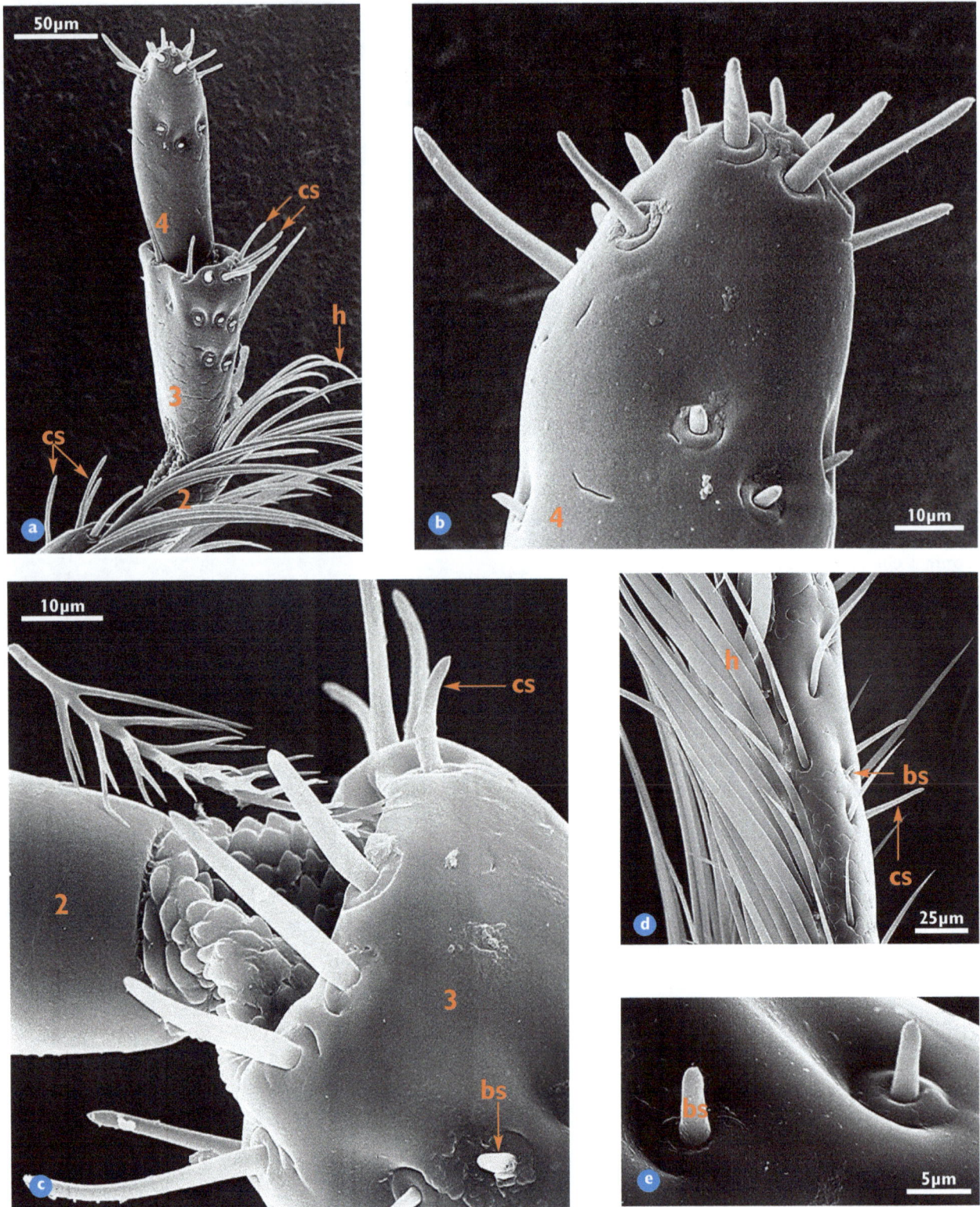

FIG. 5.30 a *Part of segment 2 and segments 3 and 4 of the labial palps. Only these three distal segments (2, 3 and 4) bear contact chemoreceptors. A ring of sensillae surround the tip of the distal-most segment while a few shorter pegs are found in the lateral walls. Segment 3 has long sensillae chaeticae (cs) on its outer distal rim and short pegs on the outer lateral surface. The second segment has sensillae chaeticae and basiconic sensillae (see d) along its lateral margin. The basiconic sensillae on the second segment are not visible here, being hidden among long cuticular hairs (h) but can be seen in d. b The sensillae at the tip of the distal-most labial segment are of varying lengths. Five neurons innervate each sensilla[4]. c Sensillae chaeticae (cs) surround the outer side of the distal rim of segment 3. Basiconic sensillae (bs) occur in cuticular depressions in the lateral wall (see also e). d Basiconic sensillae (bs) and sensillae chaeticae (cs) on the lateral margin of the second segment. e Basiconic sensillae (bs) on segment 3.*

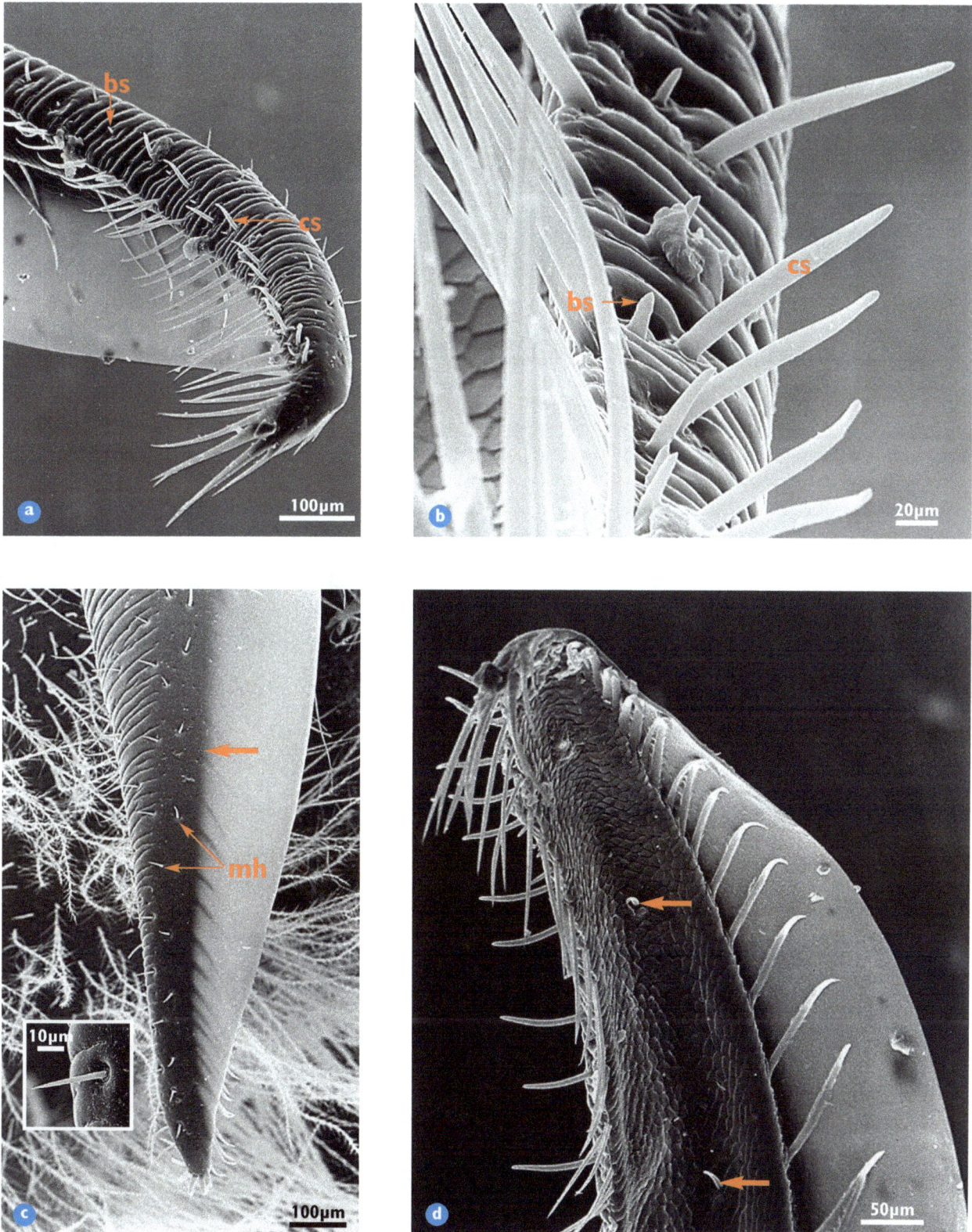

FIG. 5.31 a *Sensillae chaeticae (cs) and basiconic sensillae (bs) are situated on the lateral margin of the distal two-thirds of the galea.*
b *Basiconic sensillae (bs) roughly alternate with sensillae chaeticae (cs).* **c** *The outer surface of the galea showing the central ridge (arrow). Short, sharp-pointed mechanoreceptors (mh) (see also inset), are found on the inner surface of the ridge.* **d** *The inner surface of the galea bears a few contact chemoreceptors along its length (arrows) with a larger number at the base, near the reduced maxillary palp (see also fig. 5.25c).*

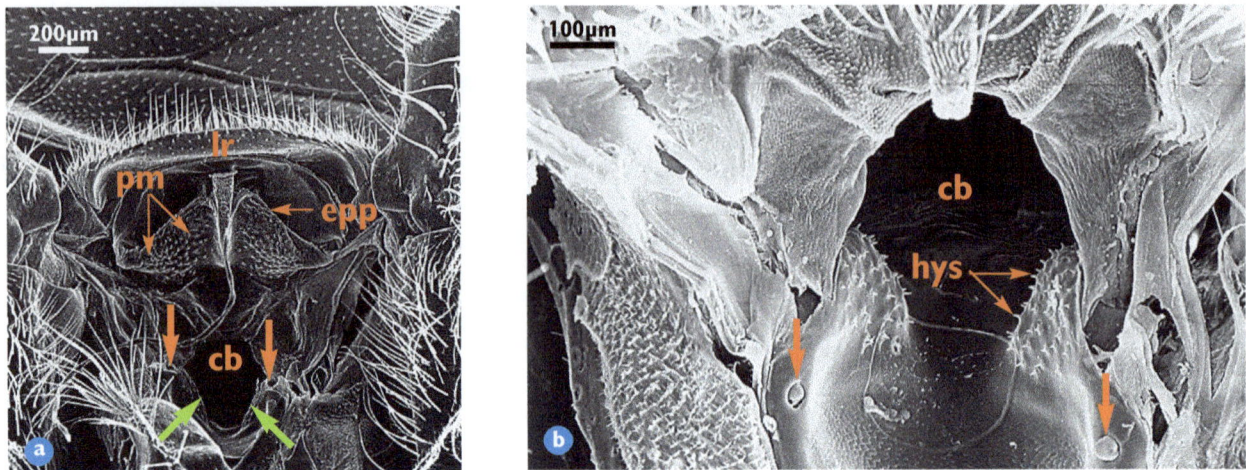

FIG. 5.32 a *The oral cavity opened to show the upper lip or labrum (lr) and the epipharyngeal pads (epp) beneath it. The floor of the cavity has been pulled away to expose the entrance to the cibarium (cb). The openings of the hypopharyngeal glands lie on either side of this aperture (orange arrows); the hypopharyngeal chemoreceptive sensillae are not visible but their site is indicated by green arrows. Very small pegs (pm) covered in debris in the living bee, are found on the surface of the epipharyngeal pads. These are said to be innervated by one neuron and are thought to be mechanoreceptive in function[19].* b *The entrance to the cibarium (cb) showing the two groups of 50–60 hypopharyngeal sensillae (hys) and the openings of the hypopharyngeal glands (arrows).* c *The hypopharyngeal sensillae (hys) resemble small basiconic sensillae with apical papillae (arrow). Their structure resembles that of cibarial receptors in other insects that are known to function as contact chemoreceptors. They are assumed to have the same function in the bee, sampling material passing in or out of the cibarium.*

response. Very few electrophysiological recordings have been made from individual sensillae, a difficult operation due to the very small size of these structures. In successful recordings, the responses of the individual cells within the sensilla can usually be distinguished so that we know the response characteristics of a few of the taste receptor cells on the labial palps and on the galeal lobes.

The **nectar** collected by bees consists of an aqueous solution of sugar whose concentration can vary as widely as 5–80%, although it is more usually around 35–45%. The disaccharide, sucrose, is the predominant or sole sugar present in some nectars, but other nectars contain almost equal proportions of sucrose and the monosaccharides, glucose and fructose, while others contain mostly glucose and fructose. There may be small traces of other sugars present, for example, maltose,

melibiose, raffinose and melezitose. In addition, there are very small quantities, in some cases mere traces, of nitrogenous compounds, organic acids, lipids, minerals, vitamins, and aromatic compounds, but these substances together comprise less than 2% of the nectar. Honeydew, the secretion of sap-sucking insects such as aphids, also collected by bees, has a similar composition to nectar but may also contain trehalose and the trisaccharide, melezitose. The nectar or the honeydew is carried back to the nest or hive in the bee's honey sac (crop) and transferred to nest workers for ripening into honey.

Enzymes secreted by the hypopharyngeal glands, including invertase (α-glucosidase) and glucose oxidase, are added to the nectar. Invertase converts the sucrose into its component sugars, glucose and fructose. This has important consequences for the production and storage of honey since converting sucrose into glucose and fructose at hive temperatures makes it possible to produce a very concentrated solution of sugars, usually over 80%. At the temperature of the hive (30°C), the solubility of glucose in a solution of fructose increases markedly if the concentration of the fructose is raised above 1.5 g per g of water[22]. As a result of the increased fructose concentration, one gram of water can carry 1.25 g of glucose, twice as much as a dilute solution of fructose can dissolve, enabling the production of a highly concentrated food for storage. Sucrose does not have this high solubility. Apart from the high energy value of the stored product (3040 kcal/kg[12]), fully ripened honey is protected from microbial attack by its low water content. The hygroscopic environment thus created, rapidly dehydrates most bacteria and yeasts, and allows the capped honey to remain stored in the comb without deterioration. The freshly collected nectar is quite dilute and would provide an ideal growth medium for micro-organisms; however, in honey with a sugar concentration of between 23% and 30%, glucose oxidase catalyses the oxidation of glucose to form gluconic acid and hydrogen peroxide, which together act as an antibacterial agent. The activity of glucose oxidase is gradually reduced as the glucose concentration increases, ceasing when conversion is complete[7,22].

In foraging tests where bees were presented with some 34 different sugars or substances closely related to sugars, 30 of which tasted sweet to man, the bees accepted only seven sugars[20]. When bees are supplied with decreasing concentrations of the individual compounds in stepwise order, a limiting value, termed the acceptance threshold, is reached, below which the solution is no longer collected. The seven sugars, in order of their apparent sweetness to the bee determined in this manner, were: sucrose = maltose > trehalose = glucose = fructose > α-methyl glucoside > melezitose. Five of these sugars are present in appreciable quantities in nectar and honeydew: sucrose, glucose and fructose in nectar, and all three, together with trehalose and melezitose, in honeydew. Maltose and α-methyl glucoside are apparently sweet to the bee although they are not normally part of its diet. Nevertheless, bees are able to use all of these seven sugars in their metabolic processes. Most of the sugars that were rejected by bees are found to be of little or no nutritive value to them[13].

When foragers were further tested with 19 non-sugar compounds that tasted sweet to man, including artificial sweeteners, none of them proved acceptable to the bees. If bees do not collect a particular test solution it may be because the bee cannot taste it, or it may not be sweet enough or it may be repellent to the bee. If the test substance is added to a weak solution of sucrose that is only just acceptable to the bee and makes the sucrose solution more acceptable, it would appear that the test substance was just not sweet enough to be collected. However, if the sucrose solution is now not acceptable, then the test solution may be repellent to the bee. If the test substance makes no difference to the bee's response to the sucrose solution then it is assumed that this substance does not taste sweet to the bee. Using these criteria, it was found that none of the 19 compounds appeared sweet to the bee. Artificial sweeteners, such as saccharin, are not sweet to bees: indeed, if presented at concentrations that leave a bitter after-taste to man, they are found to be repellent. Not all insects have the same response spectrum to sugars; the Lepidoptera, for example, respond to a rather wider range.

Electrophysiological recordings made from individual contact chemoreceptors on the labial palps[24] (fig. 5.30), and on the galeae[23] (fig. 5.31), have demonstrated the presence of receptor cells that respond to stimulation by sucrose, glucose and fructose. Other sugars were not tested. The size of the response of the sugar-sensitive cells was linearly related to the logarithm of the concentration of the sugar solutions, with sucrose generating the largest response in the cells examined. Dose-related responses mean that the cells are able to signal information about the relative concentrations of stimulating sugars. Sugar receptor cells often have more than one kind of acceptor site for sugar molecules on their membranes, and there is some evidence that this is true for bee sugar receptor cells. Pyranose sugars with a six-membered ring, such as glucose, react with the so-called 'pyranose-site' while fructose reacts with a second class of acceptor site, the 'furanose-site', which binds sugars with a five-membered ring[15]. If the bee is stimulated with a mixture of glucose and fructose the response is additive. However, a mixture of sucrose and glucose is not additive but smaller than would be expected due to the fact that the glucose and sucrose molecules are competing for the pyranose acceptor sites. The presence of more than one class of acceptor site does not mean that two sugars can be discriminated by the cell; it merely increases the spectrum of its response to sugars. Stimulation of any acceptor site on the membrane will result in depolarization of the cell and the generation of nerve impulses, and these will be the same whichever class of site is stimulated.

Only one of the cells in each galeal sensilla tested was a sugar-best cell. A second cell was found to be sensitive to salt solutions, in particular to sodium chloride and potassium chloride. Behavioural tests have shown that bees are sensitive to bitter substances such as quinine, rejecting sucrose solutions with quinine added; however, no electrophysiological tests have been carried out using bitter substances in the bee. The galeal and labial sensillae tested all showed the presence of a mechanoreceptive cell, stimulated when the sensillae were bent in any direction. No water-best cells have, to date, been found in the bee but very few of the mouthpart sensillae have been tested.

Based on behavioural studies, the sense of taste appears to be around 300 times less sensitive than the olfactory sense in both bees and man, but bees and other insects differ from man in that their taste thresholds can apparently vary according to the state of their nutrition, i.e. a starved insect will have a lower threshold[13]. Bees normally gather nectars containing between 10% and 70% sugar. When forage is good, bees have been found to have a relatively high behavioural threshold for sugars, e.g. 34% for sucrose[13]. As we have seen, the sugar content of nectar can vary between 5% and 80% while honey is stored at a concentration of around 80%; thus it would be a waste of energy for bees to collect very dilute nectar which would require considerable concentration in the hive. Nectar is sampled by the mouthpart receptors before it is collected and, under good foraging conditions, only nectars with a fairly high sugar concentration are accepted. In times of sparse forage, the behavioural threshold for collecting nectar can fall to as low as 4% sucrose concentration[8,20,25]. A feedback system between the foragers and the house bees allows a forager to monitor its nectar's concentration in relation to that of other foragers. A group of bees specializing in receiving nectar unloads the foragers as they return. These bees will experience nectar from a variety of forage patches and are in a position to compare the sweetness of the various nectar loads, and so can adjust their behaviour towards the forager that they are unloading accordingly. Receiving bees solicit nectar from the forager, who regurgitates a drop between its mouthparts and the receiver extends its proboscis to sample it. A forager with a relatively concentrated nectar will be rapidly unloaded (in less than 40 seconds), and such foragers are stimulated to dance and recruit other bees to their forage source. Conversely, bees with a relatively dilute nectar, in comparison to other nectars coming into the hive, will be unloaded slowly and may have to offer their nectar to several receivers. These bees are not stimulated to recruit other bees and may themselves be recruited to forage in other areas. Since the response of the receivers is related to the abundance of forage, if there is a heavy flow of nectar into the nest, the receivers will be more selective in their choice of bees to unload rapidly, whereas if the flow falls away, foragers bearing more dilute nectar will become more acceptable. In this way, the colony as a whole keeps abreast of the changes in nectar production over a wide area of forage and is flexible in its response. However, should overheating occur in the hive, then the demand for very dilute nectar or water by the bees cooling the brood area will result in the receivers unloading bees with dilute nectar or water more rapidly than those with concentrated nectar. Foragers will then be recruited into fetching water or watery nectar until the crisis has passed[8,25]. This flexibility of response requires both the ability to taste the nectar and signal information about its concentration to the central nervous system, as well as plasticity in the behavioural thresholds at which sugars are acceptable, both of which properties are present in the bee.

The role of the tarsal and antennal taste receptors in feeding are not well understood in the bee. Stimulation of the tarsi or the antennae with a sugar solution results in extension of the proboscis, as it does in many insects that feed on sugars. In some insects, for example, the blowfly, there is a hierarchy of behaviour, with the proboscis being lowered when the tarsal receptors have located a source of sugar, or water in the case of a thirsty fly. When the hairs on the blowfly's proboscis make contact with the food, the lobes of the proboscis are spread; the food is then sampled by the gustatory papillae on the outspread lobes and, if suitable, it is sucked up into the oesophagus. However, in the bee, the proboscis is very often extended before it lands on the flower and it is certainly not dependent on tarsal stimulation. It is not clear what the stimulus is for this extension, although floral odours alone can result in proboscis extension, and possibly colour may play a part. The antennal and tarsal receptors do have a very much lower threshold for sugars, as low as 0.06%, but their role in the process of nectar collection remains unknown.

Physiological factors governing the regulation of meals are quite well understood in a number of insects whose food collection is purely to satisfy their own needs, for example, the fly and the locust. In the fly, sugar solution is passed into the crop which becomes distended and so stimulates the stretch receptors in the crop nerve net. Stretching of the foregut is also monitored. Stretch receptor activity provides negative feedback which, via the central nervous system, increases the behavioural acceptance threshold for sugars and so terminates feeding. The gradual emptying of the crop and the passage of slugs of sugar back into the foregut and thereafter into the midgut, results in decreased stretch receptor activity which ultimately reduces the acceptance thresholds and increases locomotor activity to find food[26]. The physiological factors involved in terminating nectar collecting have not been examined in the worker bee. On returning to the nest, the bee is unloaded quite rapidly, although this does depend upon the quality of its honey sac contents. Emptying the honey sac presumably stimulates it to recommence foraging but again the mechanisms involved are not known. Many studies of the factors influencing the efficiency of foraging have been made, but these address neither the underlying physiological mechanisms controlling the collection and storage of food, nor the secretion of food for the larvae, queen and young workers and drones. Much work remains to be done on the physiology of feeding in bees.

5. Feeding:
5.3 Collecting the pollen

In addition to nectar, **pollen** is essential for honey bee nutrition, providing a source of protein for both larvae and adults. Pollens contain 6–28% protein, together with up to 20% lipids, including sterols, which are also important for nutrition[7,27]. Sterols are essential for the synthesis of cholesterol by the bee. Other components are present in variable amounts, including starch, sugar, various minerals and vitamins.

Detailed studies of foragers show that, although most bees in a colony collect nectar, some bees (up to 25%) collect pollen exclusively, and others (around 17%) both pollen and nectar[7]. Pollen is not stored in large amounts and the numbers foraging for pollen are adjusted according to colony requirements: brood rearing in particular stimulates pollen collection, possibly by the presence of brood pheromone[8]. Figure 5.33 shows a pollen collector leaving a flower.

Pollen is produced by the anthers of the flower, situated at the outer end of the stamens. When the pollen is ripe the anther wall splits open revealing the pollen grains. Bees collect pollen in a number of ways, depending on the flower structure. They may bite the anthers to release the pollen and then pull them towards their body with their forelegs to trap the pollen grains on the hairs covering the thorax and legs. Bees may run along the length of a flower spike shaking the pollen onto their body hairs; alternatively, they may open up closed flowers with the forelegs and reach in to take the pollen with their mouthparts and forelegs. Bees collecting nectar may also collect pollen on various parts of the body as they seek the nectaries[11]. The rough, and sometimes sticky, pollen grains readily become attached to the long, slender hairs that cover much of the bee's body. The hairs are subdivided — so that they have a 'feathery' appearance (fig. 5.34) — which makes it easier for pollen grains to become trapped.

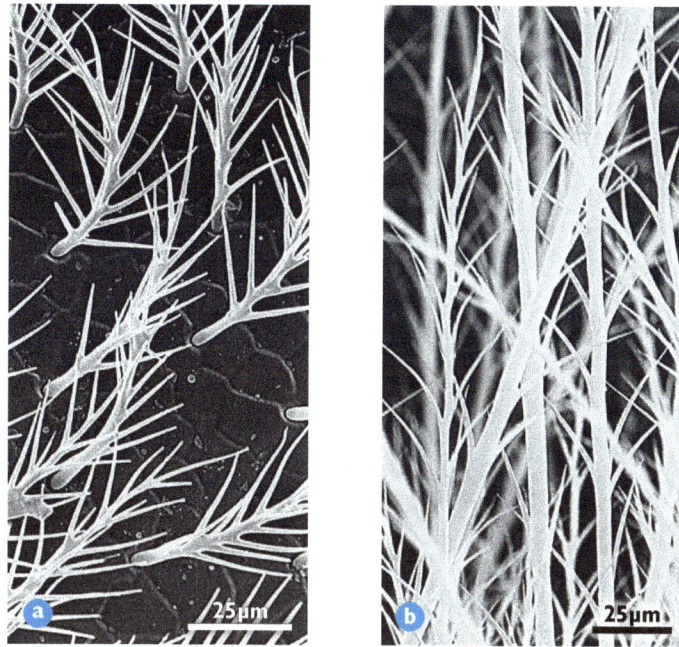

FIG. 5.34 a and **b** *Examples of plumose, or branching, hairs on the body of the bee.*

Bees can collect substantial amounts of pollen — loads up to 30 mg have been reported — and they may visit many flowers to amass this load.

How do they manage to transport the pollen back to the hive? To answer this question, some knowledge of the structure of the legs is essential. The three pairs of legs each have six segments: the **coxa**, which articulates with the thorax; the **trochanter**; the **femur**; the **tibia**; the **tarsus**; and the **pretarsus** (figs 5.35, 5.36, 5.37). The coxae of the three legs are articulated at slightly different angles to the thorax so that in action, the legs are radially distributed from the sides of the body[11]. The forelegs face forward and swing directly forwards and backwards; the middle legs face forward but at an angle to the body; and the hindlegs are attached to the posterior part of the thorax and are normally directed backward. The legs terminate in the pretarsus, which bears two claws, or ungues, and a soft pad, the **arolium** (figs 5.26, 5.35, see also fig. 8.12). The claws and the pad assist the bees in walking on a variety of surfaces, both horizontal and vertical, the

FIG. 5.33 *A pollen collector returns to the hive with massive loads of pollen packed in its corbiculae, or pollen baskets.*

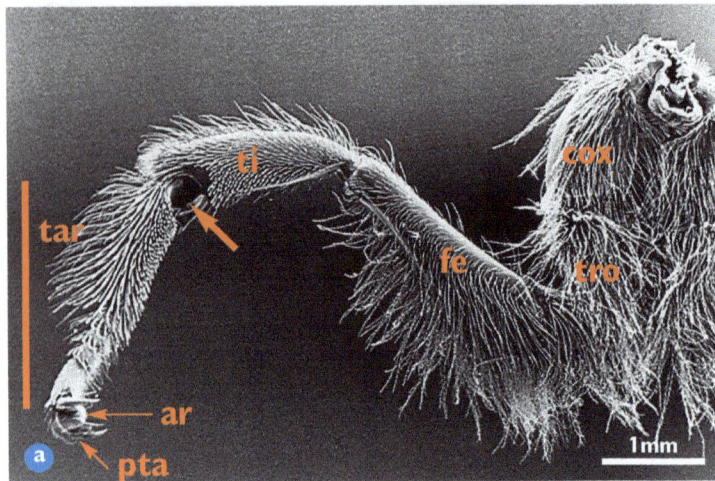

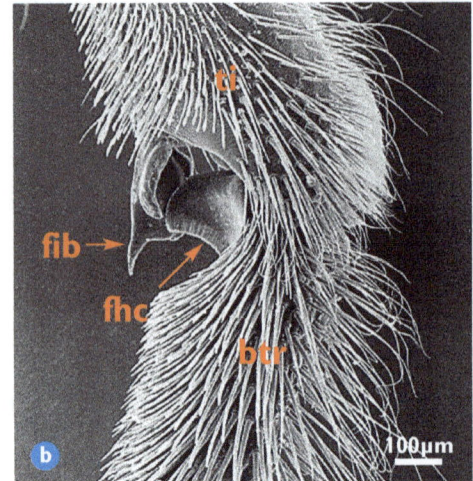

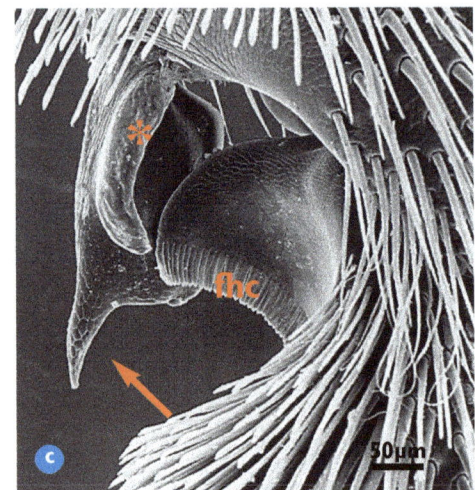

FIG. 5.35 a *The first leg of a worker bee. The segments of the leg comprise the coxa (cox); the trochanter (tro); the femur (fe); the tibia (ti); the tarsus (tar) and the pretarsus (pta). The tarsus has five tarsomeres, a basitarsus, covered with long hairs used to collect pollen, and four small tarsomeres. The distally situated pretarsus bears the claws and a soft pad, the arolium (ar) (see also fig. 5.26). Note the position of the antennal cleaner (arrow).* **b** *The basitarsus (btr) bears the antennal cleaner on its inner, proximal surface. The cleaner consists of a deep notch with a comb of fine hairs (fhc) on its outer surface. A large flattened spur, the fibula (fib) projects downward across the notch from the inner, distal margin of the tibia (ti).* **c** *To clean the antenna, the foreleg is raised and passed over the antenna so that the antennal base slips into the notch. The tarsus is then flexed in the direction of the arrow, so that the antenna is brought up against the fibula and securely held in the cleaner. The antenna is now drawn through the cleaner, the fine hairs of the tarsal comb (fhc) cleaning its outer surface, the thin accessory lobe of the fibula (star) scraping the inner surface. Antennal cleaners are also present on the front legs of the queen and the drone.*

sharp-pointed claws allowing the bee to cling to the surface. If the surface is too smooth for the claws to grasp, the arolium, which is normally folded between the claws, is brought down into a horizontal position, where the pad is unfolded and pressed flat against the surface so that it acts as a suction pad[2] (for description of claw and arolium action see fig. 8.11).

The tarsus of each leg has five subdivisions, the **tarsomeres** (fig. 5.26a): the first one, known as the **basitarsus**, is much larger than the others and forms part of the pollen-collecting apparatus. The basitarsus of the foreleg is covered with long hairs and is used to brush over the head and the front of the body to collect pollen grains. While collecting, the proboscis may be extended, and either honey or nectar (if

the bee is gathering nectar as well as pollen) regurgitated. The forelegs are passed over the proboscis and become sticky with the extruded fluid. This helps to attach and bind together the pollen grains as the forelegs sweep the mouthparts and head. In addition, the basitarsus of the foreleg bears the **antennal cleaner**, a notch containing a comb of fine hairs at the top of the segment (fig. 5.26a; 5.35b, c). The bottom end of the tibia bears a jointed spur, the **fibula**, which overlaps the notch. When the bee needs to clean debris from its antennal sense organs, it extends the foreleg forward and passes it over the antenna. The antenna slips into the notch and the tarsus is flexed so that the fibula is brought across the notch opening. The enclosed antenna is then drawn through spine is present on the distal end of the tibia (fig. 5.36b) and it has been suggested that this spine is used for detaching wax from the wax mirrors or for handling propolis, although these functions have been disputed[11].

The inner surface of the basitarsus of the hindlegs possesses nine transverse rows of long spines, set at an angle of 45° (fig. 5.37b). Although the bee can comb pollen from the abdomen with these pollen brushes, their chief function is to collect the pollen from the middle legs and transfer it to the concave region of the outer tibia of the hindlegs for storage during flight. Pollen transfer to this region, known as the **corbicula**, or **pollen basket**, is achieved during flight by a structure at the tibio-tarsal joint, the **pollen press** (fig. 5.37a). Collecting from the middle legs is achieved by grasping them one at a time between the hindlegs and then drawing them forwards, so that the pollen is scraped off onto the hind basitarsi. When packing the pollen into the corbicula, the bee hovers in front of the flower, extending the hindlegs downward, then brings them together and moves them up and down past each other. The pollen press comes into action at this point. Where the tibia and the basitarsus articulate together, there is a deep notch. The tibial lip of this notch has a row of wide, pointed spines on its inner surface forming a **rastellum**, or **rake** (fig. 5.37c), while the tarsal lip of the notch is widened to form a shallow concave shelf, the **auricle**, bordered by a fringe of long hairs and limited at its posterior end by a ridge. The surface of the auricle is covered in small spicules of cuticle. As the inner surfaces of the hindlegs are rubbed up and down against each other, the spines of each rastellum rake the pollen out of the pollen brushes on the opposing leg. The pollen falls off onto the auricular shelf and is retained there by the surrounding fringe of hairs (figs 5.37c, 5.38).

When the tarsus is flexed upwards, the pollen in the auricular shelf is pressed against the opposing tibial surface (hence the term pollen press), and squeezed onto

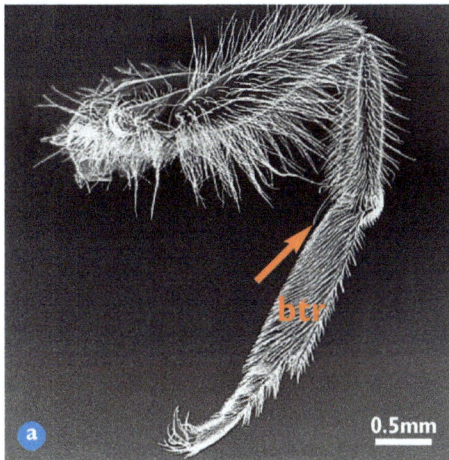

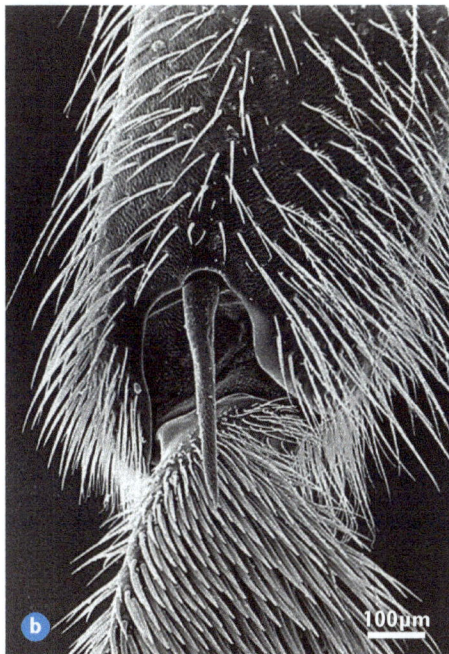

FIG. 5.36 a *The basitarsus (btr) of the middle leg is broad and flattened and covered with long hairs. It brushes the thorax with these hairs to collect pollen grains. Arrow indicates the position of the tibial spine.* **b** *The spine of the middle pair of legs is situated distally on the inner surface of the tibia.*

the notch, being brushed clear of dust and other particles as it goes.

The basitarsus of the middle pair of legs is broad and flattened, and the inner surface is covered with hairs (fig. 5.36a) with which the bee brushes the thorax to remove pollen grains. The pollen gathered by the forelegs is also scraped off onto the basitarsus of the middle legs. A single large

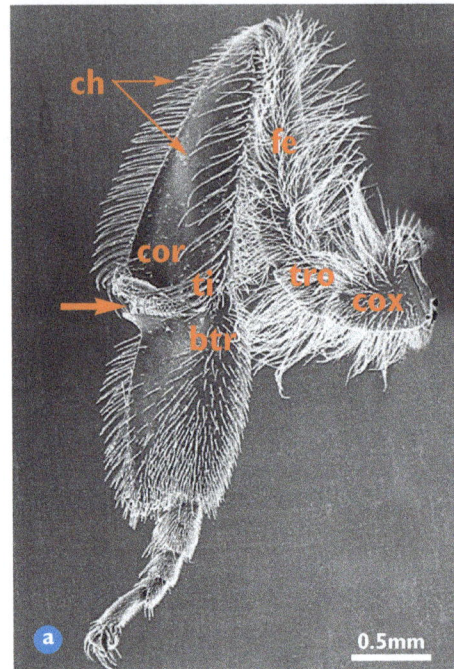

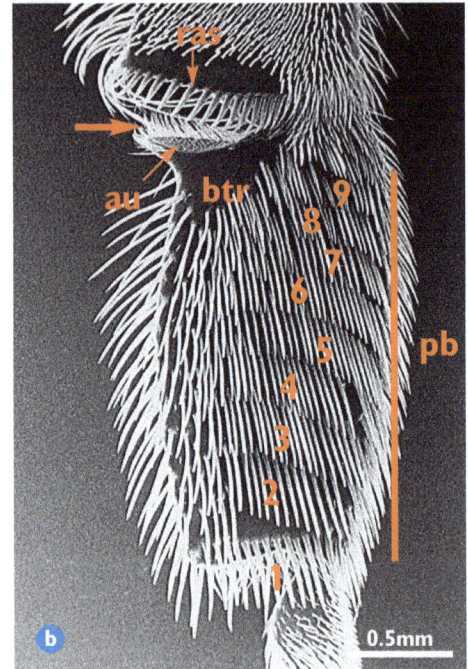

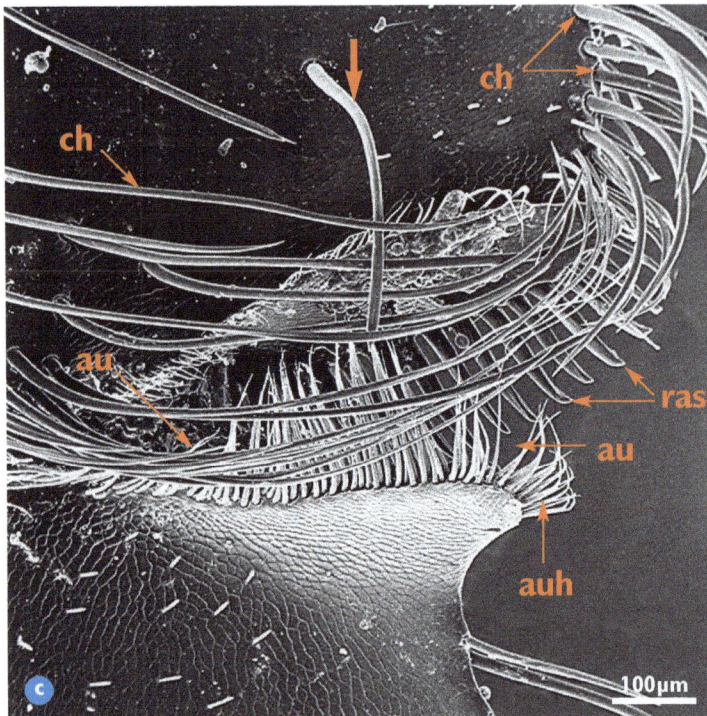

FIG. 5.37 a *The outer surface of the worker hindleg showing the broad, concave surface of the tibia, known as the corbicula, or pollen basket (cor). The corbicula is fringed on either side with long hairs (ch). An* arrow *shows the position of the pollen press developed between the distal end of the tibia (ti) and the proximal end of the basitarsus (btr). Femur (fe); trochanter (tro); coxa (cox).* **b** *The inner surface of the basitarsus (btr) of the hindleg carries nine rows of spines forming the pollen brush (pb). A row of stiff spines on the inner edge of the shallow depression on the tibia forms the rastellum (ras), or pollen rake. A shallow shelf, the auricle (au), on the inner end of the basitarsus opposes the tibial notch to complete the pollen press. The auricle bears a fringe of hairs around its outer surface (*arrow*).* **c** *The pollen press and the distal end of the corbicula viewed from the outer surface of the leg. Pollen is raked from the basitarsal pollen brush on the inner surface of the opposing leg by the rastellum (ras). The pollen falls onto the auricular shelf (au) where it is retained by the fringe of hairs (auh) around the outer surface of the auricle. The tibio-tarsal joint, or pollen press, is then closed by flexing the tarsus. The rastellum prevents the pollen from emerging on the inner surface of the tibia so that it emerges on the outer surface and collects in the corbicula. The long hairs fringing the corbicula (ch) help to catch and hold the pollen. A long stiff hair (*arrow*) in the centre of the corbicula helps to anchor the pollen mass.*

the outer side of the leg where it is caught by the hairs surrounding the corbicula. By repetitive action of the rastellum and the pollen press on each leg, pollen loads are gradually built up on each corbicula. A single long hair, or spine, near the bottom of the corbicula is said to help in retaining the pollen mass (fig. 5.37c), and the initial addition of a little honey as the grains are combed from the front of the head helps to hold the mass together. The middle legs may also help to pat the pollen into a compact mass[2,11].

When both corbiculae are loaded, the bee returns to the hive (fig. 5.33). It then stretches across a cell, lowers its hindlegs into it and prises off the pollen load with the basitarsi of the middle legs[11]. It normally displays no further interest in the load. Honey bee colonies show temporal division of labour, with bees progressing through cell cleaning and capping; to tending the queen and the brood; food handling, building and cleaning the comb; guarding and ventilating the hive and foraging[25]. Those workers engaged in food handling pack the pollen into the cells using their mandibles and forelegs and also treat the pollen for storage to prevent germination and bacterial activity. This is done by adding a phytocidal acid, whose composition is not known, apparently produced in either the hypopharyngeal or mandibular gland and thought to be related to 10-hydroxy-2-decenoic acid[7,27]. Enzymes promoting initial digestion, together with a little honey are also added. Enzyme activity prevents anaerobic metabolism and fermentation[7]. The processed pollen is often referred to as 'bee bread'.

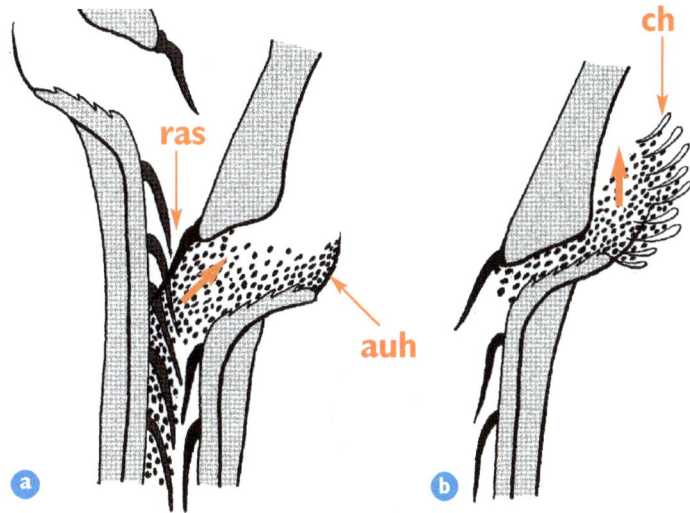

FIG. 5.38 A longitudinal section through the pollen press area of the hindlegs, viewed from the rear, to show the action of the pollen press. When packing pollen into the press, the bee hovers, extending the hindlegs downwards and moving them up and down past each other. a As the right leg is moved down past the left leg, the rastellum (ras) of the right leg scrapes pollen from the basitarsal pollen brush of the left leg. The pollen falls onto the auricle (arrow) where it is retained by the fringe of hairs on the rim of the auricle (auh). b The tarsus is flexed, closing the press, and pollen is forced up and into the corbicula on the outer surface of the tibia (arrow). Long hairs on the margin of the corbicula (ch) help to retain the pollen which is then packed into a compact mass by the middle legs. Redrawn from Dade[11].

6. Respiration:

how does the bee breathe?

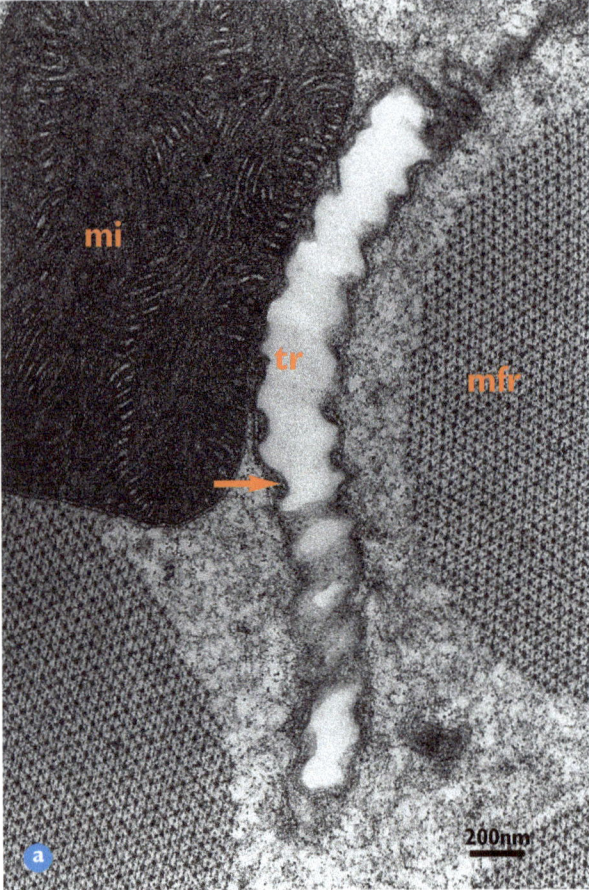

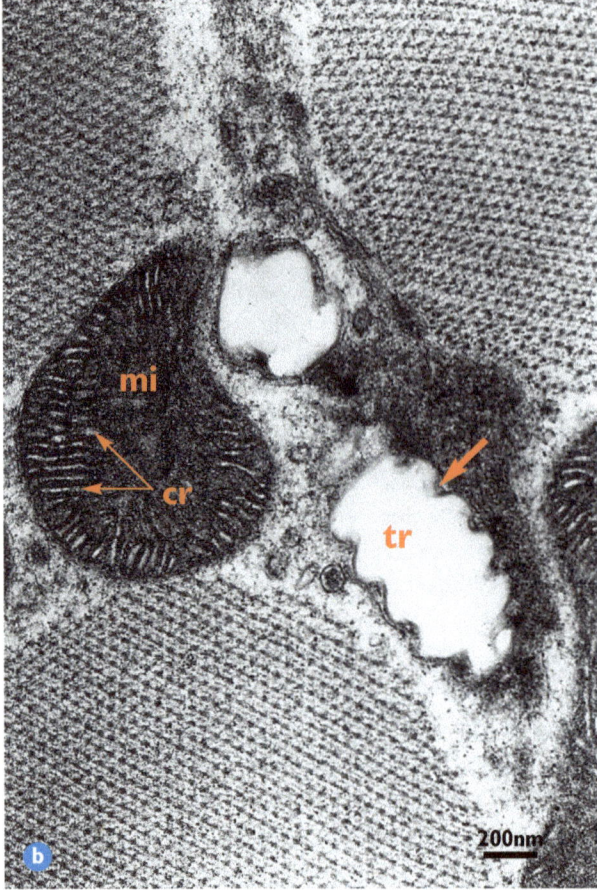

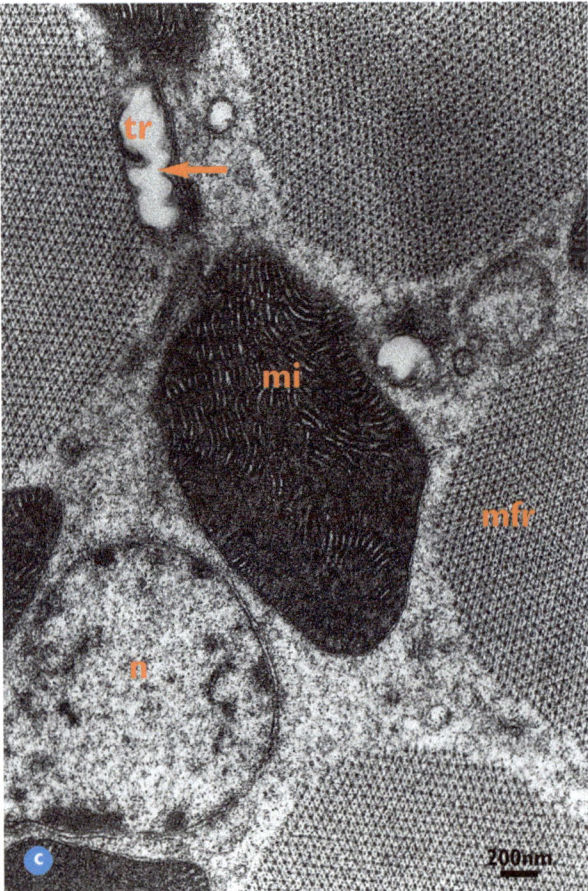

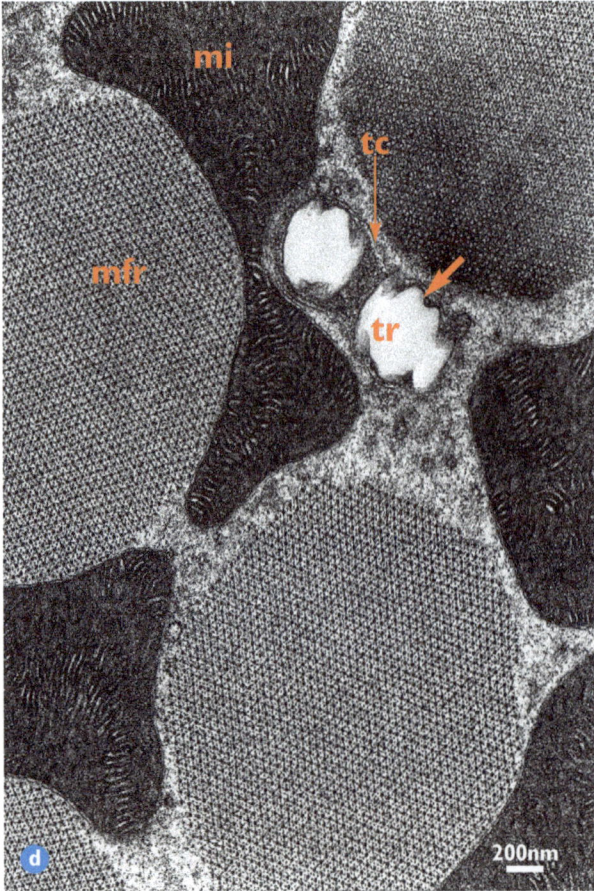

All of the activities of the living animal require the continuous expenditure of energy. Energy is derived from the oxidation of food molecules by a precise and orderly chain of reactions within the cells of the body and is stored there in chemical form until required. A supply of oxygen to each cell is essential for energy production and, since carbon dioxide is a waste product of energy metabolism, each cell also needs to get rid of carbon dioxide. Energy is produced in a similar manner in the cells of all animals, but the site of gas exchange with the environment, and the way in which oxygen is transported to the cells and carbon dioxide is removed, is not the same in all animal groups. Gaseous exchange with the atmosphere requires a membrane permeable to oxygen and carbon dioxide, having a large surface area in relation to the volume of tissue to be served. However, such requirements could result in excessive water loss since a membrane permeable to oxygen is usually permeable to water. Most active terrestrial animals show adaptations designed to overcome this, such as the development of lungs — internal cavities containing the respiratory membrane, with a single restricted entrance and exit for air.

Oxygen transport could take place by diffusion through the tissue fluids to the cells but its diffusion rate in a watery medium is very slow. An animal whose oxygen needs can be met by diffusion from the exterior through its tissue fluids has to be less than one millimetre thick. Some flatworms, living in a moist atmosphere and exchanging gas over the whole body surface, meet both the requirements for gas exchange with the exterior and the size limits for oxygen diffusion in tissue fluids. Larger animals require special adaptations. In man and other mammals, the continuous demand for oxygen by the tissues of a relatively large organism is met by carrying the oxygen from the lungs to the tissues within a closed circulatory system. The watery medium, plasma, has a lower capacitance for oxygen than air, i.e. at the same partial pressure, the maximal quantity of oxygen contained in a given volume of water is much lower than that contained in the same volume of air, hence the evolution of respiratory pigments to increase the oxygen carrying capacity of the blood. Within the highly vascularized lungs, oxygen combines with the respiratory pigment, haemoglobin, in the red blood cells. The blood is pumped rapidly around the body and the oxygen is given up from the haemoglobin in areas where oxygen tension is low. All areas of the body are plentifully supplied with fine capillaries so that no cell is further than three or four cells away from a capillary. Thus oxygen has only a very short distance to diffuse through the tissue fluids. Many control mechanisms operate on the circulatory system to ensure that the oxygen demands of a given region of the body are satisfied at any particular time of need, for example, blood flow to the skeletal muscles is increased if the animal is alarmed — the 'fight or flight' response.

FIG. 6.1 *Transverse transmission electron microscope (TEM) sections through the flight muscle of the worker bee showing that oxygen is piped directly into the muscle. The finest endings of the tracheal system, the tracheoles (tr) indent the muscle fibre, i.e. the muscle cell, and form close associations with the mitochondria (mi) and the myofibrils (mfr). Even in these finest of tracheal endings the spiral or annular rings of the taenidia (arrow) are visible. In a note the close association of the tracheole with the mitochondrion leaving a very short distance for oxygen to diffuse. In the myofibril, the regular hexagonal arrangement of the thick and thin filaments can be seen. In b note the cristae (cr) of the mitochondria, formed by the prolific folding of the inner mitochondrial membrane. c shows that tracheoles are in close apposition to the myofibrils as well as the mitochondria. Nucleus of the muscle fibre (n) (see chapter 7 for further details of flight muscle). d The myofibrils of the flight muscle fibre, the most active tissue known, are surrounded by mitochondria. The tracheoblast cell (tc) that surrounds each tracheole, forms a very thin layer of cytoplasm at the finer ending of the tracheole.*

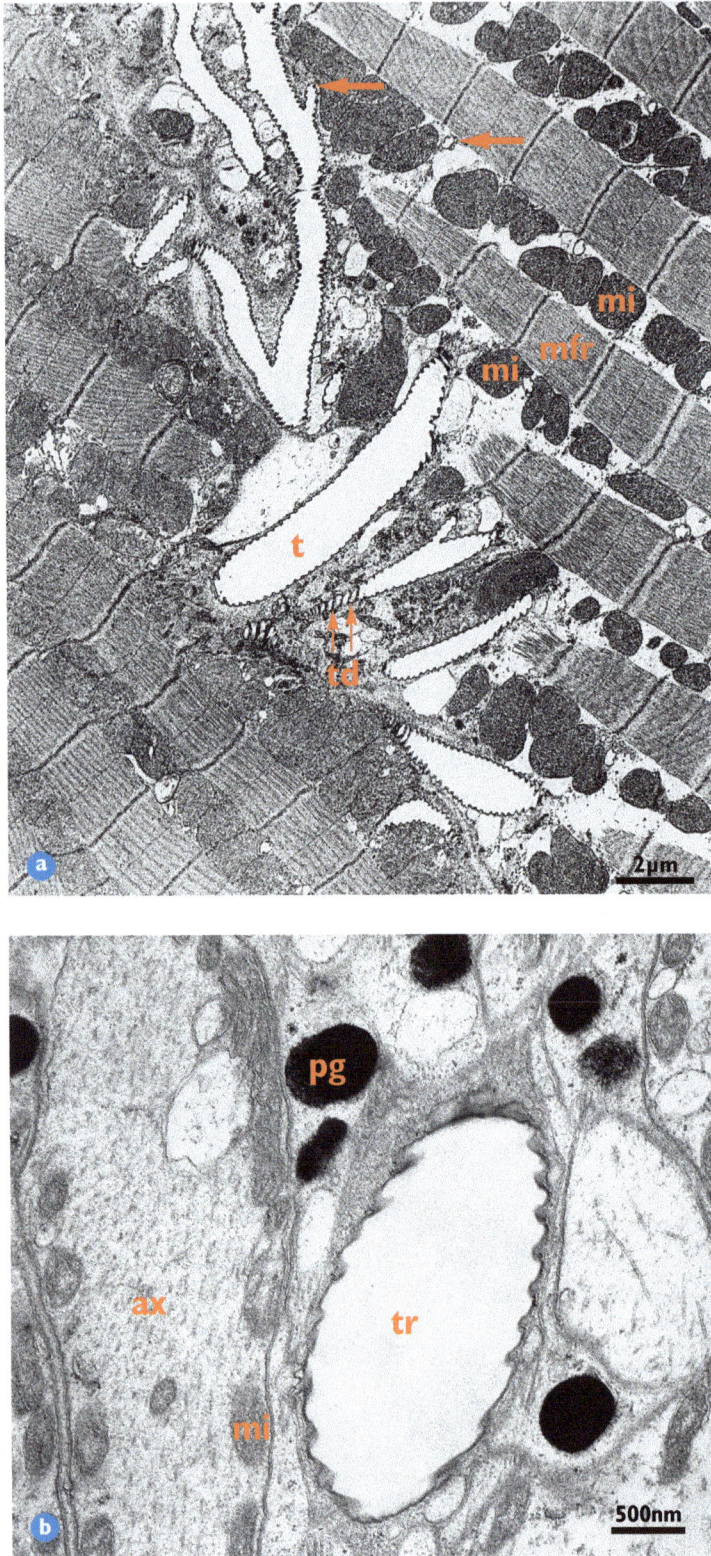

FIG. 6.2 a *A TEM section through a flight muscle of a bee showing finer endings of the trachea (t) subdividing to form the fine tracheoles (arrows) that penetrate the muscle fibre. Even the finer tracheoles are strengthened by taenidia (td). Note the column of mitochondria (mi) that occurs between the myofibrils of the muscle tissue (mfr).* **b** *A TEM section through the base of the compound eye, showing a tracheole (tr) in close proximity to a retinal cell axon (ax). Pigment granules (pg); mitochondrion (mi).*

Respiration in insects

Insects have developed an entirely different system, piping oxygen directly to the tissues in air-filled tubes known as **tracheae**. There are two advantages in using air as the transport medium to the tissues rather than a fluid transport system: its greater capacitance for oxygen and the fact that the rate of diffusion of oxygen through air is several hundred thousand times faster than through water. Thus, this system provides efficient oxygen transport from the atmosphere throughout the body into every tissue. The final diffusion pathway from the tracheal system to the site of use in the cell, which must take place in tissue fluids, is very short, usually a matter of microns.

The tracheal tubes are formed by the invagination (infolding) of the outer tissue layer of the body wall, the **ectoderm**. This means that they, too, are lined with a thin layer of cuticle produced by the ectoderm to cover the surface of the body. The cuticle lining the tracheae differs from the surface cuticle in that it lacks the wax waterproofing layer, is not hardened (sclerotized), and lacks the ingredient chitin in the finer tubes. This thin cuticular lining is known as the **intima**. The cuticular intima of the tracheae is thrown into thickened folds, called **taenidia** (figs 6.2, 6.3), which spiral along the tubes making them resistant to lateral compression, while allowing some longitudinal extension and compression. This is a common device used in flexible hoses. Where chitin microfibrils are present in the intima, they are orientated to provide mechanical strength, being tangentially arranged on the ridges of the taenidia to assist in resisting lateral compression, and longitudinally arranged in the valleys between taenidia to resist overextension.

The external openings of the tracheal system are called **spiracles**. They are located laterally on the thorax and abdomen. The bee possesses the maximum number found in insects, two pairs of spiracles on the thorax and eight on the abdomen (figs

6.5 and 6.10). However, in the bee, the first abdominal segment, the propodeum, is fused to the third thoracic segment so that the first abdominal spiracle appears to be a third thoracic spiracle, and the eighth abdominal spiracle is in the sting chamber.

The basic tracheal plan of insects consists of short tracheal trunks opening from the spiracles to join the main lateral tracheal trunks running longitudinally on each side of the body. These longitudinal trunks are interconnected by ventral trunks in each thoracic and abdominal segment. In some insects, there may also be ventral and dorsal longitudinal trunks. Numerous smaller secondary trunks arise from these primary trunks and ramify throughout the body giving off even smaller, tertiary branches. These, in turn, branch to form the very finest of the tubes, known as **tracheoles**, that taper from a diameter of around 1.0 μm, to end blindly at a diameter of around 0.1–0.2 μm (fig. 6.4). The tracheoles supply all the tissues of the body. Each tracheole lies within a cell, the **tracheoblast**, which forms a thin sheath around it (fig. 6.1d). The tracheoblast cells containing the tracheoles come very close to the cells that they are supplying, and their cell

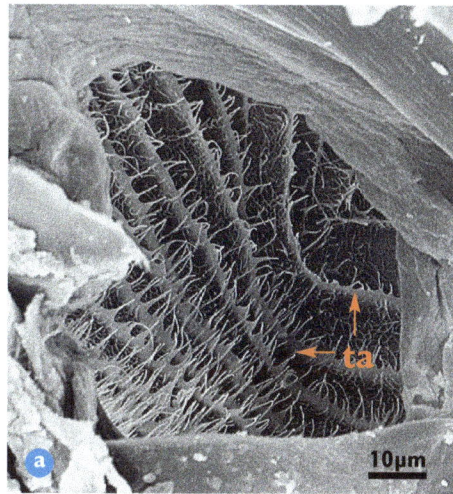

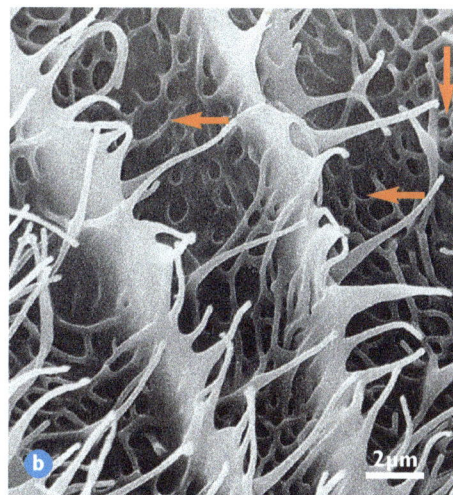

FIG. 6.3 a Looking inside the tracheal trunk opening from the first thoracic spiracle of the worker bee. The strengthening taenidial rings (ta) are clearly visible and are seen to bear small hair-like processes. This is quite a common feature of insect tracheae. **b** The cuticular lining of the trachea is often thrown into complex patterns (arrows) between the taenidial folds.

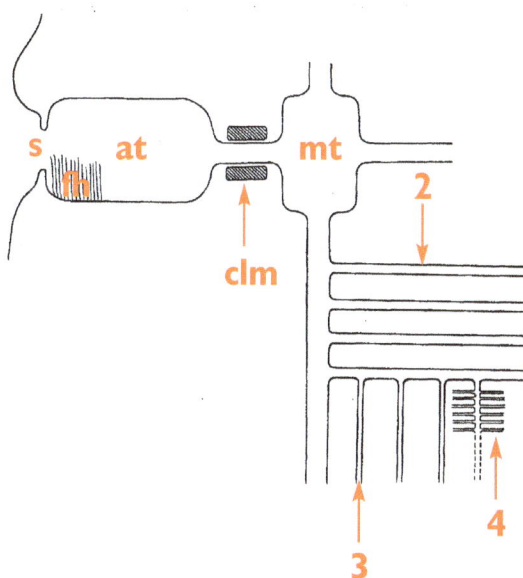

FIG. 6.4 Schematic illustration of the basic insect tracheal system. The tracheal system opens onto the surface at the spiracle (s). Beneath the spiracle there is often a cavity, the atrium (at) which may contain filtering hairs (fh). After the atrium there may be a simple system of cuticular levers and muscles (clm) which serve to open and close the entrance to the tracheal system. Beyond this, the tracheal system opens into a main trunk (mt) which runs longitudinally through the body. From this main trunk arise a number of branches which supply dorsal and ventral regions of the body. Transverse branches supply central organs and tissues and may form connectives (2, 3 and 4) across the body. The main tracheal branches subdivide to form finer and finer branches, the secondary and tertiary branches (arrows 2 and 3) and the tracheoles (arrow 4). Redrawn after Bursell[13].

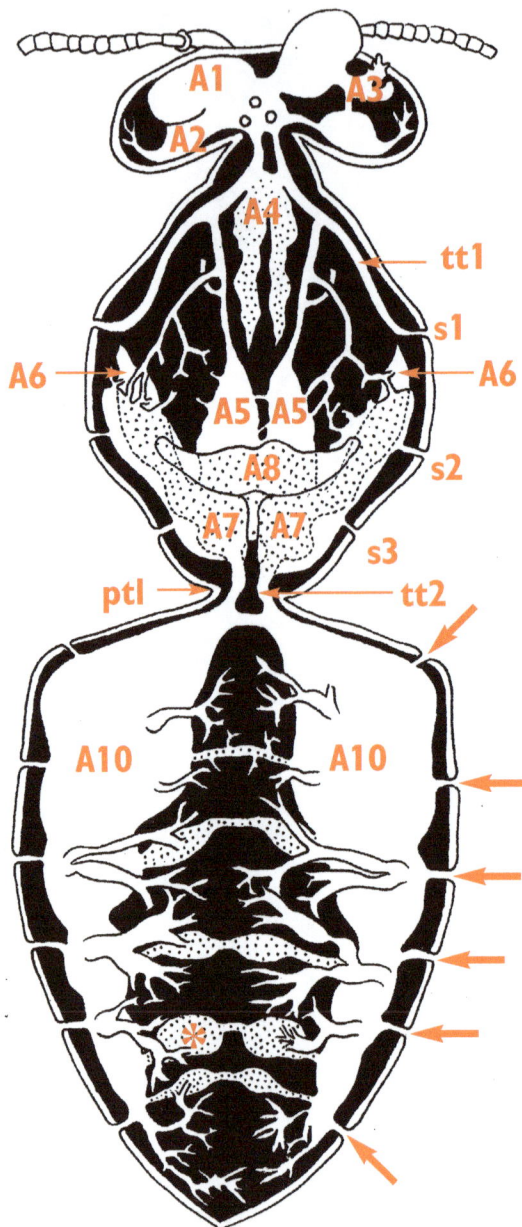

FIG. 6.5 *In the adult bee, the longitudinal tracheal trunks are expanded into large air sacs. The large tracheal trunks (tt1) arising from the first spiracle (s1) pass forward into the head and expand into three air sacs covering the top of the brain (A1), supplying the compound eyes and optic lobes (A2) and the underside of the brain and mouthparts (A3) (part of A1 has been moved forward on the right, to reveal the underlying A3). From these air sacs numerous tracheae penetrate the tissues and subdivide to form fine tracheoles. Posteriorly running branches also arise from the first spiracular tracheal trunks, giving rise to a complex of interconnected air sacs (A4, A5, A6, A7 and A8) supplying the thorax. Spiracles 2 and 3 (s2, s3) also open into these air sacs. Two large tracheal trunks (tt2) run from the thoracic complex through the petiole (ptl) into the abdomen where they expand into two enormous lateral air sacs (A10). Two small air sacs (A9) lie just inside the abdomen (not visible here). Abdominal spiracles (arrows) open into these sacs which are connected to each other via transverse commissures (star). The first abdominal spiracles (s3) are situated on the propodeum; the eighth abdominal spiracles are hidden within the sting chamber. Tracheae from these sacs ramify among the abdominal tissues. Stippled areas lie ventrally. Redrawn from Snodgrass[1].*

membranes may even fuse with those of the recipient cells or indent their surface (fig. 6.1). Oxygen from the tracheole thus has only a very short distance to diffuse in a watery medium to enter the mitochondria of the recipient cell, where it will take part in energy producing reactions. When an insect is opened up, all these air-filled tubes show up as a silvery network covering all internal organs.

The larval tracheal system in the bee is very similar to the basic insect tracheal plan but in adult insects, modifications are often seen. In many insects, the tracheae widen at intervals and lose their taenidia to form compressible **air sacs**. These may be small expansions of a trachea or, in good fliers, they may extend into very large sacs, occupying a considerable volume of the body space when full of air. The adult bee has a very elaborate development of air sacs (fig. 6.5); indeed, many of the tracheae in the bee lack well developed taenidia and chitin microfibrils, and appear more like air sacs in structure. At the front of the thorax, two large tracheal trunks arise from the first spiracles and pass forward, through the neck into the head, where they expand into three air sacs covering the brain, optic lobes and frontal region of the head[1]. The nervous system requires a plentiful supply of oxygen and the air sacs covering the brain and optic ganglia give off large numbers of tracheae that subdivide forming fine tracheoles. The tracheoles branch to all parts of the brain, giving a particularly rich supply to the neuropile areas, i.e. the areas concerned with synaptic integration. The two main tracheal trunks also give off posteriorly-running branches that connect with a complex system of air sacs supplying the legs and the flight muscles of the thorax. Basically, there are central-, intermediate- and laterally-lying sacs, ensuring that the large dorsoventral and dorsolongitudinal flight muscles are surrounded by air sacs, from which tracheal branches are given off at very frequent intervals to enter the muscles. All of these air sacs are interconnected and supplied by the first three spiracles. Two large tracheal tubes leave the thoracic air sac complex and traverse the petiole into the abdomen, where they expand into two enormous lateral air sacs that extend through the first five abdominal segments behind the petiole. The air sacs are connected to the abdominal spiracles on each side, and to each other via transverse commissures. They give off tracheae that branch around the body wall, heart and other organs in each body segment.

The oxygen supply to the flight muscle of insects has received particular attention in view of the fact that it is the most active tissue known in nature. The **mitochondria** are the crucial site of energy metabolism in the flight muscles. These organelles can occupy 30–40% of the mass of the muscle (fig. 6.1), and an adequate supply of oxygen to them is essential during flight. The tracheal supply to the flight muscles consists of a primary tracheal supply of abundant, large tracheae, which may be expanded into air sacs, as is the case in the bee. These air sacs are wrapped around the muscle fibres. Very many short tracheae are given off from the air sacs: these are the secondary tracheae of the flight system. Almost immediately in the bee, they break up into an enormous number of tracheoles, the tertiary system, which enter the muscle transversely and branch repeatedly[2]. These branches enter the muscle every few microns along its entire length. The muscle membrane is also indented at intervals by transverse tubules (the T system), which form part of the excitatory mechanism of the muscle. Some tracheoles enter the transverse tubules and so gain access to the interior of the muscle. Tracheole diameters decrease from 0.1–0.2 μm to around 0.05–0.08 μm where they meet the mitochondria. Every mitochondrion in bee flight muscle is either near or in contact with tracheoles (fig. 6.1) so that diffusion through the body fluid has all but been eliminated in this very active tissue.

How does enough oxygen reach the respiring cells?

We have seen that fine tubes containing air in contact with the exterior reach very close to each individual cell. Calculations based on the dimensions of the tracheal systems of sample insects have shown that gaseous diffusion alone can meet the respiratory needs of small insects, or of larger insects when they are inactive[3]. However, insects can switch very rapidly from an inactive state to a very active one. In the bee, there is a 50-fold increase in oxygen consumption when it switches from rest to flight. This compares with a 24-fold increase in humans switching from rest to very strenuous activity. Oxygen consumption in bee muscle during free hovering flight has been measured at between 79 ml and 94 ml oxygen per gram of muscle per hour. This is a higher rate than found in many insects[4], and no oxygen debt is built up during activity.

Is simple gaseous diffusion sufficient to meet the needs of the active insect?

Recent studies show that most insects, including the bee, ventilate the tracheae and air sacs when they are active and, in many cases, even when they are at rest, resulting in convective gas exchange[5]. This mass flow of air, coupled with the abundant tracheal supply to the flight muscles, ensures an adequate supply of oxygen during flight.

Ventilation is most often achieved by pumping movements of the abdomen. These may take the form of dorsoventral compression and expansion of the abdominal segments or, as in the bee, a rapid lengthening and contraction of the abdomen accompanied by a slight dorsoventral expansion and compression. Pumping movements act on compressible regions of the tracheal system, such as the air sacs and any tracheae oval in shape or not strengthened by taenidial rings. Air sacs remote from the pump can be affected through displacement of the haemolymph, or body fluid, by the pumping movements. If the spiracles remain open, tidal ventilation will occur, i.e. the ventilation movements will simply pump air in and out of the tracheal system through the same spiracles. Since insects are very vulnerable to dessication, they employ a number of structural adaptations to reduce water loss, including covering the surface cuticle with a waxy layer and withdrawing their respiratory surfaces inside the body. This leaves the apertures into the respiratory system as the major source of water loss from the body. In the majority of insects, spiracular closing mechanisms have been developed which can seal off the openings into the respiratory system except when gas exchange is taking place. The mechanisms of spiracular control must be able to balance the conflicting interests of efficient gas exchange with reduction of transpiratory water loss.

In many quiescent insects, respiratory cycles may be established in which the spiracles remain closed at first and no gas exchange takes place. Oxygen is used up and the carbon dioxide (CO_2) produced is stored temporarily as hydrogen carbonate (bicarbonate) in the tissues so that a negative pressure develops in the tracheae. In the second phase of the respiratory cycle, the spiracles make small opening and closing movements (often referred to as 'fluttering'), allowing air to be sucked in by the negative tracheal pressure. Finally, the spiracles open wide and there is a massive release of CO_2 accompanied by ventilatory pumping in some insects. Since the air is sucked in by negative pressure during the fluttering periods, this type of respiratory cycle is known as **passive suction ventilation**[6], and it makes an important contribution to conserving water in the insect even though there are ventilatory pumping movements for a short part of the cycle.

When an insect becomes active, particularly when it begins to fly, then the demand for gas exchange comes to outweigh the necessity to conserve water. Spiracles are opened, notably in the thorax, and ventilatory pumping movements commence. In some insects, autoventilation also occurs as a side effect of the flight movements, as deformations of the thorax by muscular contraction may compress nearby tracheae. However, this has been thought not to be the case in the bee since the extent of the muscle contraction in flight is very small, although it has recently been suggested that autoventilation may have some role in the bumble bee[6].

In many insects, the movements of the spiracular valves can be synchronized with the ventilation movements to produce a directed flow of air through the insect. Most commonly, this is a directed flow of air through the primary tracheae of the thorax during flight. Air is drawn in at the front of the thorax and expelled through the third thoracic or first abdominal spiracle, or through the last abdominal spiracles.

What do we know about respiratory movements in the honey bee?

Although the energetics of flight, including oxygen consumption, have been extensively studied, the respiratory movements of the bee have received relatively little attention. Most of the spiracles are small, and the valves of those on the abdomen are hidden within an outer atrium, so that observations or measurements of spiracular movements are very difficult and virtually impossible to achieve during tethered flight.

The spiracles of the bee illustrate a number of the opening and closing devices seen in insects. The first spiracle, lying just below the anterior angle of the mesothoracic wing base, is hidden beneath a flat lobe projecting from the rear edge of the pronotum (fig. 6.6). The edge of the lobe is lined with a dense array of cuticular hairs, which also help to conceal the deep depression in the cuticle leading to the spiracle entrance. When the lobe is removed, the depression can be seen to lead forward to the intersegmental membrane between the prothorax and the

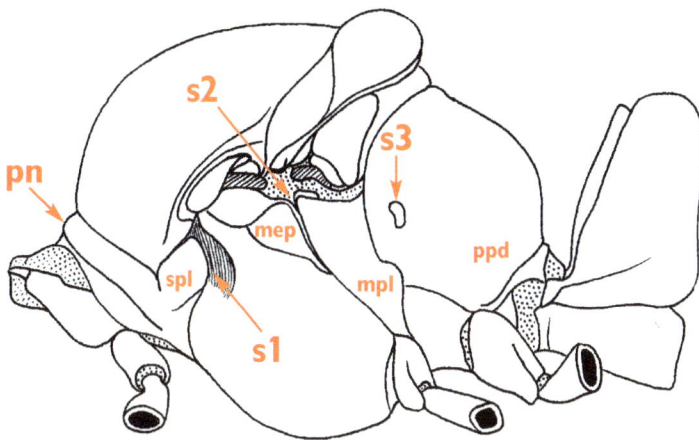

FIG. 6.6 *Side view of the thorax of a worker bee to show the locations of the spiracles. The first spiracle (s1) lies in the intersegmental membrane between the prothorax and the mesothorax. It is not visible from the exterior, being covered by a flat lobe projecting from the rear margin of the prothorax (spl). The spiracular opening is further protected by a fringe of hairs on the spiracular lobe. The spiracular lobe has been pulled aside and the hairs removed. The second small spiracle (s2) is hidden in the membranous fold (stippled) beneath the wing articulation, between the mesepimeron (mep) and the metapleuron (mpl) plates. The third spiracle (s3) lies in the propodeum (ppd), that part of the thoracic box derived from the abdomen. This is the largest spiracle and is readily visible from the exterior. Pronotum (pn). After Snodgrass[1].*

mesothorax. The oval-shaped spiracle lies in this membrane, its longest diameter being around 0.14 mm[1]. A plate, or **operculum,** is attached to the upper half of the spiracle and this can be pulled down to close the spiracle by the action of a small muscle attached to an inward projection of the plate and inserted into the ventral wall of the mesothorax (fig. 6.7d). When the muscle contracts, the plate is pulled down to the lower rim of the spiracle: however, there is nothing to hold it firm at the lower rim, which may account for the fact that this spiracle is a point of entry for tracheal mites. When the closer muscle relaxes, the elasticity of the cuticular plate causes it to move upwards again.

The second thoracic spiracle is extremely small and hidden in a membranous fold between two thoracic plates, the **mesepimeron** and the **metapleuron** (fig. 6.8). Like the first thoracic spiracle, this one opens directly into the trachea, but appears to have no closing device. Solely on the basis of its size and position, the second spiracle is regarded as contributing little to gaseous exchange in the honey bee.

The largest pair of spiracles are located on either side of the propodeal segment, the first abdominal segment that has become incorporated into the thorax. The spiracles are long (0.23 mm), oval-shaped openings

situated laterally on the tergal plate of the propodeum (figs 6.6, 6.9). Each spiracle is surrounded by a thick cuticular rim enclosing a shallow atrium from which an aperture leads into the tracheal trunk. The spiracle can be closed by a membranous valve with a strengthened cuticular margin that projects forward from the rear edge of the cuticular rim and is caught in a groove under the front lip of the rim[1]. The margin of the valve is extended into two lobes which provide attachment for muscles. A large closer muscle runs between these lobes (fig. 6.9b, c) which, when contracted, springs the margin of the valve forwards so that it catches in the groove under the front rim, closing the spiracular aperture. The spiracle is opened by the action of a second muscle, which runs between the lower lobe of the valve and the ventral wall of the propodeum. Contraction of this muscle flattens the margin of the valve so that it is no longer lodged in the frontal groove and thus is pulled away from the aperture.

The abdominal spiracles lie in the lateral walls of the tergal plates of the first six segments, i.e. segments II–VII (fig. 6.10). In the worker and the queen, segments VIII, IX and X are all withdrawn into segment VII so that the last pair of spiracles, belonging to segment VIII, are concealed within the sting chamber. The abdominal

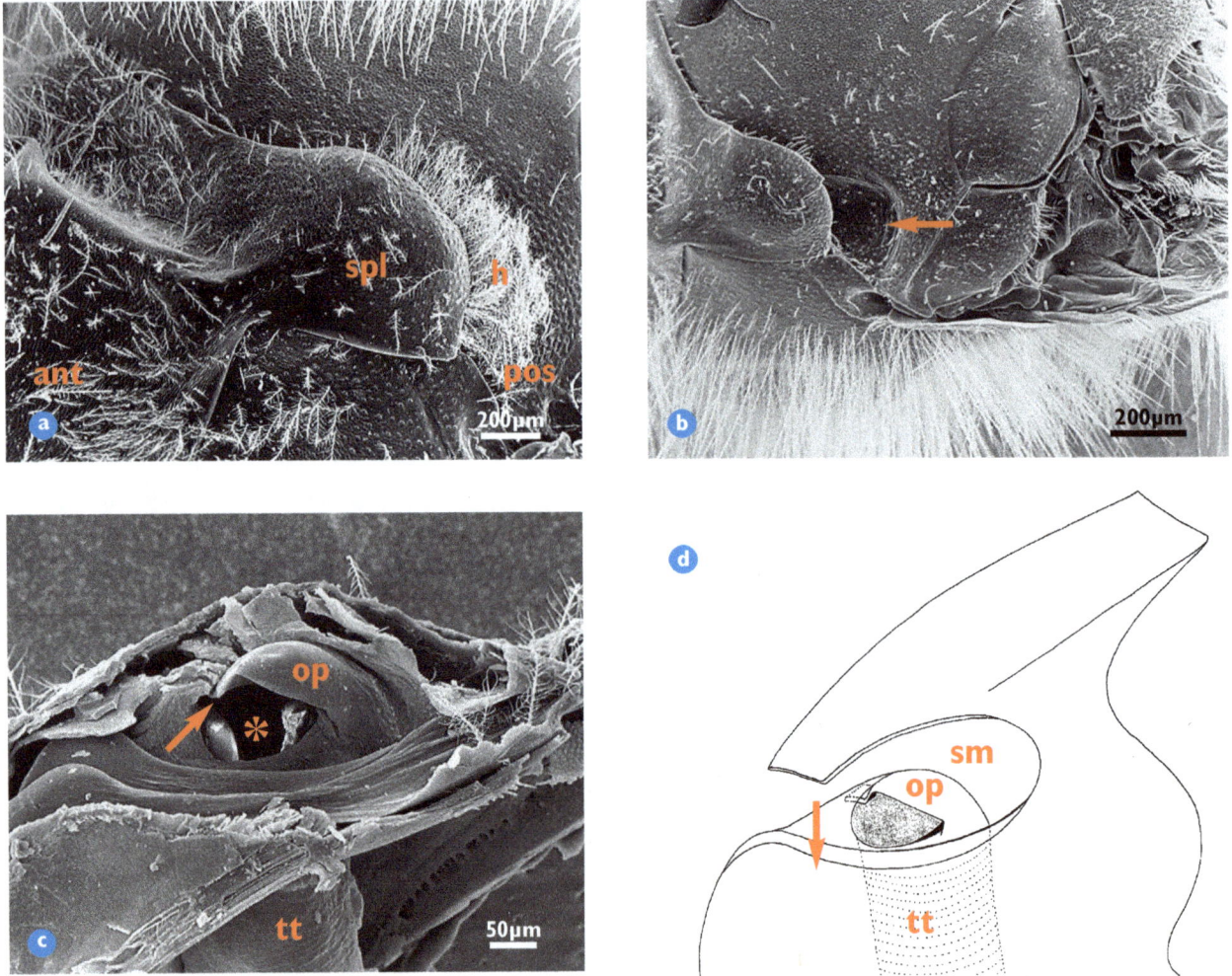

FIG. 6.7 a *The first spiracle is hidden under a spiracular lobe (spl) whose edge is ringed with fine hairs (h). Some of these hairs together with the surrounding thoracic hairs, have been removed for clarity. Anterior (ant); posterior (pos).* **b** *The fringe of hairs has been removed from the spiracular lobe and the lobe itself displaced to one side to reveal the deep depression in the cuticle (arrow) which leads forward to the spiracular opening.* **c** *The surrounding cuticle has been cut away to show the spiracular opening (star) at the end of the depression. The first spiracle has an external closing mechanism, the operculum (op) a dome-shaped sclerotized plate that partially covers the opening. A small arm projects internally under the edge of the membrane surrounding the spiracle (arrow) to provide attachment for the closer muscle. The spiracle opens directly into the tracheal trunk (tt); no atrium is present here.* **d** *The closing mechanism of the first spiracle. The closer muscle runs downwards from the inwardly projecting arm (arrow) of the operculum (op) to the mesothoracic wall. Contraction of this muscle pulls the domed operculum down over the spiracular opening. When the muscle relaxes, the elasticity of the cuticular operculum causes it to spring back into shape. Intersegmental membrane containing the spiracle (sm); tracheal trunk (tt).*

spiracles are all similar in structure, consisting of oval slits around 0.06 mm long[1], although rather larger in the drone. Each spiracle opens into an atrial chamber, rather deeper than that of the propodeal spiracle, with the tracheal opening at its inner end. The atrium contains cuticular hairs (fig. 6.11b, c), a common occurrence in insects. These hairs are believed to act as a filter, keeping particles from entering the tracheal system. They are also said to cut down the rate of transpiration from the trachea and thus aid in conserving water. The tracheal entrance can be closed by a valve in a somewhat similar manner to that of the propodeal spiracle. The valve is a fold of membrane with a strengthened margin, that hangs down from the top of the tracheal entrance. Within the fold of membrane, a closer muscle runs from one side of the margin to the other. When this muscle contracts, the lower margin of the membranous fold is drawn down against the bottom of the tracheal entrance and the trachea is closed (fig. 6.11c). An opener muscle runs from a ventral process on the margin of the valve to the wall of the sternum. When the closer muscle relaxes and the opener contracts, the membranous valve is drawn away from the entrance of the trachea. These movements are not visible outside the spiracle which makes it difficult to examine the respiratory movements of the bee in any detail.

The abdominal ventilation movements can be seen very clearly when the bee alights from flight. The abdominal segments contain a dorsal tergal plate and a ventral sternal plate. The rear of each tergal plate overlaps the front edge of the succeeding plate and the sternal plates show the same arrangement. The overlapping plates are

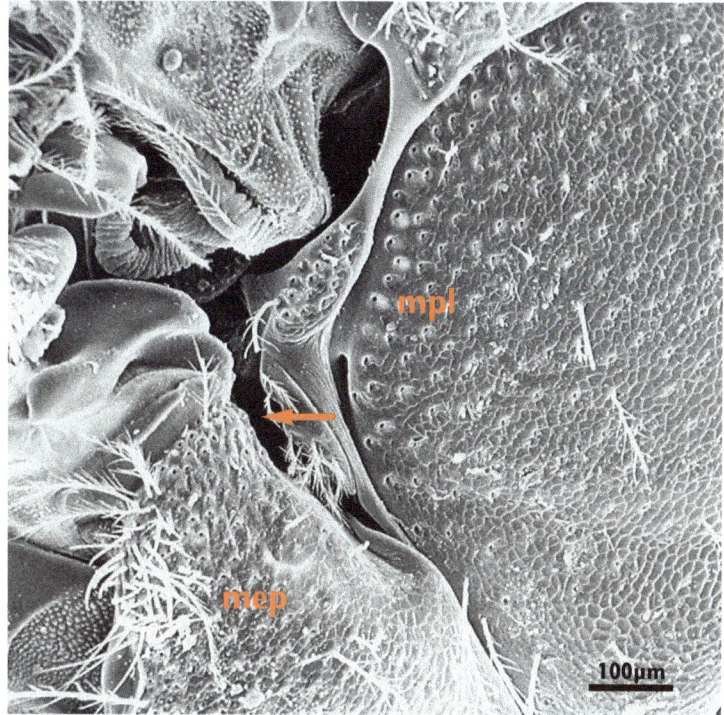

FIG. 6.8 *The small second spiracle lies in the deep membranous fold (arrow) between the upper ends of two mesothoracic plates, the mesepimeron (mep) and the metapleuron (mpl). Thoracic hairs have been removed to expose the fold.*

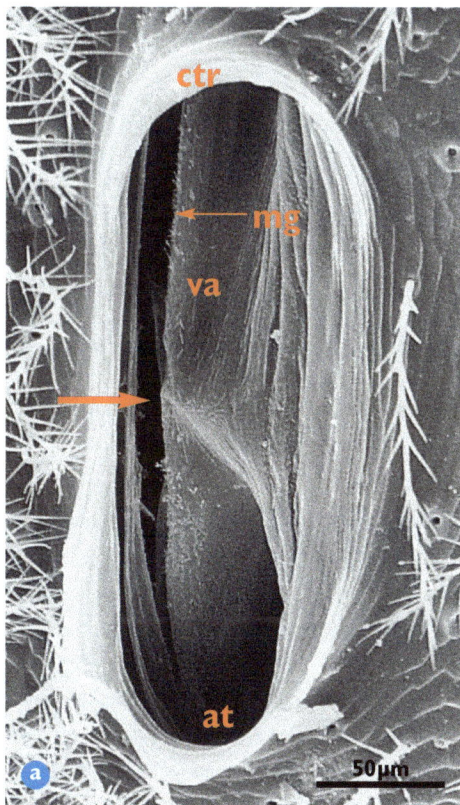

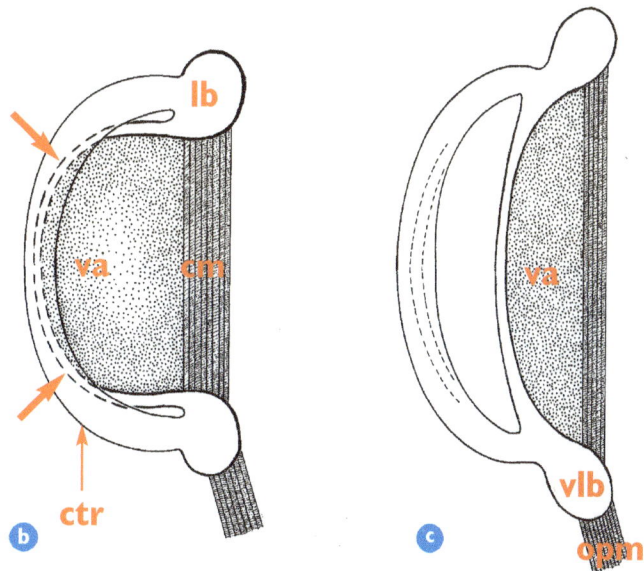

FIG. 6.9 a *The third spiracle, lying on the propodeal segment of the thorax, has a long oval aperture around 0.23 mm in length and 0.06 mm at its widest diameter[1]. The spiracular opening is surrounded by a cuticular rim (ctr) which forms a shallow atrium (at) containing no filtering apparatus. The tracheal aperture in this example is closed to a narrow slit (arrow) by a large valve (va). The valve consists of a soft, membranous fold with a strongly sclerotized margin (mg).* b *Interior view of a closed propodeal spiracle. The membranous valve (va) has internally projecting lobes (lb). The closer muscle (cm) runs between these lobes and, when it contracts, the valve becomes more outwardly convex in shape and its hardened (sclerotized) outer margin fits into a groove (arrows) in the cuticular rim (ctr) thus closing the aperture.* c *Interior view of the open spiracle. An opener muscle (opm) runs from the ventral lobe (vlb) of the valve (va) to the wall of the propodeal segment of the thorax. When this muscle contracts and the closer muscle relaxes, the valve is flattened and drawn away from the groove thus opening the spiracle. The degree to which the valve is flattened can be varied by the action of the opener muscle.* Redrawn from Snodgrass[1].

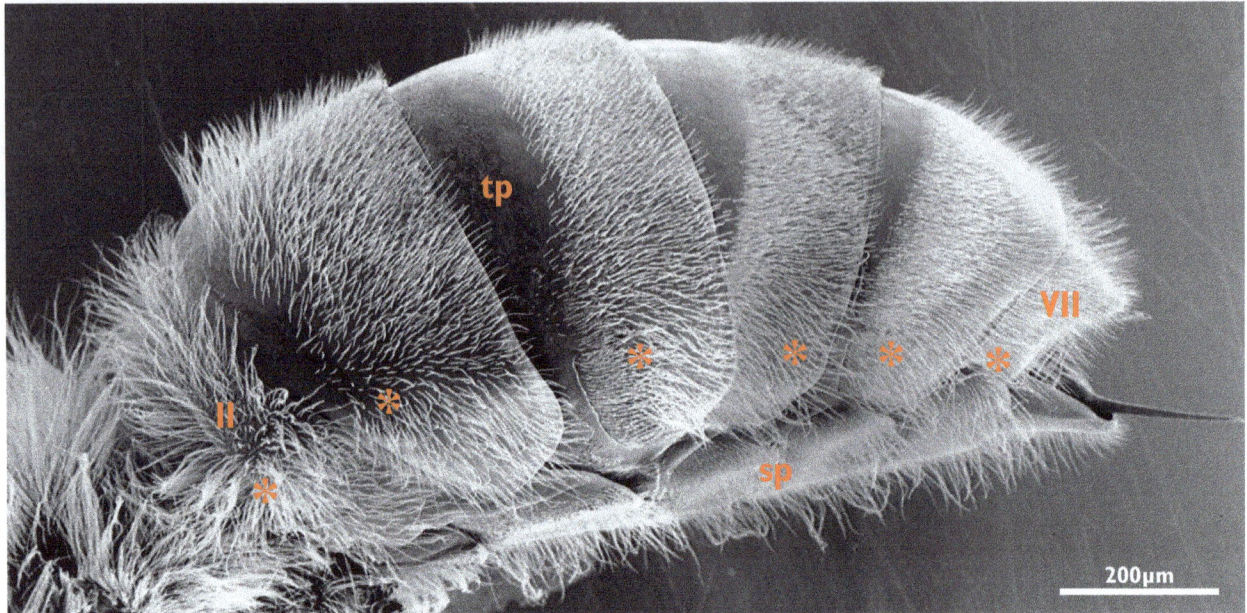

FIG. 6.10 *Side view of the abdomen of a worker bee showing the position of the spiracles (*stars*) on the left side. The paired spiracle openings are located one on each side of the tergal plates (tp). The first pair of spiracles is located on abdominal segment II, since segment I is incorporated into the thorax to form the propodeum. The last pair of spiracles is found on the spiracular plates of segment VIII, which are concealed in the sting chamber within segment VII. The other spiracular openings are concealed among the hairs covering the abdomen. Sternal plate (sp).*

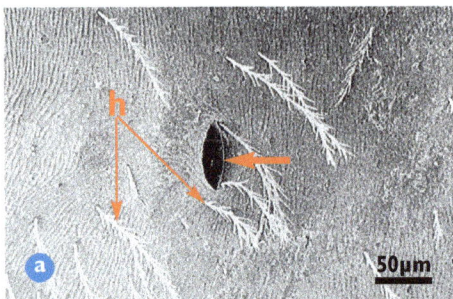

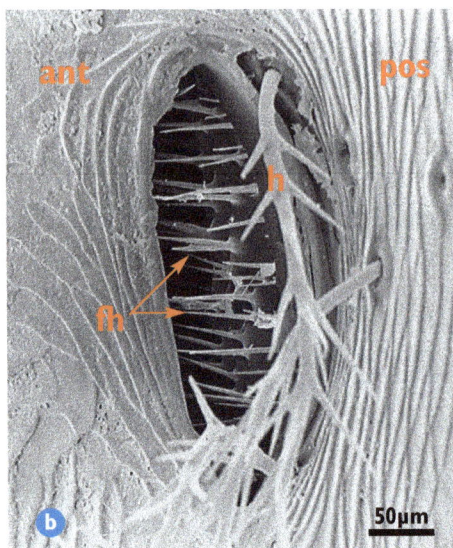

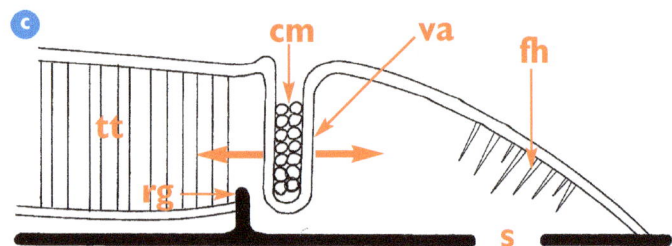

FIG. 6.11 **a** *The first abdominal spiracle (*arrow*) is revealed after most of the hairs (h) on the tergal plate have been cleared away.* **b** *The view through the spiracular aperture reveals the filtering hairs (fh) at the entrance to the atrium. Some of the large body hairs (h) remain around the entrance. Anterior (ant); posterior (pos).* **c** *A longitudinal section through the first abdominal spiracle. The abdominal spiracles are opened and closed by a similar mechanism to that of the propodeal spiracle. A domed-shaped membranous valve (va) is pushed against an internal cuticular ridge (rg) at the entrance to the tracheal trunk (tt) when the closer muscle (cm) contracts. When this muscle relaxes and the opener muscle contracts, the valve is flattened and lifted away from the ridge: the valve is shown in this position. Arrows show the directions of movement of the valve as it opens and closes. (The opener muscle, attached between one lobe of the valve and the body wall, is not shown in this section.) Filtering hairs (fh) in atrium; spiracular opening (s). Redrawn after Snodgrass[1].*

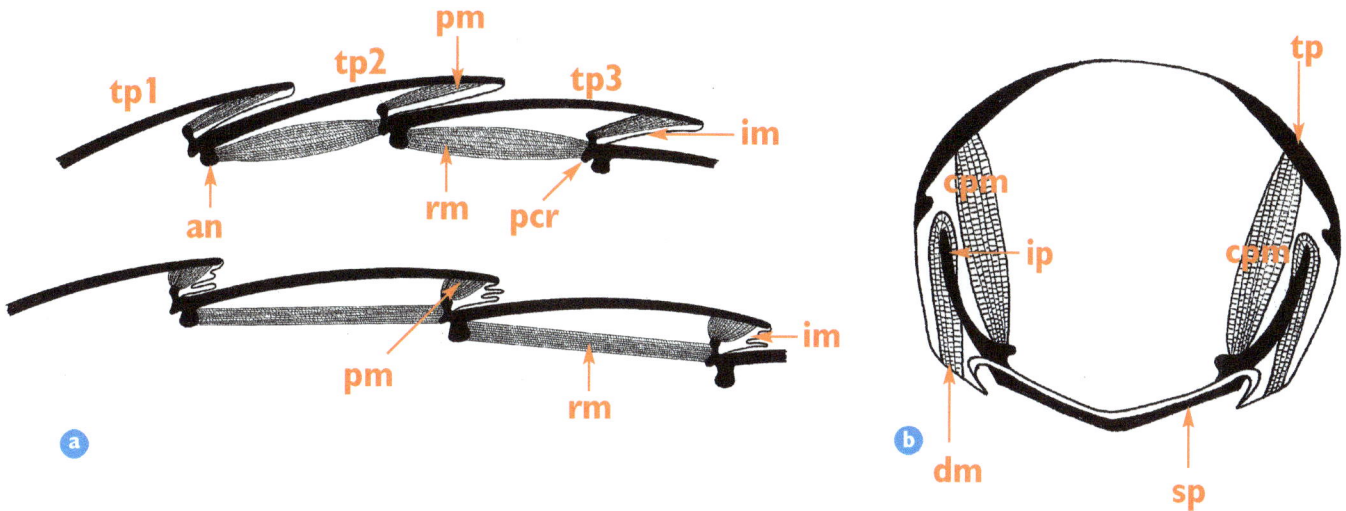

FIG. 6.12 a *Ventilation movements in the bee involve a rapid lengthening and shortening of the abdomen. Lengthening is accompanied by a slight dorsoventral expansion and shortening by abdominal compression. Each segment of the abdomen is bounded externally by a dorsal tergal plate which extends laterally over each side of the abdomen and a smaller, ventral sternal plate (see fig. 6.10). The tergal and sternal plates of each segment overlap those of the succeeding segment and alteration in abdominal length is achieved by the movement of these plates over one another. This diagram illustrates the way in which movement of the tergal plates (tp1, tp2 and tp3) is accomplished. Movement of the sternal plates is achieved in the same manner. A flexible intersegmental membrane (im) permits movement of the plates relative to one another. The abdomen is shortened when there is maximum overlap of the plates (upper diagram). This is achieved by the action of four pairs of retractor muscles in each segment. On each side of the body a dorsal retractor muscle (rm) runs between the internal thickened margin of one tergal plate (the antecostal ridge (an)) and the anterior margin of the succeeding plate (the precostal ridge (pcr)). A second retractor muscle runs from the lateral tergal wall to the anterior margin of the succeeding plate (not shown here). The sternal muscles show a similar disposition. The abdomen is lengthened when there is minimal overlap of the plates, accomplished by the action of two pairs of protractor muscles (pm) in each segment. These muscles are attached to the rear margin of one tergal or sternal plate and the anterior margin of the succeeding plate. Their contraction causes the plates to slide apart.* **b** *The dorsoventral compression and extension that accompanies shortening and lengthening, respectively, is achieved by the action of two pairs of compressor muscles and one pair of dilator muscles per segment. The two compressors on each side run from the lateral wall of the tergal plate (tp) to an internal ridge of the sternal plate (sp). Only one pair (cpm) is shown in this transverse section through a segment. Contraction of these muscles pulls the tergal and sternal plates towards each other and, together with the action of the retractor muscles, reduces the volume of the abdomen. This action compresses the abdominal air sacs. The dilator muscles (dm) are attached ventrally on the tergal wall and extend dorsally to the upper end of a long internal projection (ip), or apodeme of the sternal plate. Contraction of these muscles forces the tergal and sternal plates apart and, together with the lengthening of the abdomen, increases the volume of the abdomen.*

joined by a flexible membrane and this allows them to slide over one another so that the abdomen may be lengthened or shortened. The normal, or shortened, position of the abdomen is achieved by the action of two pairs of retractor muscles between the tergum and the sternum of each segment. The first pair consists of a large, median dorsal muscle on each side of the tergum which runs from the thickened internal margin at the front of the plate to the anterior margin of the succeeding plate. The second pair consists of a smaller, lateral muscle on either side running from the anterior part of the tergum to the anterior strengthening ridge on the succeeding plate. Two pairs of muscles with similar dispositions form the retractors of the sternum. When all these muscles are fully contracted, there is maximum overlap of the plates (fig. 6.12a). The abdomen is lengthened by the contraction of the reversed protractor muscles of the tergum and sternum. These short muscles lie in the infolded membranes of the tergum and sternum, where the plates overlap. They are attached to the rear of one plate and to the front of the succeeding plate (fig. 6.12a, lower diagram). When these muscles contract, the membrane between the plates is thrown into a fold and the posterior plate slides out from under the anterior plate, resulting in a lessening of the overlap between plates and a consequent lengthening of the abdomen[1].

The shortening and lengthening of the abdomen is accompanied by dorsoventral expansion and contraction. This is achieved by three muscles running between the tergum and the sternum either side of each segment. Two of the three muscles run directly between the tergum and sternum, and these pull the plates towards each other, compressing the abdomen when they contract. This movement accompanies the shortening of the abdomen. The third muscle is attached to the tergum at its most ventral position with its other end attached more dorsally to the upper end of a long lateral internal extension of the sternum (fig. 6.12a, lower diagram). Contraction of this muscle separates the tergal and sternal plates and so expands the abdomen dorsoventrally. This movement accompanies the lengthening of the abdomen in the ventilation movements. The activity of these muscles, of course, is not solely concerned with the respiratory movements of the insect; there are other instances when flexibility in the abdomen is required. There may be differential degrees of contraction of these muscles between segments in order to contribute, with other muscles, to a variety of abdominal movements, for example, bending of the abdomen when the queen is laying eggs, or expansion when the worker is carrying a large nectar load.

Respiration in the adult worker bee has been examined under a variety of conditions. Oxygen consumption[7,8] and CO_2 production[9] have been measured; ventilation and/or spiracular movements recorded; and the movement of air currents around the bee in a gas chamber observed[13]. However, isolated observations, made under very different conditions, do not give a complete picture of respiration in the bee over its whole range of activity, although there is enough information to show that in the bee, as in other insects, respiratory movements vary with the insect's state of activity. At temperatures below 12°C, a chill coma or generalized muscular paralysis develops[7], and no ventilatory movements are made by the abdomen. In this state, gas exchange in the bee occurs by means of continuous diffusion of oxygen in and of CO_2 out through the spiracles, although some CO_2 will dissolve in tissue and body fluids and be lost through the general cuticular surface[9]. The first thoracic spiracle is assumed to be open continuously while passive diffusion is occurring, the very small second spiracle is always open and the abdominal spiracles may be open, partially if not fully. Above 12°C, the resting bee switches from continuous diffusion to a discontinuous, convective ventilation regime[9] and this has been observed up to a temperature of 15°C. The spiracles are closed during the major part of the cycle but are opened at irregular intervals, when short bursts of abdominal pumping movements are observed. These occur at a very low frequency, around 30 cycles/hour, and each short burst of pumping is accompanied by the emission of CO_2. Fluttering movements of the spiracles, during which air is sucked in, were not clearly visible in these bees, so possibly air entered only during the period of abdominal ventilation.

Above 15°C bees start to become active, and above 28°C the flight muscles can operate normally. The respiratory movements change again as the bee becomes active. Abdominal ventilation bursts gradually cease to be intermittent as the frequency of pumping movements increases to between 100 and 200 cycles/minute. If the bee is resting or moderately active, tidal ventilation appears to take place, with air moving in and out of the first thoracic spiracle and possibly the second spiracle[10]. The third, or propodeal, spiracle remains closed. Movement into and out of the abdominal spiracles is probably slight under resting conditions but, as activity increases, the beginnings of a directed air flow are seen, with a weakly-directed air flow through the thorax and into the abdomen. Extreme activity, such as that seen in flight, will increase CO_2 production. Under artificial CO_2 stimulation, the action of the spiracles, in conjunction with the abdominal movements, produces a directed flow of air through the thorax which is assumed to mimic the actual respiratory movements during flight[10]. Air is drawn in through the first pair of spiracles when the abdomen expands; these spiracles then close, and the propodeal spiracles open when the abdomen contracts and air is forced out of them. This creates a rapid flow of air through the thorax of the flying insect. Air is also thought to be drawn in through the abdominal spiracles when the abdomen expands and then, with the abdominal spiracles closed, the air is forced out of the propodeal spiracles as the abdomen contracts. A directed stream of air in two directions through the body is unusual but has been reported in other insects. Thorax-specific oxygen consumption rates increase with the onset of foraging to increase flight range and decrease energy costs (see chapter 7).

Spiracular movements are controlled largely by the central nervous system. Opener and closer muscles are innervated by motor neurons whose cell bodies lie in the ganglia of the ventral nerve cord; those of the closer motor neurons are situated in the ganglion anterior to the segment in which the spiracle is located. Activity patterns in these motor neurons may be modulated centrally as a result of oxygen deficiency in the tissues, or by the state of the insect's water balance[11]. Elevated levels of CO_2 have their main effect peripherally, acting directly on spiracular muscles. The presence of two control systems allows the insect to exchange gases efficiently, both at rest and when active, while minimizing water loss. The importance of reducing water loss is indicated by the fact that there is an increased tolerance for intra-tracheal CO_2 in dehydrated insects, while the direct action of CO_2 on the spiracular muscles ensures that suffocation will not result from measures to resist dessication.

Ventilation movements are produced by rhythmical sequences of nerve impulses to the appropriate muscles of a segment from their motor neurons in the abdominal ganglia of the ventral nerve cord. Overall control is provided by a pacemaker which overrides the rhythms of the other ganglia. Where located, it lies either in the third abdominal ganglion or last thoracic ganglion, but its site has not yet been established in the bee. The activity of the pacemaker has been found to be modulated by centres sensitive to CO_2 concentration within the brain and thoracic ganglia of some insects. The motor patterns of the ventilatory centres drive the spiracular activity patterns of the ganglia when ventilation occurs so that spiracular movement is co-ordinated with ventilation movements.

Tracheal mites

The parasitic mite, *Acarapis woodi*, invades the young adult bee via the first pair of spiracles. Arriving on the surface of a new host, it is said to be attracted to the region of the first thoracic spiracle, first by the vibration of the wing bases, and then by the intermittent bursts of air coming out of the spiracles[12]. Although the propodeal spiracles are large enough to admit the mites, they do not appear to enter by this route. Only young bees are susceptible to infestation, being most vulnerable when they are newly emerged from their brood cells. Susceptibility then decreases until the bees are around nine days old, after which they are seldom invaded. Once inside the bee, the mites infest the large trachea leading away from the spiracle, although they have also been found in the head and abdominal air sacs. They feed by piercing the tracheal walls of the host with their mouthparts and sucking up the host's haemolymph. Mites leave older bees via the first spiracle again, climb a hair on the thorax and then grasp a hair on a passing bee with their forelegs.

Since the tracheae leading from the first thoracic spiracles are so important for the supply of air to the thorax, it was long supposed that the presence of numerous mites in these tracheae would impair the bee's ability to fly. However, bees severely infected with mites appear to forage in a normal manner and the same proportion of infected bees are found among those flying and hive bees in infected colonies[12].

7. Flight:
wings, aerodynamics, sensory control and metabolism

FIG. 7.1 *Wings of a worker honey bee, left pair from above.*

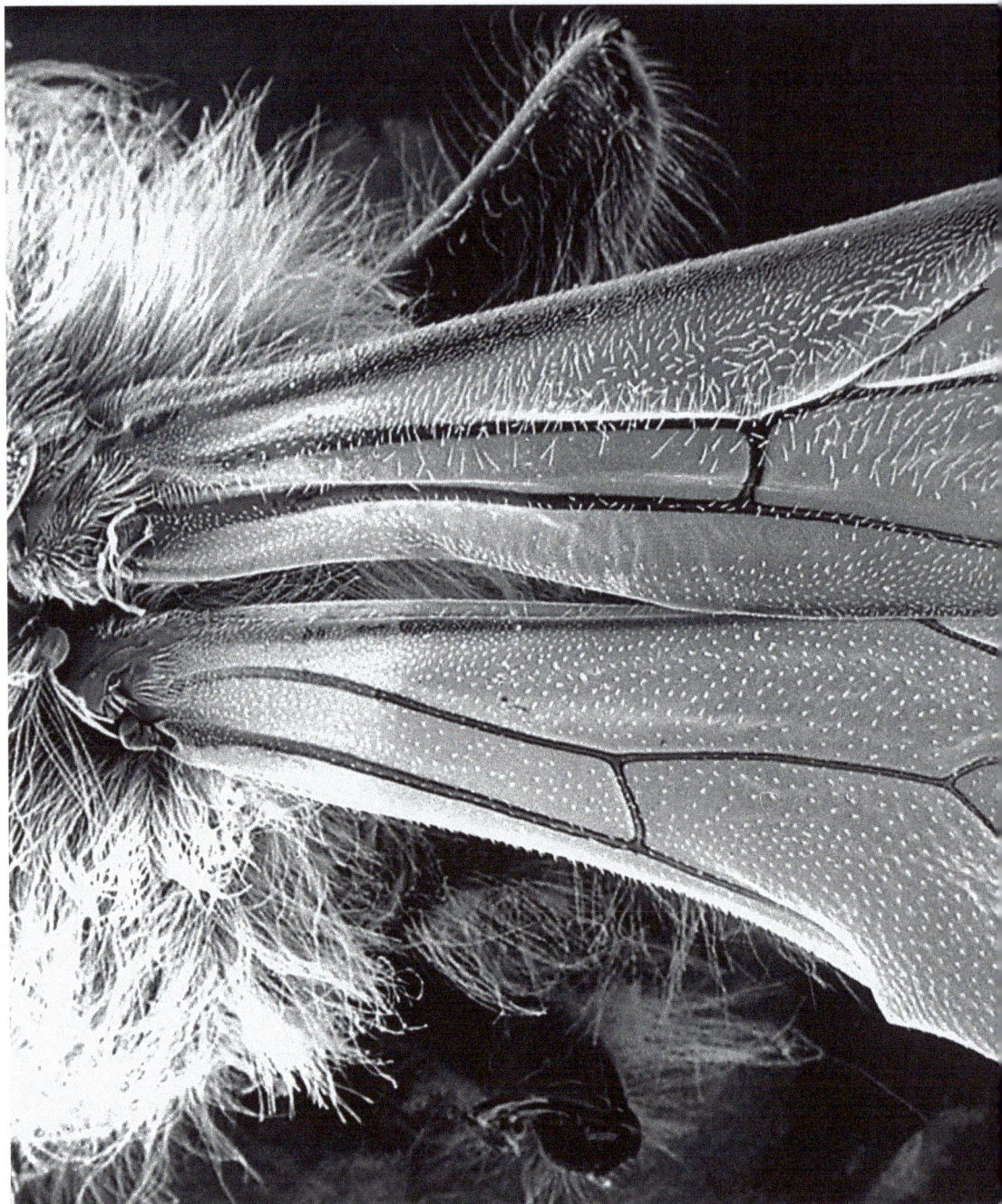

The ability to fly enables honey bees to survive on a specialized diet of nectar, pollen, tree sap and water, which they have to collect from a multitude of different sources dispersed in both space and time. Flight enables them to forage effectively over a wide area and at greater distances from the hive.

The fossil record indicates that insects conquered the air around 350 million years ago, 200 million years before the first dinosaurs and birds evolved the power of flight. The importance of flight in the evolutionary success of insects is clear, for about 1.2 million species of insect possess wings (fig.7.1).

How did wings evolve?

The earliest known flying insects first appear in the fossil record from Carboniferous swamp forest deposits. Even then some of them looked remarkably similar to today's dragonflies, some were like cockroaches and others more like mayflies.

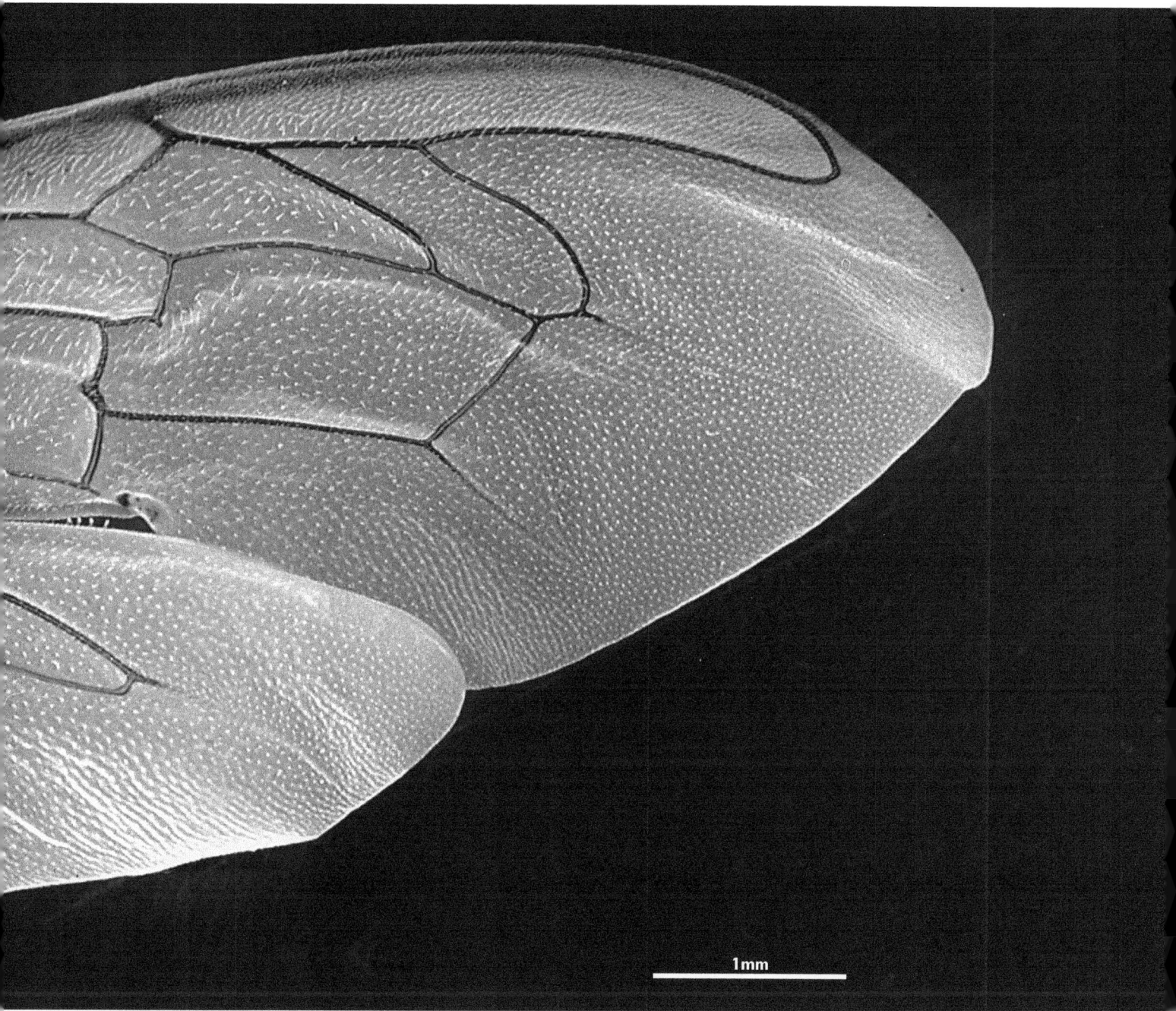

1mm

These fossil insects do not tell us how wings arose in the first place because they were already in possession of fully formed and functional wings. It is estimated that the ancestors of these early insects were perfecting the wing formula for 20 million years before we first see them as fossils. We can only speculate about the way insect flight evolved, based on the evidence from fossils, from studying modern insects and from a consideration of theoretical bio-mechanics. There has been much debate between entomologists about the basis of such an important evolutionary event.

The debate revolves around two important questions. Firstly, there is the anatomical problem. From what part of the ancestral insect were the wings derived? Were the earliest wings originally part of the legs, or were they new outgrowths from the body? Secondly, what were the precursor 'wings' used for? It is most unlikely that an ancient insect suddenly appeared with a funct-ioning set of wings. They most likely developed for some other function and these 'proto-wings' later evolved into structures suitable as aerofoils. Maybe this function was connected with locomotion — enabling a little extra distance to be gained from gliding in air or floating on water. Or perhaps it was some completely different function like protection, defence, respiration or temperature regulation. Form and function are inextricably linked during evolution, so what proto-wings might have been used for depends heavily on exactly what they were like structurally. Could they move or were they rigid flaps? Where on the body were they?

It was believed for a long time that early wings developed as rigid extensions of the body. The evidence for this comes from a series of fossil insect nymphs from the late Carboniferous, 250 million years ago, which have several pairs of wing or lobe-like structures jutting directly out from the thorax and abdomen (fig. 7.2).

However, since the 1970s, an idea first suggested in the 19th century has gained more and more support, championed by the palaeoentomologist Kukalová-Peck[1]. She suggested that the early mayfly-like fossils may be misleading and in fact early wings were extensions from the legs. The aquatic nymphs of modern mayflies possess tracheal gills, sail-like extensions of the coxae, the section of the leg closest to the body. Some of the tracheal gills function as ventilators, wafting water past the gills, others seem to be protective plates covering the gills. The idea that prehistoric wings developed from this kind of leg extension, and not directly from the thorax, has a number of important points in its favour.

Firstly, the legs and wings of modern insects develop from the same place in the embryo, while the tissue that makes up the thorax has a quite different developmental pathway. Many scientists believe we can learn about the ancient evolution of modern structures from the way in which they develop in the embryo. Secondly, most modern insect wings have certain kinds of sensory receptors on their surface that are also found on legs, but not on thoraxes[2]. A third point is that if proto-wings were extensions of the leg, not the thorax, then they would have been articulated with the body and could conceivably have been movable, and voluntarily controlled at an early stage. This is important, because it greatly improves the chances of proto-wings being actively controlled structures serving a useful function prior to their evolution into wings.

There have been numerous suggestions over the years as to what exactly that earlier function might have been. They could have been ventilators or gill protectors, much like the lobes carried by mayfly nymphs, but we do not know for sure whether the earliest insects with wings were aquatic. The aquatic stage in the life cycle of modern mayflies seems to be a secondary development, for they appear to have evolved from terrestrial insects that returned to the water. In this context, it seems more likely that the gill-related function in mayflies developed after the original function of the proto-wings was lost. The Russian entomologist Brodsky has argued that because the original function

FIG. 7.2 *A protereismatid mayfly nymph reconstructed from fossil remains from Lower Permian deposits from Oklahoma, USA. Lateral paranotal extensions were present (arrows) on the thoracic and abdominal segments.* Redrawn from Kukalová-Peck[1].

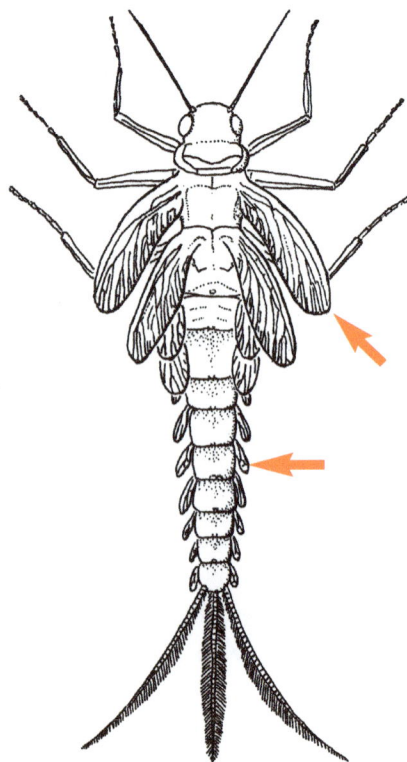

was lost when the structures became useful as wings, it cannot have been that important[3]. Perhaps it was something very specific to a particular insect species, like egg holding or signalling to the opposite sex?

Another idea that has received a lot of attention is that the proto-wings had a thermoregulatory function. Like the ears of an elephant, they could have been used to dissipate heat from the body. In 1985, two American scientists, Kingsolver and Koehl, used resin models of early insects to investigate this hypothesis. They added proto-wings of various lengths to several sizes of model insect and tested the effects on thermal uptake and aerodynamic forces[4]. They found that small wing-like structures were effective heat-exchangers but as the size of the 'wings' increased, the thermal advantage from the increased length decreased, until at a certain size, there was no further advantage in getting any bigger. It is at exactly this size that the structures begin to have useful aerodynamic properties! This is possibly a very neat piece of theoretical evidence for wings having developed from thermoregulatory structures.

All of these ideas assume that the very small proto-wings could not possibly have had any advantage in early attempts at flight. As we shall see in the following section, modern insect aerodynamic theory tells us that if the wings could be flapped, they would produce far greater aerodynamic forces than we might predict for their size, so perhaps we are over-complicating the story, and proto-wings were always potentially available for promoting some enhanced locomotor performance in water or air.

Even if proto-wings were aerodynamically useful structures, they would not have enabled the sophisticated flight skills we observe in modern insects. However, if these early insects were getting into the air by some other means, then there are various ways in which even relatively undeveloped wings might lend a slight aero-

dynamic advantage to the insect that possessed them. The first fliers could have been gliders, climbing to the tops of plants to feed and then jumping down or across to other plants nearby. A structure like a plate or lobe could then reduce the rate of descent by providing extra lift and carrying the insect a greater distance along the ground. One of the earliest known fossil winged insects has a wing shape similar to that of a butterfly, which would have supported gliding. Another possibility is that insects were jumping into the air, perhaps to escape from predators. However, this seems unlikely because the oldest winged insects did not have legs adapted for jumping[5]. A third idea is that winged insects evolved from very small insects that were simply blown into the air and floated around on the wind.

If the ancestors of winged insects lived in water, they could have used their proto-wings and legs like the fins of a fish. Recently a species of stonefly was discovered in Madagascar that was flightless but had very short wing-like structures, but it used these not for flying, but for skimming over the water surface[6]. It is believed that these insects keep their legs in contact with the surface, and that the 'paddles' propel it forward over the water, rather than supporting its weight in the air, and enable it to move faster.

Although many of these fossil insect lines developed wing-like structures and subsequently lost them, they do serve to demonstrate ways in which the ancestors of modern flying insects might have used their proto-wings. The main obstacle to understanding the evolution of insect flight is that, in the absence of adult insects from the early fossil record, we can only guess what the immediate ancestors of the first flying insects looked like. If they were completely terrestrial, without an aquatic stage in their life cycle, then some of the explanations proposed above are completely ruled out. Also we do not have any idea how big they were. It is unlikely that the passive drifting model could have led to insects the size of the giant dragonflies that appear in the later fossil record. It is possible that convergent evolution was at work, and flight evolved in different ways, some of which led to less successful solutions and which subsequently disappeared.

Aerodynamics, and how flight is possible

You will undoubtedly have heard the claim that a bumble bee should not be able to fly, given its weight and the size of its wings and muscles. Until about 15 years ago, scientists found it difficult to explain how insects could fly, for according to the laws of conventional steady-state aerodynamics, their wings are simply not large enough to get their bodies airborne. However, the reality is that they can support their own weight in the air, and some wasps have been shown to be able to lift up to nine times their own body weight in prey[7]. Even now, the interaction of basic aerodynamic principles and the complex mechanisms of dynamic force generation are still being unravelled and described.

However, before considering the recent developments in insect flight aerodynamics we should consider some general aerodynamic principles that have been understood and used by engineers for more than a century, leading to such developments as the jumbo jet and supersonic military aircraft.

The most important forces in aerodynamics are drag and lift. When an insect moves through the air the component of force acting on it parallel to its direction of movement and resisting it is called drag. **Lift** is defined as the force perpendicular to drag generated by the insect. During steady horizontal flight, lift must exactly equal the weight of the flying insect but, at take-off and in ascending flight, lift must exceed this. Another group of forces known as acceleration forces do not appear to be important in insect aerodynamics[8].

How is lift generated?

In the steady-state condition there are two principles involved. The first is that gases move from regions of high pressure to regions of lower pressure, for example wind is air moving from an area of high pressure to an area of low pressure over the surface of our planet. The second, known as Bernoulli's principle, states that air pressure is related to the speed at which the air is moving. Fast moving air is at a lower pressure than that of slow moving air. Thus if air can be induced to move faster over the upper surface of a 'wing' than beneath it there will be a net upward force on the 'wing' as air tries to move from the high pressure area below it to the lower pressure area above it.

Wings, known in engineering terms as aerofoils, are designed so that the air moving over their upper surface travels a greater distance than the air moving underneath, and is therefore forced to travel faster.

Figure 7.3 shows the flow of air around a typical aircraft wing in flight. Any object inclined to the horizontal, or cambered as in the diagram, will behave as an aerofoil and generate lift when there is an airflow around it. Air approaching from the front is forced to split into two streams, which pass above and below the wing to rejoin at the trailing edge. The inclined surface causes the split to occur just below the leading edge, and the air that takes the 'high road' has to round the bend first. At the trailing edge, the two flows of air rejoin just behind the wing rather than following its exact shape because of the viscosity of the air. The air moving over the top has a greater distance to cover so moves faster, and generates a lower pressure above the wing than that produced by the air moving below it, resulting in an upward force, lift, acting on the under surface of the wing.

The magnitude of the lift generated is directly proportional to three important factors: the speed of the air moving over the wing, the area of the wing, and the angle between the wing and the direction

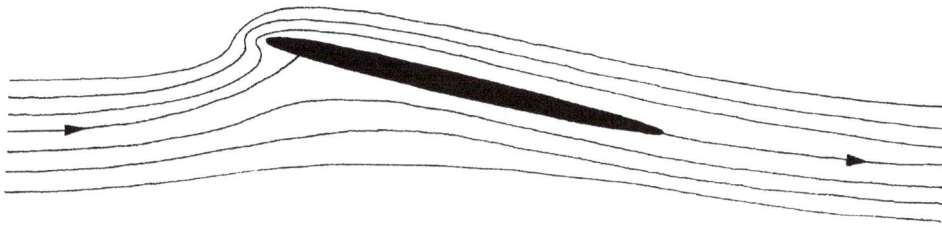

FIG. 7.3 *Diagram showing viscous flow around an aerofoil (black). Arrows on flow lines indicate direction of flow.* Redrawn from Barnard and Philpott[9].

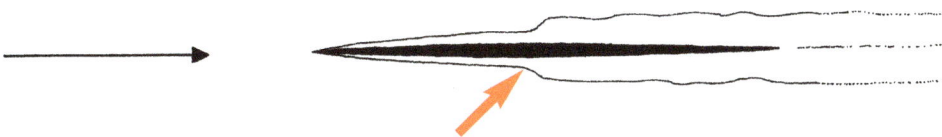

FIG. 7.4 *Diagram showing the theoretical position of the transition point* (orange arrow)*, where the flow ceases to be laminar, for a thin flat plate at zero angle of attack. Black arrow indicates direction of flow.* Redrawn from Barnard and Philpott[9].

of air flow, called the angle of attack. In theory, the faster the wing moves and the larger its area, the more lift it creates. Increasing the angle of attack, up to certain limits, also increases the lift. The wing area and the angle of attack of an aircraft's wings is increased by deployment of flaps on the trailing edge of the wings at take-off and landing to produce enough lift at relatively low speed to lift the machine off the ground. Once cruising at high speed, the angle of attack can be reduced so the flaps are retracted.

All these effects are complicated by another very important force in aerodynamics called drag. **Drag** is the force resisting movement. Its magnitude depends upon the shape and surface roughness of the moving object and, like lift, is dependent on its speed and the angle of attack of the wing. It results largely from friction between the air and the surface of the flying object, so size and shape of the wing are very important if the amount of lift relative to the amount of drag (the lift to drag ratio) is to be maximized.

Much of the effect of drag occurs in the very thin layer of air over the surface of the wing, known as the boundary layer. At the very surface of a moving wing the air is stationary but at increasing distances from its surface, depending upon the density and viscosity of the air and the 'roughness' of the wing, its speed will increase. The friction caused by 'carrying' the slower moving air within the boundary layer results in

the friction drag of the wing. The thickness of the boundary layer increases with distance from the leading edge of the wing and at a certain point it becomes so thick that the flow over the wing surface ceases to be smooth, or laminar, and becomes turbulent. This point, where the boundary layer changes to turbulent flow, is called the transition point (fig. 7.4).

Exactly where the transition point occurs also depends on the density and viscosity of the air, the speed of travel and the width of the wing. The transition is important because the type of flow in the boundary layer makes a big difference to the overall drag effect. Laminar flow produces less drag than turbulent flow. As was noted earlier, the angle of attack can be increased to increase lift. However, even with laminar flow there comes a point above which a further increase in angle of attack results in the air flow separating from the wing contours leaving a large area of turbulence. This condition is referred to as dynamic stall. Its effect is not only to enormously increase drag, but also to disrupt the aerodynamic flow that is sucking the air over the top surface of the wing faster than it is travelling beneath it. This disrupts the production of lift and can have disastrous consequences for an aircraft.

At this point we can consider an effect of scale on the aerodynamics of wings. A small wing will have a transition point that is relatively much further back from the leading edge of the wing than a large wing, thus making it much more susceptible to flow separation and consequent dynamic stall. Any feature that promotes turbulence in the boundary layer will bring the transition point forward and thus reduce the risk of stall. Scales or hairs on the surface of insect wings may be important for this reason, as may their very thin section.

Why flight by animals is more complicated

Until now we have been considering what is known as 'steady-state' aerodynamics. The wing under consideration has a fixed angle of attack, and a fixed area, and the only thing that changes is the speed of the air over the wing that results from propulsion through the air. However, insects, bats and birds flap their wings. The angle of attack of the wing changes constantly as does the speed of the air over the wing surface and its direction relative to the wing. Even the size and shape of the wing can be altered during flight. This makes the calculation of the aerodynamic forces considerably more complicated. It is possible to approximate them separately for a number of different positions in the flapping cycle and then to average them over the full cycle, but certainly in the case of insects this is not fully adequate or appropriate. The study of insect flight is now almost entirely concerned with understanding 'unsteady' aerodynamic forces generated by the flapping motion itself.

Why do flying animals have to flap their wings?
Observation shows that birds and large butterflies, for example, can stay airborne without flapping, by gliding. As the generation of lift relies on air moving over the wing, gliding flight is possible in certain circumstances because either lift is sacrificed for forward descending movement or because the air flow is itself directed upwards as it is in thermals or in up-currents near cliffs, etc. Normally, in order to maintain forward movement and to maintain height, a propulsive force, thrust, is required, and the thrust generated must exceed the drag created by the movement.

Flight is different from ground-based forms of locomotion, like walking or running where the backwardly-directed action force applied to the ground results in an equal and opposite reaction force generated by the limb's friction with the ground surface resulting in forward displacement of the animal. There is no physical contact between the animal and the ground when it is flying through the air. In aircraft, the thrust is created using the energy from the engines to suck air into the engine and then to force it out behind. Animals' wings create aerodynamic thrust and lift forces by flapping. Flapping involves both translation, up and down movement of the wing, and rotation around the long axis of the wing at the top and bottom of the wing stroke.

Figure 7.5 shows the aerodynamic forces generated during the flapping cycle of a bumble bee's wings. On both the upstroke and the downstroke, the lift force is not totally perpendicular to the forward direction of flight (right to left across the page). This is because lift acts perpendicular to the direction of airflow over the wing, which in turn depends on the way the wing itself is moving through the air. On the downstroke the wing moves forwards and downwards and on the upstroke it moves upwards and backwards. Its leading edge is rotated forward (pronated) (fig. 7.19) and tilted downwards on the downstroke and rotated backwards (supinated) (fig. 7.19) and tilted upwards on the upstroke. On the downstroke the direction of the resultant lift force will be angled upwards and forwards and on the upstroke it is angled upwards and backwards. In forward flight the backward rotation of the wing on the upstroke causes the under surface of the wing to face the direction of airflow and this tends to reduce the backwardly-directed component of the force. In this way, during forward flight lift is largely produced during the downstroke and thrust during the upstroke.

The amount of thrust generated depends on how much the resultant lift force is tilted towards the direction of flight. If it is

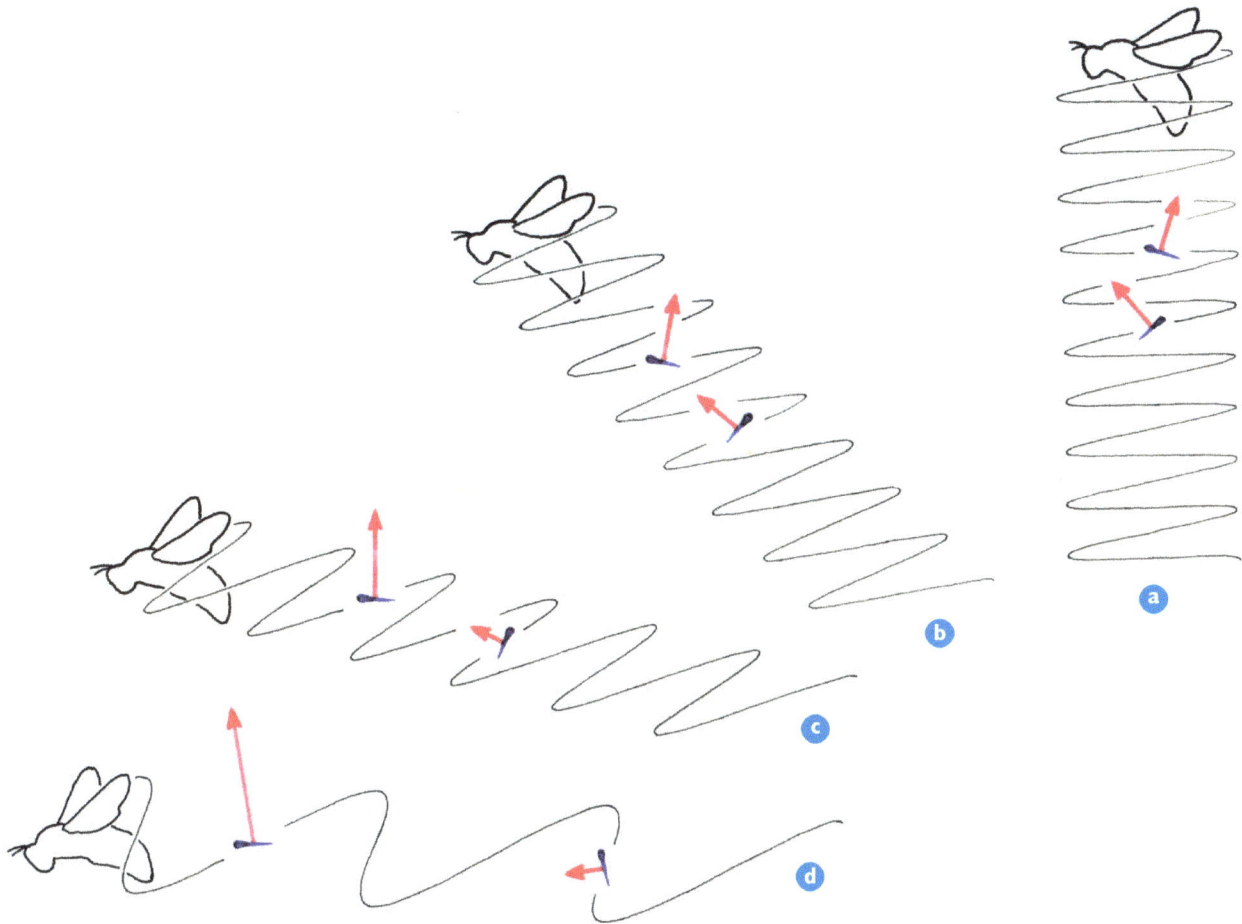

FIG. 7.5 *Diagram showing the path of wingtip (black line) of a bumble bee during* **a** *hovering,* **b** *forward flight at 1m/s,* **c** *forward flight at 2.5 m/s,* **d** *forward flight at 4.5 m/s. The wing is shown in blue (leading edge thickened) for midpoints in the up- and downstroke, and the magnitude and direction of the lift force is represented by a* red arrow. Redrawn from Ellington[10].

inclined heavily forwards, a large proportion of it will appear as thrust rather than lift. The resultant thrust and lift is averaged over the whole wingbeat cycle, which in a flying honey bee is completed more than 200 times every second. In general, insects do not increase their wingbeat frequency nor the wing stroke amplitude when changing speed.

The precise direction of the overall aerodynamic force depends on the stroke plane, i.e. the plane through which the wing moves relative to the body in one complete cycle (fig. 7.5). Increasing the stroke plane angle results in faster flight because a greater proportion of the force appears as thrust and some additional thrust is generated on the upstroke as well as on the downstroke.

Insect flight is still more complicated!

We have seen how a bee might achieve its flight manoeuvres with subtle changes in the flapping action of its wings, but the process is still more complicated as the flapping action creates further problems. Apart from causing a headache for scientists trying to calculate the aerodynamic forces involved, it also takes a lot of extra energy on the part of the insect. At the top and bottom of every stroke, a potentially large amount of energy could be wasted as the wing has to be decelerated right down to a stop before being accelerated again in the opposite direction. This use of energy is minimized by storing it elastically in the cuticle of the thorax, and in the flight muscles themselves. Some insects also have a pad of a rubber-like protein called resilin at the base of the wing that aids in this storage of energy. About half of the energy required to decelerate and accelerate the wing at each stroke is stored elastically, although so far it has not been possible to measure directly the energy storage at wingbeat frequencies typical of flight[8].

Another major aerodynamic consideration is the small size of insects. As explained above, size affects the transition point in the wing boundary layer, making them much more susceptible than large aircraft to dynamic stall. In addition, the boundary layer itself is relatively much thicker in an insect wing than that of an aircraft, so its effect, in terms of drag, is much greater. The effect of the viscosity of the air is considerable for insect-sized fliers that in relative terms are swimming through syrup. Combined with low air speed, this tends to give insects a very low lift to drag ratio.

Bees are at a particular disadvantage because they have an extremely high wing loading — that is, the ratio of body weight to wing area. Since the steady-state aerodynamic forces generated are proportional to the area of a wing, it would seem reasonable to expect heavier insects to have larger wings, as they have more weight to support and all insects would have approximately the same wing loading. In fact, the wing loading of a typical bumble bee worker is 15 N/m^2, which is an order of magnitude higher than wing loadings measured for butterflies of similar size[11]. Measurements of honey bee wing loadings have recently been made[12].

How do insects stay up in the air?

The measurement of the forces generated during flapping flight is not easy, as they are created only when air moves over the surface of the wing, and when this occurs in nature the insect is normally moving through the air, thus making measurement even more difficult. However, a similar effect can be simulated experimentally by keeping the test subject in a fixed position and passing air over it. The forces generated by the insect in flight have therefore

been measured from insects tethered in wind tunnels where the airflow can be precisely controlled; if the insect is tethered to a sensitive mass balance, the amount of lift created can be measured accurately. Another method is to use scale models of wings, and such a model of the wings of the fruit fly (Drosophila melanogaster), has recently added greatly to the current understanding of non-steady-state aerodynamics[13].

In the mid 1960s, it was first suspected that the lift generated was greater than that predicted from conventional, 'steady-state' aerodynamics. American scientist Leon Bennett measured the lift force generated by a flapping model of a beetle wing, and compared it with the lift force generated by simply rotating the wing to create lift in the same way as a helicopter rotor blade that is known to obey the laws of steady-state aerodynamics[14]. He found that the flapping motion generated two to three times more lift than predicted from a wing of its shape and size. More recent calculations agree with this finding and confirm that the amount of lift required to support the weight of most insects exceeds by two or three times the maximum that could possibly be generated by the wing using steady-state aerodynamic principles[10].

There followed a considerable effort to understand where and how this extra lift was generated. It was already clear that the flapping movements and the continually changing orientation of the wings during each beat is complicated and involves much twisting and rotation of the wing. Also the fact that flying insects produce a buzzing sound or flight tone suggests that flapping produces puffs or pulses of air, and this was confirmed by measurements of air speed in the wake of a fly during flight[15].

The first 'unsteady' mechanism to be described from observation of an insect in flight was called the 'clap and fling' and was first described in a small parasitic wasp (Encarsia)[16]. The wasp is very small with a wingspan of only 1.3 mm, and beats its wings about 400 times per second during flight.

In 'clap and fling' the wings clap together at the top of the upstroke and are then pulled apart rapidly starting from the leading edge. In the split second when the leading edges are separate and the trailing edges are still together, air is sucked into the space between the wings. This creates an area of low pressure above the wings and enhances lift. By creating low pressure above the wing surface at the beginning of each stroke, the mechanism kick-starts the normal lift generation process. Clap and fling has now been demonstrated in several other small insects including thrips, fruit fly and whitefly. Some larger insects such as butterflies and locusts[17] use a similar movement, called 'clap and peel', which can produce up to 25% more lift than normal wing movements[10].

In the late 1970s, Ellington passed streams of smoke over a scale model of an insect wing and was able to show that there were vortices — rapidly spinning columns of air, like micro tornadoes — being generated around the 'wing'[18].

Vortices have long been known to be important in aerodynamics, because of the same principle that creates lift — the faster the air is moving, the lower its pressure. In a vortex, air is spinning round very fast, so the pressure is low, as in the centre of a tornado. A vortex produced on the upper surface of the wing will have the effect of increasing the lift force acting on the wing. Several kinds of vortex have now been identified in the air around a flying insect, generated by the flapping action of its wings.

Perhaps the best understood is the leading edge vortex — a spiral of spinning air that spans the front edge of the wing and flies off at its tip. Ellington's experiments have shown that the flow of air over a hovering hawkmoth's wing actually separates from the wing at its leading edge, as in a dynamic stall[19] and reattaches further down the wing. The leading edge vortex occurs in the gap where the airflow is sep-

arated from the wing. Insects such as hawkmoths, and probably also bees, make extensive use of this effect during hovering and forward flight, because it allows them to operate at high angles of attack, thus generating even more lift, without suffering the ill effects of dynamic stall.

Other vortices are generated by the trailing edge of the wings at the top and bottom of the wing stroke. They are caused by the rapid rotation of the wing edge during its change of direction. After each flapping cycle, the top and bottom vortices join together and peel off the wings in the wake of the insect in a kind of doughnut-shaped swirl[20]. The 'doughnut' has the effect of causing air behind the insect to flow backwards and downwards, which in turn leaves a lower pressure around the insect itself and again generates extra lift. It is these 'doughnuts' of turbulence flying off behind the insect that create the pulse-like changes in air flow that are largely responsible for the audible buzzing tone produced during flight.

More recently Dickinson and his co-workers have used a scaled-up model of the wings of the fruit fly beating in oil rather than air, called 'robofly'. This has enabled the measurement of the forces and their direction produced during flapping and has led to clarifying descriptions of the ways that lift can be generated during the wingbeat cycle[13]. Three mechanisms have been described (fig. 7.6). The first, called **delayed stall**, occurs during the translation phase of the downstroke when the wing has a high angle of attack and promotes the formation of the leading edge vortex that reduces pressure above the wing. As the wing approaches the bottom of the downstroke it starts to rotate, moving the leading edge backwards and generating **rotational lift** by a mechanism that has been likened to that which generates 'backspin' on a tennis ball. Rotation produces a further powerful force at the end of the downstroke that is resolved primarily as thrust that pushes the insect forward. A third lift-enhancing mechanism is available during the upstroke. Some of the energy

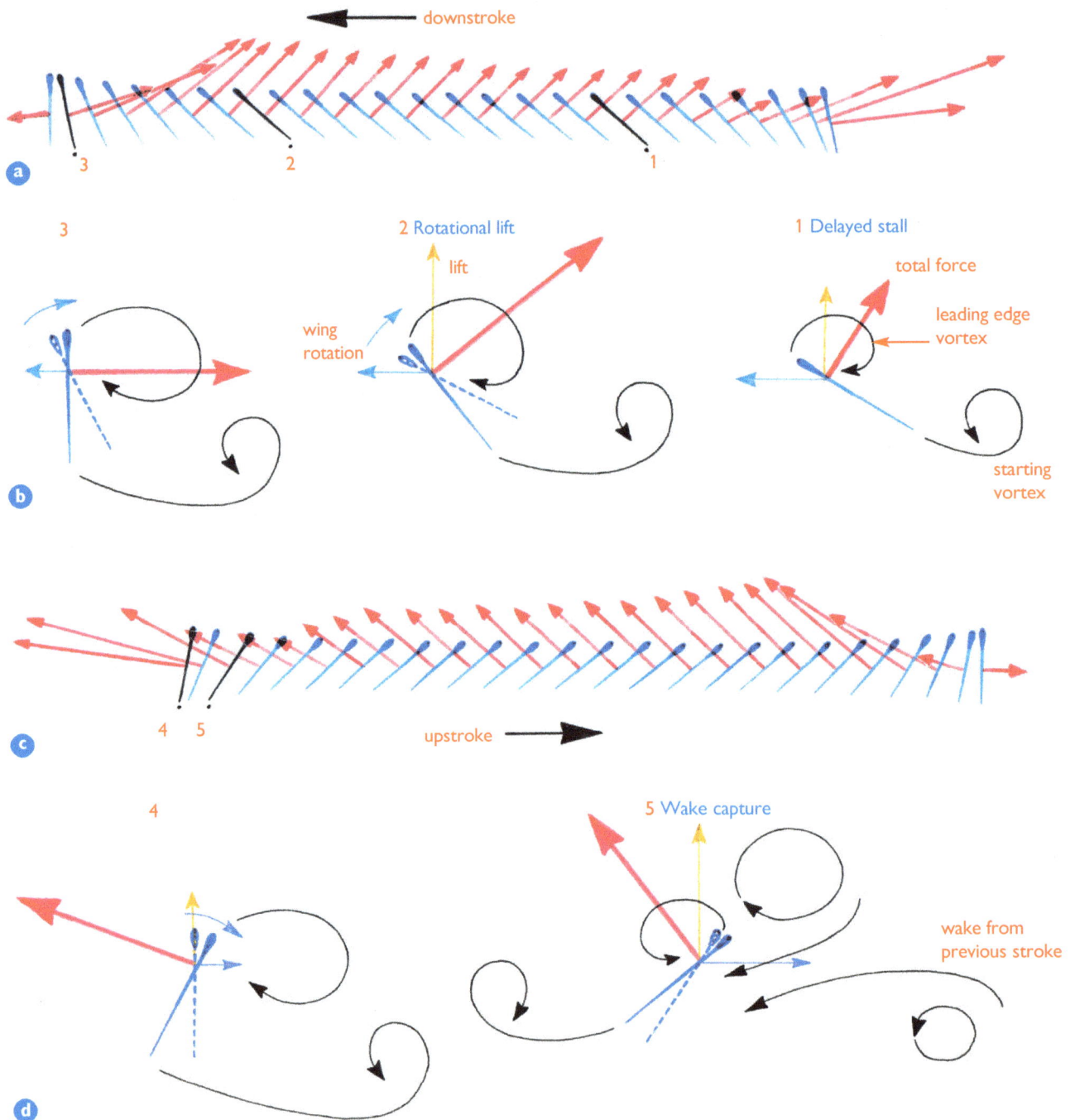

FIG. 7.6 *Diagramatic representation of wing movements and forces produced. Thick* blue *lines represent the wing seen end on and thickened to indicate the front (leading) edge. Curved* thin blue *arrows indicate the direction of wing rotation and horizontal* thin blue *arrows the movement, forwards or backwards, of the wing. Thin black lines represent air flow lines.* Red arrows *show the magnitude and direction of the total force and* thin gold arrows *the lift component of the force.* **a** *Downstroke. Successive wing positions (in* blue) *during the downstroke as the fly moves from right to left. The* red arrows *represent the magnitude and direction of the forces generated in the stroke plane.* **b** *Downstroke. During the downstroke, represented by wing position 1 in (*a) *delayed stall sets up a leading-edge vortex which reduces the pressure above the wing. As the wing rotates (rotation starts at 2 in (*a) *and is represented by the* dotted wing *and* thin blue arrow *in (*b2) *prior to the end of the downstroke, rotational lift is produced. Further propulsion is achieved at the end of the downstroke (*a3 *and* b3). **c** *Upstroke. Insect still moving from right to left, wing now moving up and back.* **d** *The insect has left swirls of air behind it (4) and as the insect rotates its wing again (*dotted blue line) *its path intersects its own wake (5) capturing energy resulting in further lift production. This is wake capture.*

Redrawn from http://www.sciam.com/exhibit/1999/062899fly/flight.html, after M Dickinson.

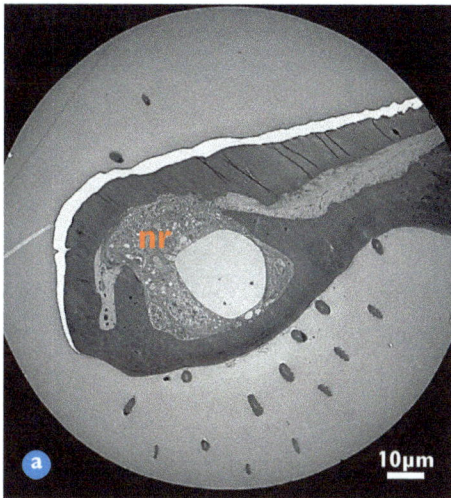

FIG. 7.7 *Light microscope section through the leading edge vein (costa) of a worker bee's wing.* **a** *Low magnification view of the lumen of the wing vein showing neurons (nr).* **b** *Higher power view of nerve tissue in the lumen of the wing vein. Cuticle (c), nerve tissue (n).*

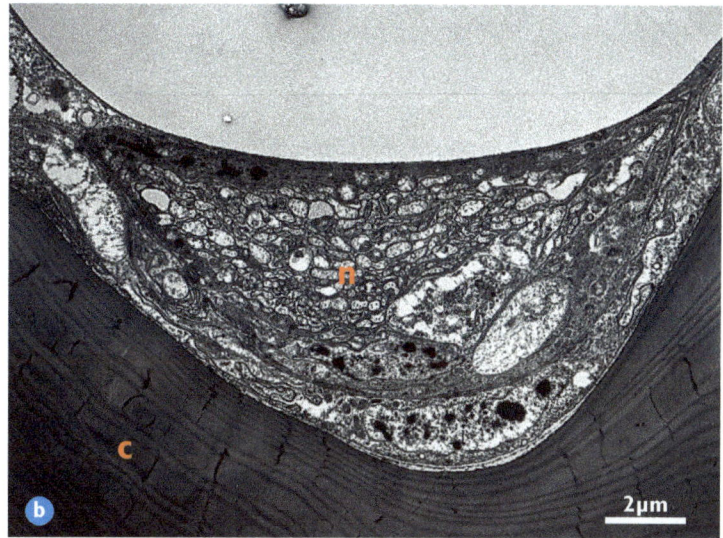

lost in the vortices shed during the downstroke can be recovered during the rotation at the start of the upstroke when the wing moves through its own wake. This mechanism has been called **wake capture**.

Further study of these vortices and non-steady mechanisms is explaining the discrepancy between conventional aerodynamics and the fact that insects can fly, but there is still a vast amount of work to do. Insect wings come in an enormous variety of shapes and sizes, and small variations can lead to fundamental differences in the nature and extent of the contribution made by the non-steady forces produced. The shape of the vortex wake has been shown to be different for different insects; in lacewings, for example, it is more of a horseshoe shape than a doughnut shape[8].

Insect flight is certainly complicated and although in the last twenty years a lot of progress has been made towards understanding it, most of the work has been carried out on a very limited number of species — hawkmoths, locusts, bumble bees and some much smaller insects such as fruit flies and parasitic wasps. Measurements of the actual forces generated, and

the precise mechanisms being used by honey bees during flight have yet to be made, but while it is clear that honey bees must make use of these mechanisms it is as yet unclear which mechanisms operate during flight.

The structure of the wings

Insect wings are thin membranes of cuticle stiffened and supported by cuticular veins. On emergence from the final nymphal skin or pupal case the wings of an adult insect are rumpled, soft and pliable. The wings are expanded by hydrostatic pressure of insect haemolymph or blood, which is pumped into the hollow veins. In the first few hours of adult life the wings dry and stiffen to their final size and shape. Some of the veins are hollow (fig. 7.7) though they do not necessarily all carry nerves[21]. Throughout life, haemolymph is moved through those that are hollow to maintain the level of cuticular hydration necessary to maintain the flexibility of the cuticle and to sustain the nerves supplying the wing's sensory structures. The veins impart both stiffness and flexibility, but also allow the wing shape and profile to alter under different conditions of inertial and aerodynamic loading[22].

There is much variation in the size and shape of insect wings and in their venation. In the honey bee the forewing is larger than the hindwing and the venation is less

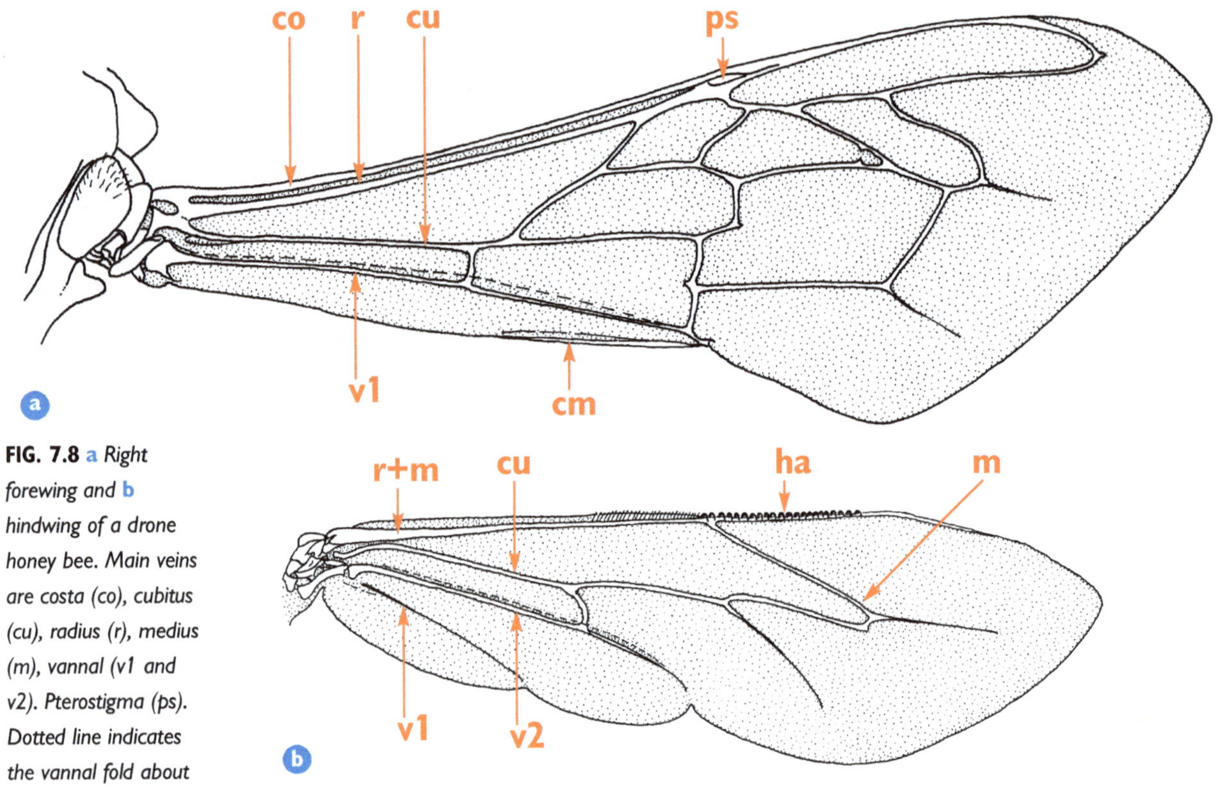

FIG. 7.8 a *Right forewing and* **b** *hindwing of a drone honey bee. Main veins are costa (co), cubitus (cu), radius (r), medius (m), vannal (v1 and v2). Pterostigma (ps). Dotted line indicates the vannal fold about which the trailing edge of each wing can bend. The coupling mechanism is indicated; coupling margin (cm) on the trailing edge of the forewing and the row of hooks (hamuli (ha)) on the leading edge of the hindwing. Redrawn from Snodgrass*[23].

complex than in some other insects (fig. 7.8).

One or two pairs of wings ?

The aerodynamic problem of having two pairs of wings beating at the same time on each side of the body has been solved in different ways. In the more primitive insects such as the Odonata (damselflies and dragonflies), both pairs of wings can be moved independently out of phase with each other and consequently both the mesothorax and the metathorax are well developed. If the hindwings are the more important pair as in the Orthoptera (locusts, grasshoppers and crickets) and the Coleoptera (beetles), then the metathorax is well developed. In the Hymenoptera (bees and wasps) and the Lepidoptera (butterflies and moths) the

FIG. 7.9 *Moth wing coupling mechanism.* **a** *Silver-striped hawkmoth (Hippotion celerio).* **b** *Wing coupling mechanism in the male hawkmoth. Hindwing bristle or frenulum (f), and hook or retinaculum (r) emanating from the radial vein of the forewing.*

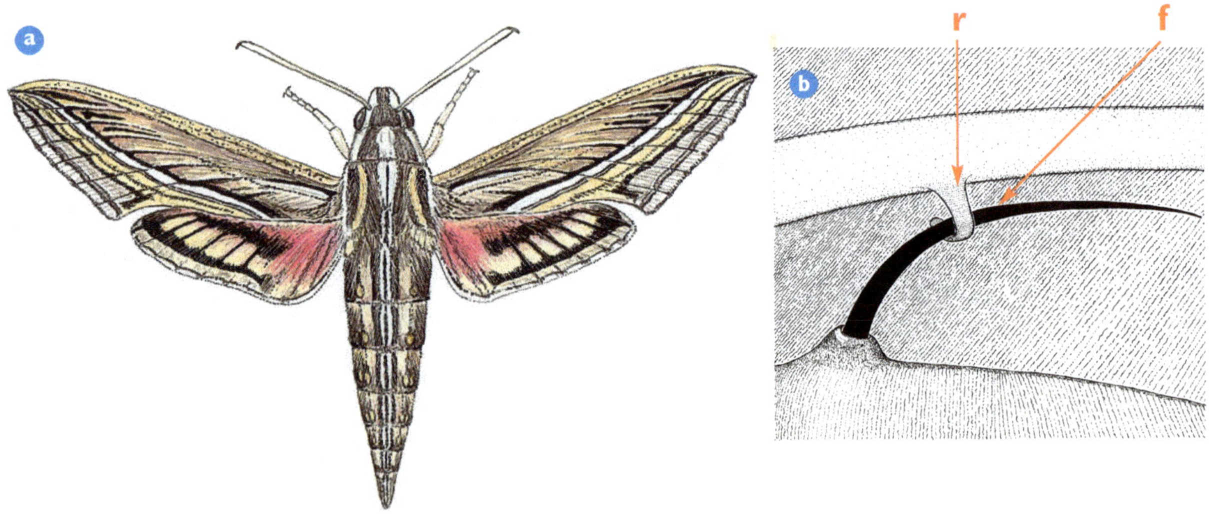

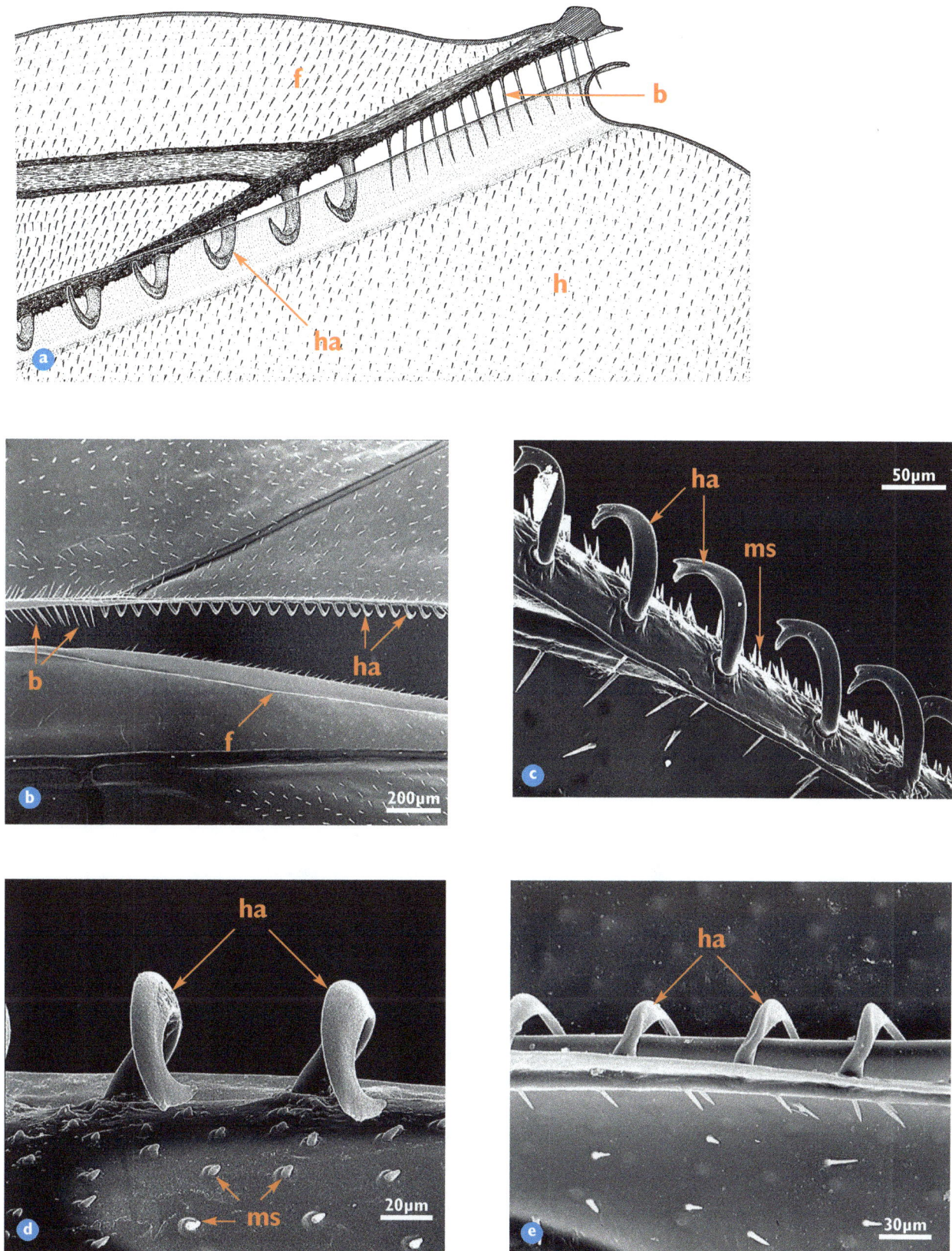

FIG. 7.10 *Honey bee wing coupling mechanism.* **a** *Diagrammatic view from below. Forewing (f), hindwing (h), hamulus (ha), large hairs or bristles, perhaps with a sensory function (b).* **b, c, d, e** *Micrographs showing detail of coupling mechanism in the honey bee.* **b** *Wings separated. Hindwing at top; hamuli (ha), bristles (b) on leading edge of hindwing. Fold in trailing edge of forewing (f).* **c** *Hamuli (ha) and small mechanoreceptor setae (ms).* **d** *Hamuli (ha) and mechanoreceptor setae (ms) viewed from above.* **e** *Hamuli (ha) coupled with fold in forewing.*

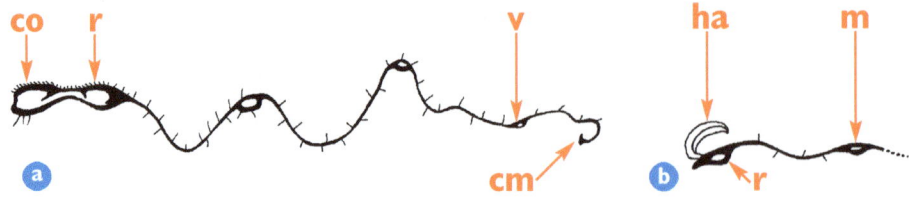

FIG. 7.11 *Diagram showing section through a worker bee's forewing* **a** *and the leading part of its hindwing* **b** *in the region of the coupling mechanism. Costa (co), radius (r), vannal (v), and medius (m) veins. Coupling margin (cm) and hamulus (ha). Redrawn from Snodgrass*[23].

mesothorax is highly developed at the expense of the metathorax, for in these insects the forewings are powered and the hindwings follow their movements more or less passively. In the bees and moths the forewings and the hindwings are mechanically linked during flight by various types of hook mechanisms (figs 7.9 and 7.10). In the honey bee the wings are uncoupled in the resting position over the back of the bee but as the forewing moves forward over the hindwing during flexion a row of hooks (hamuli) on the anterior edge of the hindwing engage in a fold on the posterior edge of the forewing, and both wings open together as a functionally single wing surface (figs 7.10 and 7.11). In the Diptera (the two-winged flies) the metathorax is reduced still further and the hindwings are reduced to a pair of club-like structures, the halteres (fig. 7.12). The halteres beat in time with the forewings and are believed to function like gyroscopes sensing turning and rotational torques whilst the insect is making complex flight manoeuvres.

How are the wings moved?

Anatomy of the thorax

Before looking at how the wings are moved we need to consider the basic skeletal structure and organization of an insect. It is generally agreed that the ancestral insect was a segmented, worm-like creature with a flexible skin. As in worms the skin, or integument, of the ancestral insect formed a sack 'inflated' by the combination of organs and hydrostatic pressure of the body fluids contained within it, giving it what is known as a hydrostatic skeleton. In most adult insects the original segmental structure has been obscured by fusion of the segments into functional groups forming the three main **tagmata** or regions of the body and hardening of parts of the cuticle. Typically the head, thorax and abdomen are comprised of six, three and 10 segments (some have 11), respectively. In adult insects the thorax usually carries three pairs of legs and these, together with the tripartite body, constitute two of the main taxonomic features of the superclass of Arthropods, the Hexapoda or Insecta. In those insects with wings they too are carried on the thorax. The head contains the mouthparts, eyes and the main sensory

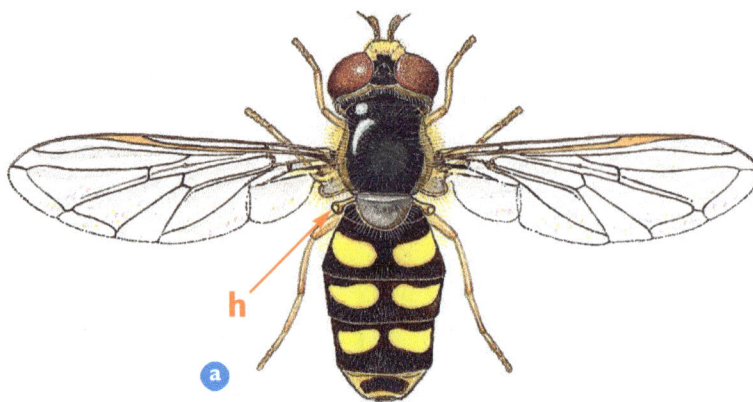

FIG. 7.12 a *Hoverfly Metasyrphus luniger.* **b** *Base of forewing (f) and modified hindwing (h) or haltere of the hoverfly.*

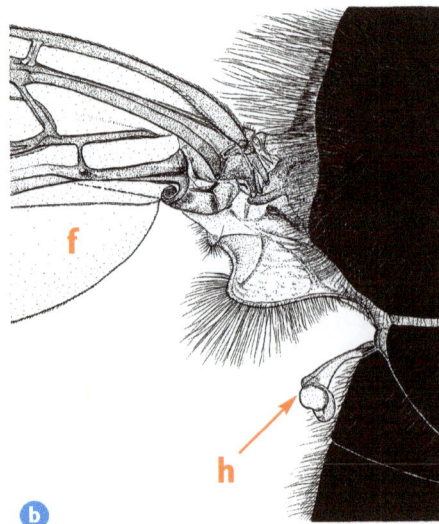

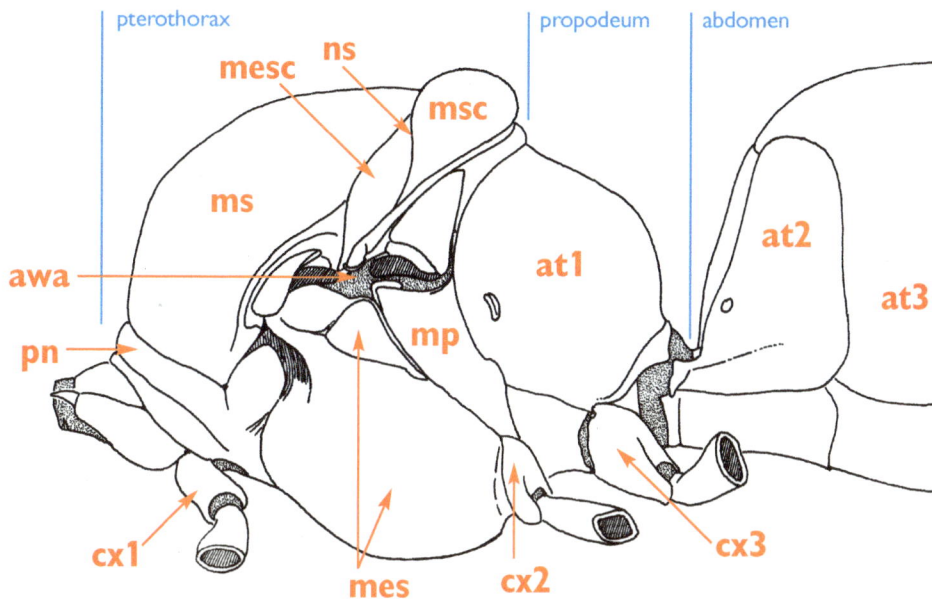

FIG. 7.13 *Side view of the bee's thorax and anterior part of its abdomen showing the major sclerites. Area of wing articulation (awa). 1st abdominal tergum (propodeum (at1)), 2nd and 3rd abdominal tergites (at2, at3), respectively. Coxae of the legs (cx1, cx2 and cx3). Mesoscutellar arm (mesc), mesoscutum (ms), mesoscutellum (msc), two parts of mesepisternum (mes), metapleuron (mp), notal suture (ns), pronotum (pn). Redrawn from Snodgrass[23].*

and nerve processing centres, the supra and suboesophageal ganglia. The thorax is the primary focus of locomotor activity and the abdomen houses the reproductive, digestive, and excretory organs. Evidence for the primitive segmental origin of insects can be seen in the repetition of neural ganglia and spiracles within the segments of the thorax and abdomen.

An important property of the exoskeleton of insects is its flexibility. The cuticle consists of polysaccharide microfibrils embedded in a matrix of protein, forming a fibrous composite that is both strong and elastically flexible[24]. The process of **sclerotization** increases the degree of binding between the fibres and the matrix resulting in hardened plates of cuticle known as **sclerites**. In those insects where sclerotization is extensive the insect is effectively enclosed in a series of hardened plates linked together by flexible cuticular membranes. A generalized segment consists of two main sclerites: a dorsal **tergum**, and a ventral **sternum** separated laterally by **pleura** and between each segment, narrow, hardened **intersegmental sclerites** formed in the folds of the intersegmental membranes. The longitudinal muscles attach to the intersegmental sclerites thus straddling the segment from front to back. In the generalized thorax the three tergal sclerites of the pro-, meso- and meta-thoracic segments form the pronotum, mesonotum

and metanotum, respectively. If wings are present they are attached to the meso- and metathoracic segments and together with the prothoracic segment the three thoracic segments are often tightly fused to each other — and in honey bees, also with the first abdominal segment (fig. 7.13) — to form the **pterothorax.**

In winged insects the intersegmental tergal sclerite of one segment has typically become fused to the front of the segment behind it, forming what is called the **postnotum**, that is the roof of the segment. The line of fusion is often visible on the dorsal surface of the thorax as the antecostal sulcus separating the anterior part (acrotergite) from the scutum. Within the pterothorax, sclerotized ridges, or **phragma**, descend from the roofing sclerites (**nota**) and serve to strengthen the roof of the wing-bearing segments and mark points of muscle attachment. These internal ridges are represented by lines, **sulci**, which demarcate areas on the external surface of the notum.

Like the tergal sclerites, those of the sternum are normally fused together to form the floor of the thorax. Internally from the sternum extend sclerotized arms, the **sternal apophyses.** In the honey bee the sclerotized elements of the side wall of the pterothorax, the pleuron, is visible as the **epimeron** and **episternum** (fig. 7.13) which

FIG. 7.14 *Right forewing of a drone, opened and flattened and viewed from above to show the orientation of the wing sclerites. The 1st, 2nd, 3rd and 4th axillary sclerites (ax1, ax2, ax3 and ax4), respectively. Axillary lever (axv). Redrawn from Snodgrass*[23].

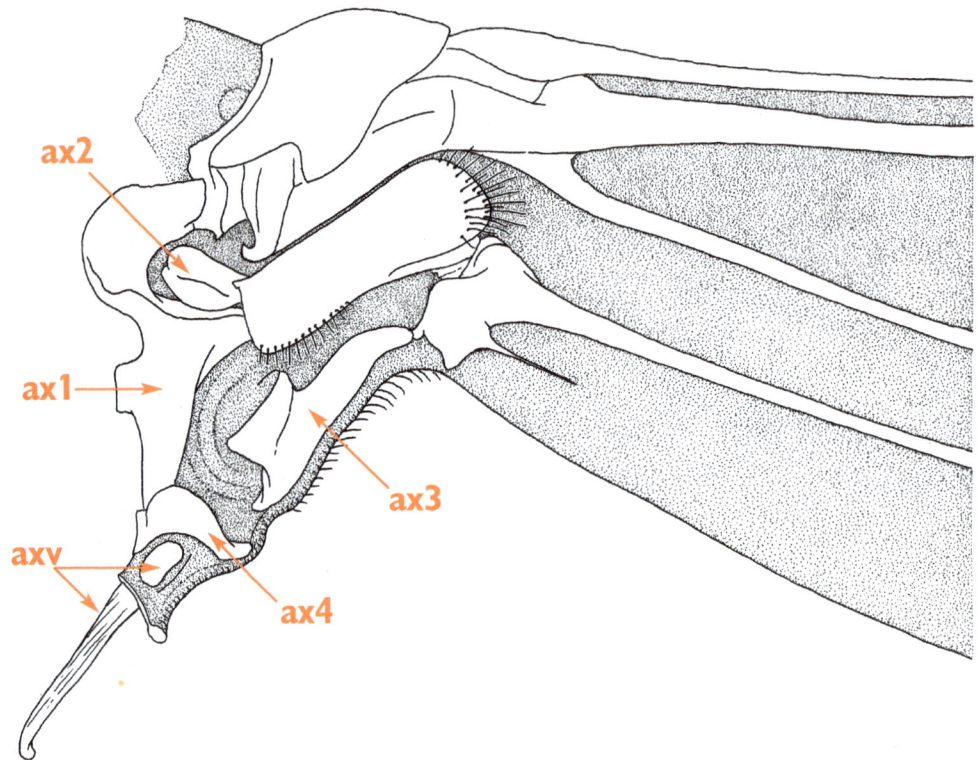

are fused to each other and to the sternum to form the sides and base of the thorax. Internally from the side walls of the thorax, pleural apophyses extend to link via pleurosternal muscles to the sternal apophyses. The pleurosternal muscles regulate and control the rigidity of the thorax and the apophyses form an internal framework supporting the ventral longitudinal muscles. The proximal segments of the legs, the coxae, connect to the thorax at points between the pleural and sternal sclerites. In the bee pterothorax (fig. 7.13), the roof is formed, from front to back, from a reduced pronotum, a greatly enlarged mesonotum divided by the scutoscutellar sulcus into an anterior scutum and posterior scutellum (strictly the mesoscutum and mesoscutellum, respectively) which are separated by a flexible joint: the **notal suture**. The suture permits upward and downward movement of the roof of the thorax and plays a vital role in the thoracic flight mechanics. The very small metanotum is fused to the tergum of the first abdominal segment, called the propodeum (fig. 7.13).

Wing articulation in the honey bee

Part of the central, lateral region of the thorax, between the nota and the pleura is a membranous area associated with the attachment of the wings. Protruding into this area are the anterior and posterior notal processes, both extensions of the scutum but with the latter fused to the scutellum. The notal processes act as the fulcra of the levers that are the wing. Within this membrane and that of the base of the forewing lie the epipleurites, two small plates, the basalare and the subalare, associated with the episternum and epimeron and four small and separate **axillary sclerites** (fig. 7.14). The articulation of the sclerites with the other elements of the thorax and with the wing base is complex; it is summarized in table 7.1 and the interested reader is referred to works by Snodgrass[23], Pringle[25] and Dade[26] for more detailed information. In the normal flight position the base of the forewing is rolled over on itself such that the posterior part is orientated downward in the vertical plane and therefore not visible from above, and this obscures the axillary sclerites to a great extent. The first axillary sclerite articulates

Table 7.1 Axillary sclerites of the honey bee forewing (after Snodgrass[23]).			
Axillary	**Proximal articulations**	**Distal articulations**	**Action**
1st axillary	*Neck* rests on anterior end of anterior notal process; *body* hinged to posterior end of wing process; *posterior end* articulates with concavity	*Head* abuts against the humeral complex in the flattened wing; *lateral projection* underlies inner edge of median plate	Anterior hinge plate of the wing
2nd axillary	*Dorsal surface* articulates with 1st axillary	*Ventral surface* articulates with wing process of metapleuron and the subalare on the upper edge of the pleuron	Pivotal sclerite of the wing base since it rests on the pleuron; it also gives the wing solid support from below
3rd axillary	Lies close against the axillary cord; *proximal end* associated with 4th axillary	*Distal end* closely connected with enlarged base of fourth vein	Flexor sclerite involved in wing flexing
4th axillary	*Inner surface* overlaps posterior end of 1st axillary and articulates with the posterior wing process of the scutum	*Outer surface* is pressed closely against the axillary cord behind the proximal end of the 3rd axillary	Posterior hinge plate of the wing
Axillary lever	*Apex* of elongate, triangular sclerite lies close against the mesal surface of the arm of the postphragma	*Dorsal angle* of base appears as a small nodule in the wing membrane immediately behind the 4th axillary	Involved in wing flexing

with the anterior notal process and acts as the anterior hinge plate of the wing. The second axillary sclerite is the pivotal sclerite of the wing base, as it is the only one resting on the pleuron and supporting it from below and articulating with both the basalare and subalare pleural sclerites. The third axillary sclerite is called the flexor sclerite by Snodgrass, as it is the only axillary sclerite that is connected to muscles and is the main skeletal element of the wing-flexing mechanism (fig. 7.15). The fourth axillary sclerite forms the posterior hinge plate by articulation with the posterior notal process. In the honey bee there is a further sclerite, called by Snodgrass the axillary lever, which is involved in forewing flexion through articulation with the third axillary sclerite. A similar arrangement of sclerites is found in association with the hindwing except that there is no fourth axillary sclerite or axillary lever in the metathorax.

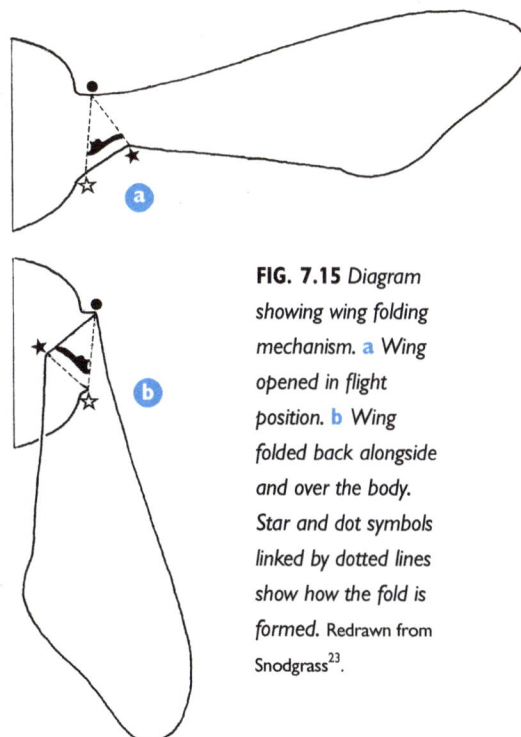

FIG. 7.15 *Diagram showing wing folding mechanism.* **a** *Wing opened in flight position.* **b** *Wing folded back alongside and over the body. Star and dot symbols linked by dotted lines show how the fold is formed.* Redrawn from Snodgrass[23].

The flight muscles

Within the honey bee thorax there are two main types of muscle that directly or indirectly move the wings (table 7.2 and fig. 7.16). The **direct flight muscles** have insertions on the plates and internal extensions of the thorax at one end and to the pleural and axillary sclerites in the wing base at the other. Contraction of the direct flight muscles therefore has a direct effect on the motion of the wing. The longitudinal **indirect flight muscles** run the full length of the thorax and have insertions on the anterior and posterior phragma, and the dorsoventral indirect flight muscles have

Muscle	Origin	Insertion	Action
Table 7.2 Major wing muscles of the honey bee (after Snodgrass[23]).			
Dorsoventral, paired, indirect, fibrillar muscle (72)	Ventral and lower lateral walls of mesothorax	Stretch upwards and anteriorly to the lateral parts of the scutum, anterior to the scutal (notal) suture	Flattens the notum (towards the sternites) which is facilitated by the opening of the notal suture. Main wing elevator
Dorsolongitudinal, paired, indirect, fibrillar muscle (71)	Posterior end of the propodeum (mesothoracic phragma)	Stretch upwards and anteriorly to small prephragma of mesothorax and strongly decurved surface of the scutum between the lateral tergosternal muscles	Arches the notum (making it dome-shaped) closing the notal suture. Main wing depressor
Accessory wing elevators (paired) (75)	Episternum of mesothorax	Tendon inserted on the margin of the posterior plate of the notum (scutellum)	Pulls directly down on edge of the notum
Basalar muscles (paired) (77)	Anterior part of mesopleuron (episternum)	Tendon inserted into inner surface of basalare	Pulls basalare inwards towards the episternum, turning the anterior margin of the descending wing
Subalar muscles (paired) (82)	Mesothoracic coxa	Tendon inserted into upper angle of subalare	Pulls subalare inwards, moving the second axillary sclerite; pulls downwards on the posterior of wing base; during upstroke rotates anterior margin of the wing up and back
Hindwing elevators (three pairs) (97, 98, 99)	Metathoracic arm of endosternum	Metanotum	Pulls metanotum downwards, turning hindwings upwards
Wing flexor muscles (76a, b, and c)	Inner surface of mesopleuron	Into large lobe of the anterior margin of the 3rd axillary sclerite	Contraction brings about flexing of the wings
Axillary lever muscle (78)	Mesothoracic arm of endosternum	Tapering distal end of axillary lever	Flexion of the forewing

Numbers in brackets refer to some of the flight muscles shown in fig. 7.18.

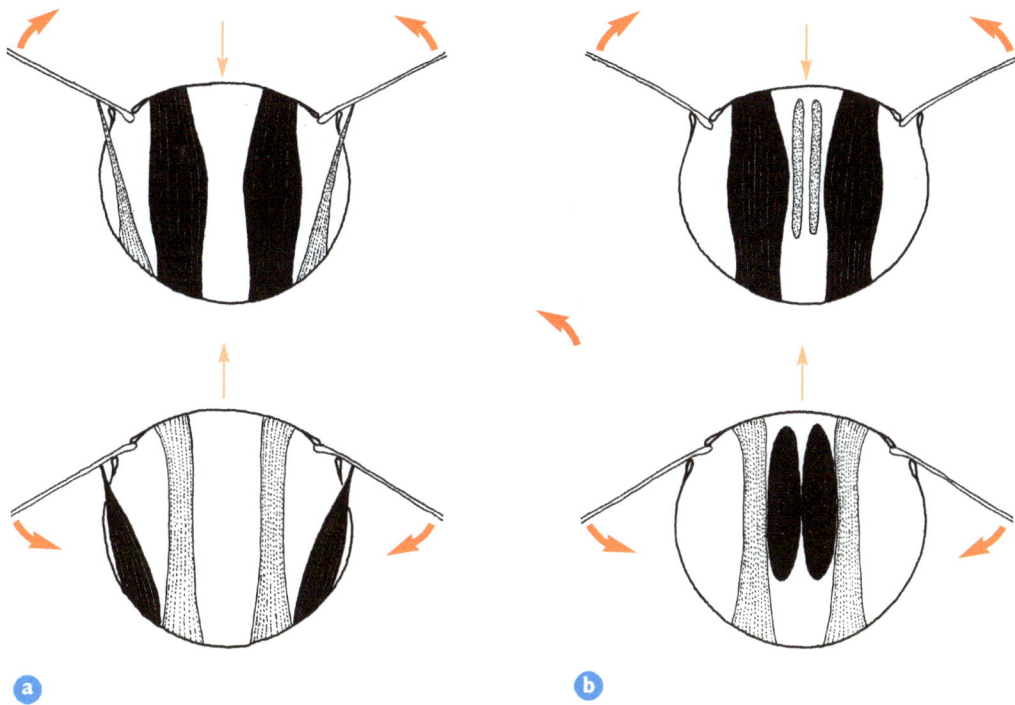

FIG. 7.16 *Transverse sections through the thorax showing the main flight muscles during the upstroke (upper pair of diagrams) and during the downstroke (lower pair).* **a** *Direct and indirect dorsoventral flight muscles and* **b** *indirect dorsoventral and longitudinal muscles. Muscles shown dark when contracted, light when relaxed and stretched. Heavy arrows show direction of wing movement. Light arrows indicate direction of movement of the roof (notum) of the thorax.* Redrawn from Snodgrass[23].

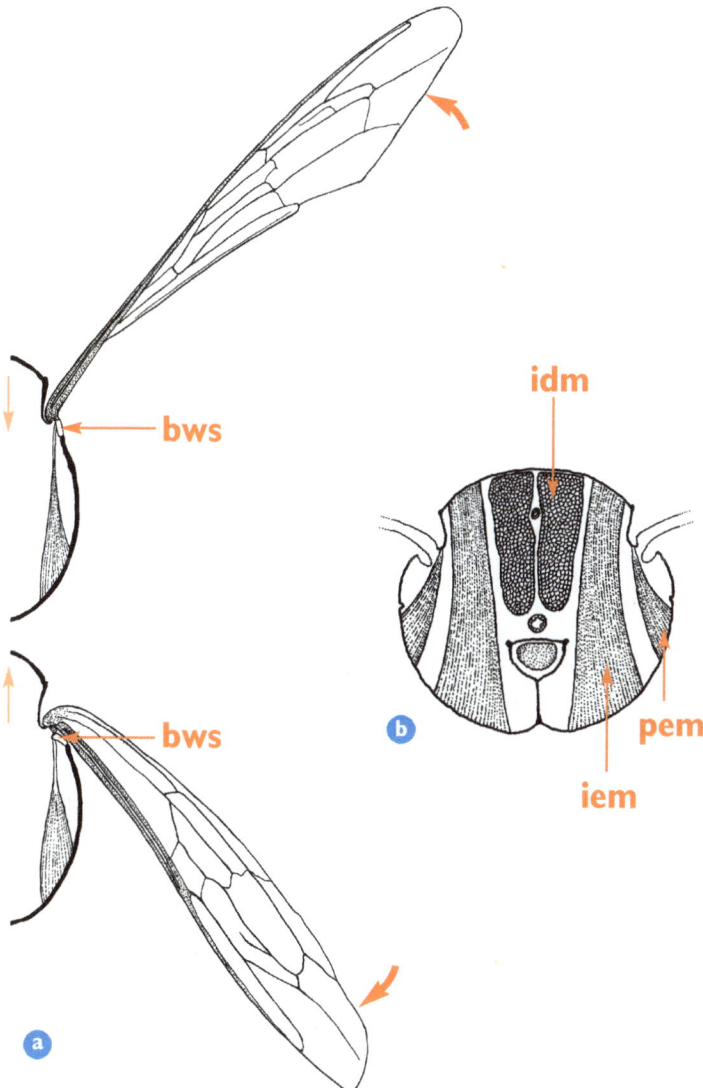

FIG. 7.17 *Diagrams showing direct and indirect flight muscles in the honey bee.* **a** *Action of basalar muscle. This muscle is a direct flight muscle as it attaches to the basalare wing sclerite (bws) and the ventrolateral wall of the thorax (episternum) and deflects the wing forward during the downstroke. Upper diagram: wing raised by action of the indirect elevator muscles (not shown, see* **b***) depressing the notum. Lower diagram: wing depressed by the indirect wing depressor muscles (not shown, see* **b***) and deflected forwards by the basalar muscle. Arrows indicate direction of movement of the notum and the wing.* **b** *Transverse section through the thorax showing the paired indirect depressor muscles (idm), the indirect elevator muscles (iem) and the pleuronotal wing elevator muscle (pem).* Redrawn from Snodgrass[23].

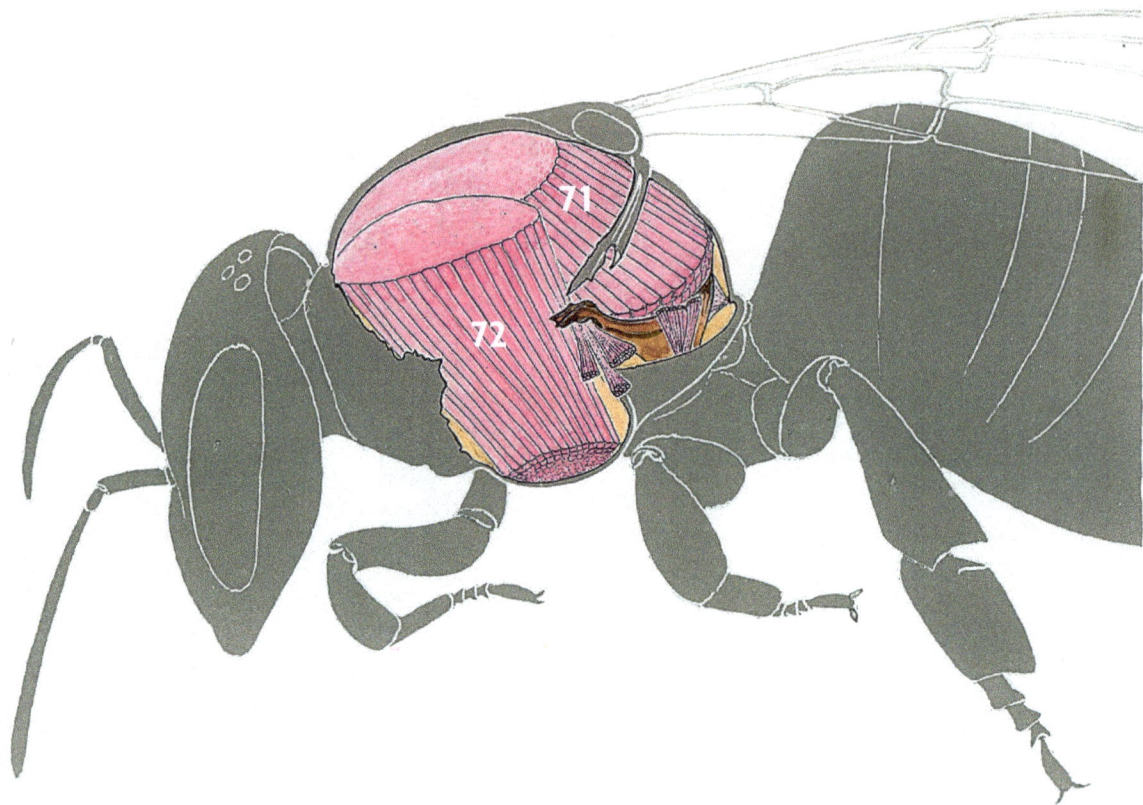

FIG. 7.18 Drawings of a progressive dissection of the thorax of a worker honey bee to reveal some of the flight muscles (numbered). **a** Left thoracic wall and part of notum removed to reveal main indirect flight muscles: 71 left indirect wing depressor; 72 left indirect wing elevator. **b** Left indirect depressor muscle 72 moved down to reveal structures behind it. Thoracic salivary gland (s). Muscles numbered after Snodgrass[23].

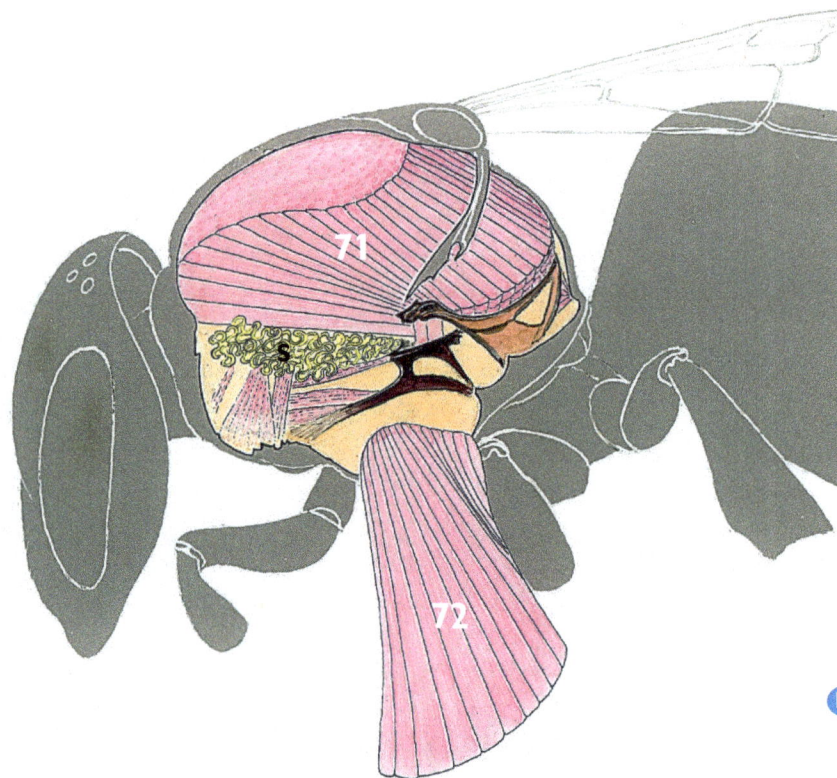

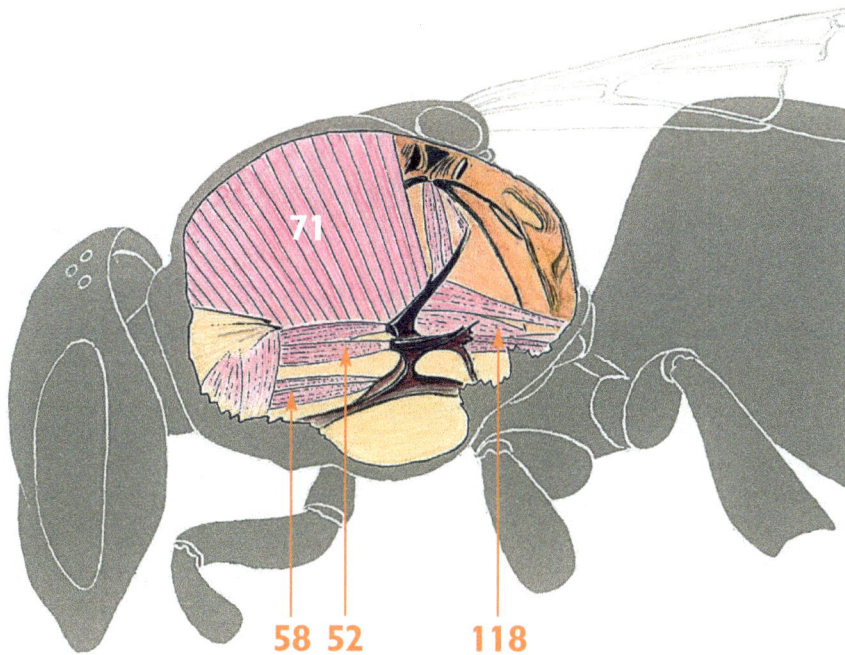

c *Thoracic salivary gland removed. Ventral intersegmental muscles exposed: 52 and 58 in the mesothorax, 118 in the metathorax. These muscles attach to the end walls of the thorax and to internal, cuticular structures shown in dark brown.*

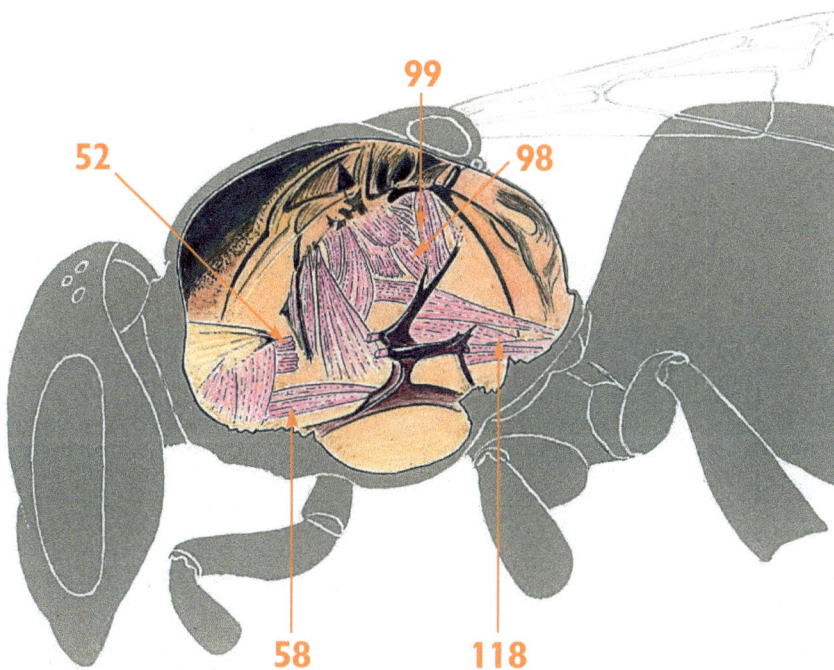

d *Further dissection to reveal small muscles attached to the internal (right) wall of the thorax; these are involved in tensioning the thorax and wing base and some of them attach to the legs.*

insertions on the notal and sternal plates and their internal projections. Contraction of the longitudinal muscles causes the notum to rise whereas contraction of the dorsoventral muscles pulls it down again. Alternate contraction of these indirect muscles deforms the thoracic box which, because of the way the thorax is structured and its articulation with the base of the wing is organized, results in wing flapping.

The roles of direct and indirect flight muscles
The relative size and importance of the direct and indirect flight muscles varies in different insects, for example in the Odonata (dragonflies) the large indirect dorsoventral muscles produce the upstroke with the direct 1st basalar subalar muscles pulling the wings down during the downstroke (fig. 7.17). In dragonflies the dorsal longitudinal muscles are reduced or even absent.

In the Hymenoptera the indirect muscles are greatly developed at the expense of the direct muscles (table 7.2 and figs 7.17 and 7.18), and the shape and placement of the axillary sclerites in the flexible membrane of the thoracic wall ensures that the wings are automatically pronated on the downstroke and supinated on the upstroke. In bees the direct muscles serve as steering muscles adjusting the rotational and other subtle movements of the wing during beating, through alteration of the mechanical configuration of the wing hinge sclerites and tensioning of the thorax.

Activation of the flight muscles In the Diptera (true flies) and Hymenoptera (bees and wasps) the activation and physiology of the direct and indirect flight muscles is very different. The direct flight muscles are individually, neurally innervated, contracting upon receipt of an appropriate nerve impulse from the flight motor system. The maximum rate of contraction and relaxation in these **neurogenic muscles** is set by the time taken to receive and act on a nerve impulse. The longitudinal and dorsoventral indirect flight muscles are stretch-activated and once set in motion,

continue to contract and relax continuously until stopped by the nervous system. Antagonistic sets of **myogenic muscles** contract rhythmically without further input from the central nervous system beating the wings of the honey bee at about 180 Hz. Myogenic muscles allow greatly increased wingbeat frequencies than would be possible in a neurogenic system which, because of the inherent delays in the neuro-muscular activation chain, are unlikely to exceed 100 Hz whereas 1000 Hz has been recorded in small flies such as gnats which possess myogenic muscles.

The wings and flight muscles are primarily used for flight, of course, but they have other uses too. With the wings folded, activation of the indirect flight muscles, albeit at a lower oscillation rate than in flight, results in what has been called 'shivering' with much of the energy appearing as heat to raise the body temperature prior to take-off in cool weather. The wings can also be used to waft Nasonov pheromone and when this occurs the wings are flexed up and down through an almost vertical stroke plane and the bee clings to the substrate. Another use is in fanning to move air about in the hive; here the stroke amplitude is reduced and the stroke plane is vertical and displaced towards the abdomen. It would appear that there is a sort of clutch and gearing mechanism, presumably controlled by tensioning different direct flight muscles, which facilitates these different activities.

Stability and manoeuvrability
Once in the air the bee has three degrees of translational freedom. It is free to move (translate) along its three orthogonal axes, forward and back along its long axis; up and down in the vertical plane and sideways (sideslip) in the horizontal plane. It can also rotate about these three axes; roll around the longitudinal axis, pitch in the vertical plane, and yaw in the horizontal plane (fig. 7.19).

With the stroke plane angled towards the horizontal, the thrust forces on the up-and

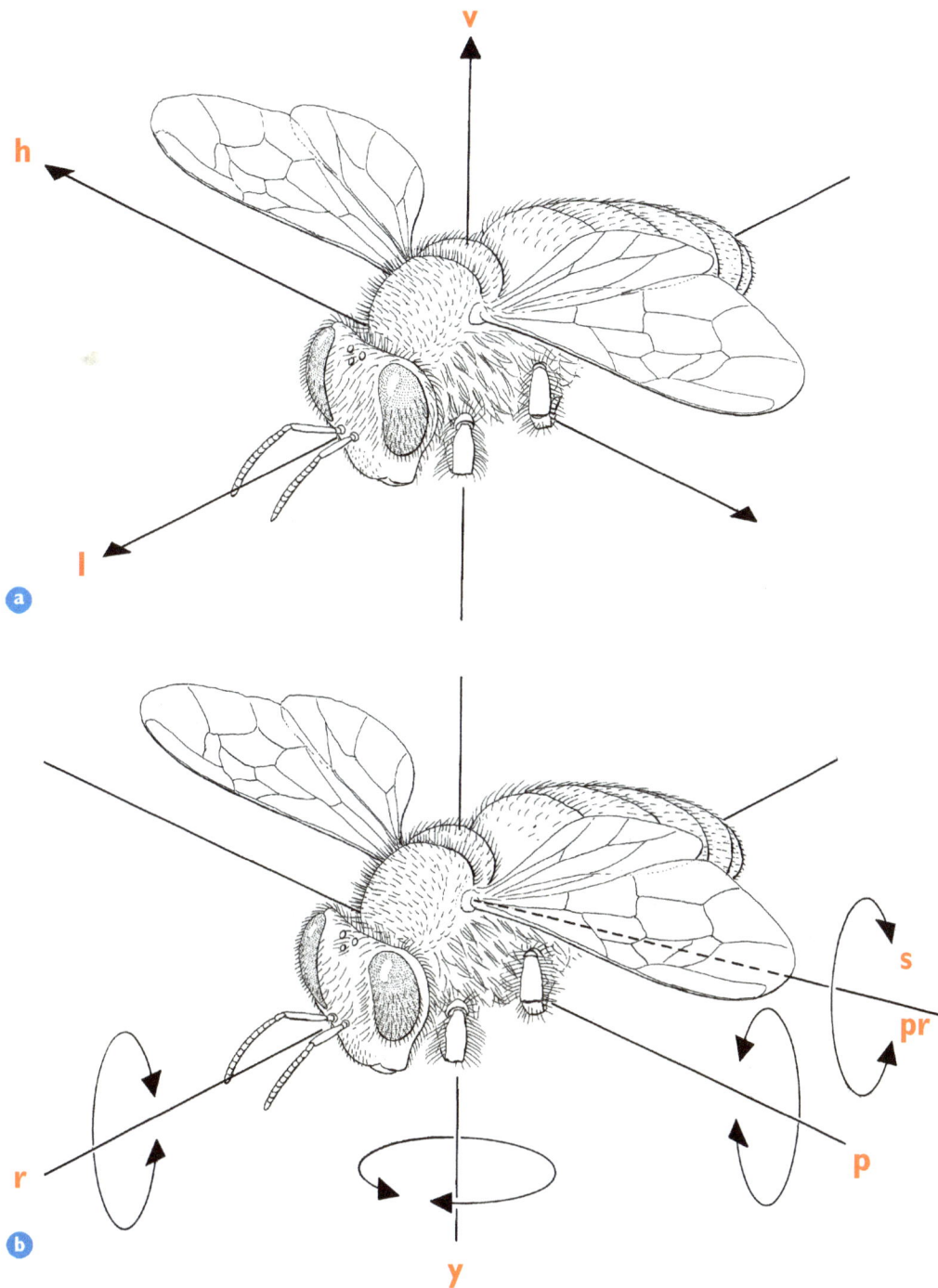

FIG. 7.19 *Diagram showing a the bee's three axes of translational movement, and b the rotation about them. a Vertical, up and down, lift (v); longitudinal, forwards and backwards, thrust (l); horizontal, lateral, side to side, sideslip (h). b Yaw (y), roll (r), pitch (p). The dotted line represents the axis about which the wing can rotate: supination (s), pronation (pr).*

downstrokes are equal and opposite, and cancel each other out. The wings are generating lift with no overall thrust, although they are still flapping. This is hovering.

A helicopter uses a similar principle to create lift without forward thrust as its circling aerofoils travel both forwards and backwards at the same time and the thrust forces cancel each other out. On the upstroke, the insect's wing rotates backwards through 120 degrees and flips upside down. As any surface inclined to the direction of airflow can act as an aerofoil, so an insect wing can function either way up.

Hovering in bees is important because they mate in flight, and often hover to inspect flowers but they are not really able to hover motionless in space like hoverflies or hawkmoths. Hawkmoths, for example, hover in front of flowers while inserting their extensible tongues into the corolla of the flower to reach the nectar on which

FIG. 7.20 *Take-off sequence: honey bee leaving the landing platform.*

they feed. Similar wing movements to those used in hovering are necessary for controlled take-off and landing manoeuvres (figs 7.20 and 7.21a, b).

The stroke plane can be varied, such that the insect will move backwards and if the two pairs of wings are following a slightly different stroke plane, this will result in the production of unequal or differentially directed forces across the body promoting a turning motion. The wing with the highest stroke plane angle will be on the outside of the turn, by the same principle that a boat will turn away from the side on which the oar is moving fastest or furthest.

The bee can initiate a roll by increasing the angle of attack of one pair of wings. Roll combined with yaw results in a banked turn. Since lift is directly proportional to angle of attack, this change will increase the lift force on one side relative to the other, causing the body to roll. Pitching the body, so that the nose moves up or down around the centre of mass is achieved by moving the entire flapping motion forwards or backwards in relation to the centre of gravity of the insect. It is clear that for effective locomotion by flight there is a need for a high degree of muscle control and co-ordination requiring a range of sensory systems of appropriate complexity and responsivity (see below).

Apart from the wings, the insect's body has a role to play in the aerodynamics of flight. The angle it is held to the horizontal can change for different types of flight. When inclined to the horizontal in forward flight,

the body behaves like an aerofoil, and can contribute up to 10% of the total lift. As an insect comes in to land, the abdomen may be held more vertically, increasing drag and reducing the forward speed. In this position the relatively heavy abdomen hangs like a pendulum bob, thus aiding stability. In larger-bodied insects like bees, this pendulum effect reduces the insect's ability to accelerate rapidly out of hovering. The ability to hover perfectly still and to make rapid sideways movements to new hovering positions is taken to an impressive extreme by hoverflies; they hold their lighter bodies horizontal during hovering. The disposition of the legs is also important and typically they are tucked up against the thorax and abdomen as soon as possible after take-off and during flight (figs 7.20 and 7.21). They may be lowered prior to landing to increase drag thus assisting with breaking prior to touch down.

The non-steady effects may also help to explain why insects in flight are so manoeuvrable. Tiny changes in the timing of separate elements and components of the wing-flapping cycle and the way in which the wings rotate, for example, can substantially alter both the magnitude and direction of the resultant forces. It has been shown in hoverflies that a change of 8% in the timing of the wing edge rotation, relative to the timing of the flap, increased the amount of lift generated by 67%[13]. In addition, the exact positioning of lines of variable flexibility in the wing itself, caused by the placement of hairs or veins or, in four-winged insects such as bees, the junction between the fore and aft pairs, could

FIG. 7.21 a Landing sequences of a heavily laden worker honey bee, shown from the side, alighting on the landing platform.

FIG. 7.21 b *Landing sequence of an unladen worker honey bee returning to the hive to alight on the landing platform. Note the legs start to extend only when the bee is very close to the platform. Touchdown is initiated by the hindlegs.*

be crucially important because they may form the axes of wing rotation[27]. The pterostigma (fig. 7.8), or dark thickened patch at the front of a bee's forewing, may also be important in this respect because it strengthens the leading edge, reducing 'flutter' and encouraging the vortex generating rotations to occur between lines the flexion and the trailing edge[28].

Sensory regulation of flight behaviour

It is clear that to take-off, to fly in a controlled and directed manner and to land again at an appropriate place involves a great deal of sensory input and subsequent filtering and processing followed by modifications to muscle activity. Fixed-wing passenger aircraft are designed to have a high degree of stability when in cruising flight, with forced deviations automatically being compensated for to keep the aircraft flying on a steady course, and active control being required primarily at landing and take-off and during course changes. For a high degree of manoeuvrability such as is required by combat aircraft, complex avionics systems are necessary to maintain dynamic stability on the edge of instability. Insects tend towards this latter case, flying on the edge of instability with continuous sensory and motor activity modulating and adjusting the output of the flight motor. The insect monitors body position and motion through sense organs known as proprioceptors, and information about the environment and objects moving within it primarily through its visual sense.

Information from the sensory system is processed by the nervous system of which the brain and the ganglia of the pterothorax are of primary importance in flight.

Vision is of great significance in flight behaviour as is clear, for example, from the fact that 80% of the volume of a dragonfly's brain consists of areas primarily associated with processing visual information. The compound eyes (chapter 2) have extensive fields of view, often extending above, below and behind the insect. During flight the visual environment appears to move past the eyes and it is this optical flow that the insect monitors. Rapid processing of motion information is facilitated by neural pathways between the brain and the thoracic ganglia[29,30]. The high flicker fusion frequency (see chapter 2, p. 31) of insects' compound eyes also facilitates the high temporal resolution of visual cues necessary during flight.

A well studied aspect of visual stability control is the optomotor response where compensatory flight responses are induced following movement of the visual field. Such responses to movement around the insect that typically induce a same-direction, following response have been extensively studied in flies[31,32,33] and locusts[34]. Such responses can be mediated by unilateral adjustments to the wing stroke amplitude or stroke plane angle. Optomotor responses to changes in translational visual flow, such as an apparent speeding-up or slowing-down of the visual environment passing beneath the flying insect, usually induces an opposite-sense response, the insect slowing-down or speeding-up to maintain a constant flow rate. Increasing or decreasing wingbeat frequency can be employed in this case[35]. Compensatory responses to translational optomotor cues enable insects to maintain a constant height of flight and to compensate for changes in ambient wind speed. By maintaining a constant ground speed, honey bees can estimate their flight distance during foraging flights[36,37,38]. However, not all translational perturbations are along the line of flight and it has been shown that insects use flow field information to compensate for drift induced by cross-winds[39]. This aspect will be discussed later in the section on foraging flights.

Goodman and her co-workers have identified a number of directionally selective, wide-field, motion-sensitive descending neurons in the brain and suboesophageal ganglion of the honey bee[40-45]. Neurons driven by either the compound eyes or ocelli following vertical, horizontal and oblique movements have been identified and mapped[46].

Insects have the neural processing capability to be able to separate translational and rotational movement of the visual field from movement of objects within the visual field, for example a drone bee in flight pursuing a queen bee has to be able to identify the moving queen against the apparently moving background and to continually adjust its course to coincide with that of the queen. To do this, bees use a continuous tracking method where the optomotor output is adjusted during tracking[47].

The dorsal ocelli (see chapter 3) also have a role to play in flight stability and are believed to function as pitch and roll sensors through sensing the horizon[48,49,50].

As we have seen in chapter 1, the antennae are well-placed and equipped to monitor airflow and hence an insect's air speed. During flight they are typically held out forward in a characteristic manner and, in those species that have been studied, mechanoreceptors in the Johnston's organ (see fig. 1.21) respond to the degree of antennal bending caused by air flowing over the head during flight[51,52]. It has also been shown, at least in the locust, that the antennae respond to the pulses of thrust produced by the beating wings, and this information is fed back to the flight motor system[53] and may help to regulate the wingbeat frequency in the locust's neurogenic flight motor. The antennae can therefore function as air speed sensors while the compound eyes monitor ground speed.

Hair plates on the head itself have been studied in locusts[54] and shown to be responsive to airflow and to possess neurons that pass through the brain to both the suboesophageal and thoracic ganglia[55]. It has been shown in locusts that asymmetric stimulation of these hair plates from a simulated cross-wind during tethered flight induces compensatory deflections of the abdomen and hindlegs[56]. The head of the honey bee is covered in hairs but it is not clear whether any have an airflow monitoring function.

Hair plates on other parts of the body and wings are undoubtedly important in monitoring the relative position of one part of the body to another and to airflow, respectively. The hair plates in the neck of the honey bee (see chapter 4), while acting as gravity sensors in the grounded bee, may also function to monitor the orientation of the head to the thorax during flight. They function in this way in flies[57], and during yaw-turns in the locust[58,59] the head moves

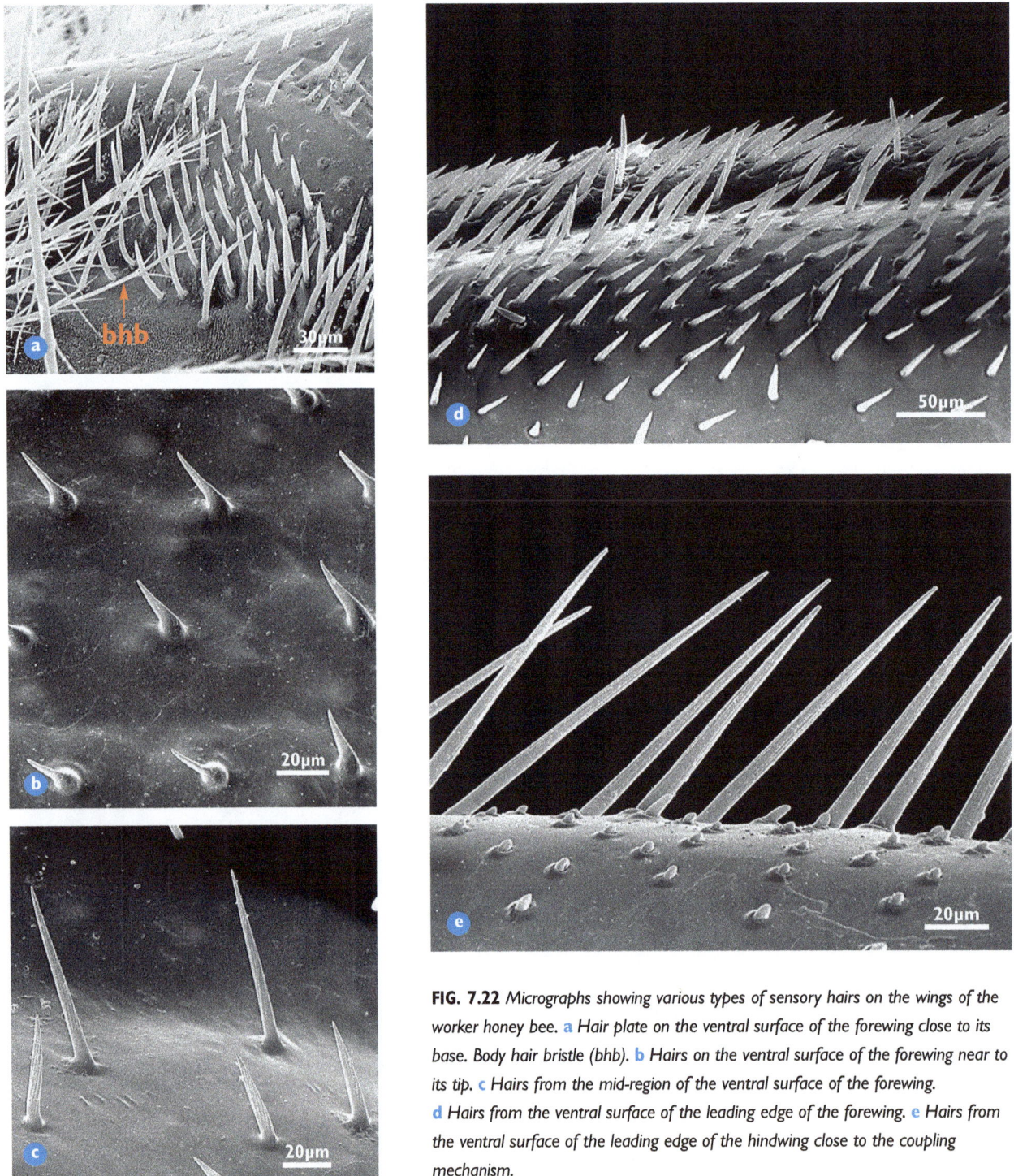

FIG. 7.22 *Micrographs showing various types of sensory hairs on the wings of the worker honey bee.* **a** *Hair plate on the ventral surface of the forewing close to its base. Body hair bristle (bhb).* **b** *Hairs on the ventral surface of the forewing near to its tip.* **c** *Hairs from the mid-region of the ventral surface of the forewing.* **d** *Hairs from the ventral surface of the leading edge of the forewing.* **e** *Hairs from the ventral surface of the leading edge of the hindwing close to the coupling mechanism.*

first followed by the thorax driven by the appropriate cross-body alteration in wing movements. Hairs are present on the wings of the honey bee (fig. 7.22) and are probably associated with monitoring air-flow over the wings.

Within insects' bodies, stretched between different parts of the body wall and in the wing base, are **chordotonal organs** or **stretch receptors**. They are typically a small group of elongated sense cells that form strands or sheets. They monitor the distance between their two points of attachment and in some instances, such as in tympanal organs, vibrations, and hence are involved in sound reception[60]. In the base of the bee's forewing there are thought to be between 50 and 62 chordotonal organs[61].

Other sense organs in the integument of the insect are the campaniform sensillae (fig. 7.23) that monitor bending and twist-ing of the cuticle. Fields of campaniform sensillae are present on the ventral surface of the base of the wing close to its leading edge (fig. 7.24). Campaniform sensillae appear as oval bell- or dome-shaped mounds, each set in a hole in the cuticle, and are typically arranged with the longer axis of their oval shape with a common ori-entation. In the locust they make direct neural connections to the thoracic gan-glia[62] and in the fly probably monitor tor-sion and degree of camber during wing rotation and translation[63,64,65]. They are present on the bee's wing in large numbers (fig.7.24) arranged in fields on both the upper and lower surface of both the forewing and hindwing[25,61].

Foraging flights

The use by the honey bee of the sun com-pass during navigation has been described in chapter 2, and of gravity in the waggle dance in chapter 4. The sun compass is a key element in the worker bee's ability to relocate a profitable food source repeat-edly and to report this knowledge to other workers in the hive during the waggle dance, as first described by von Frisch[66]. However, von Frisch recognized that the

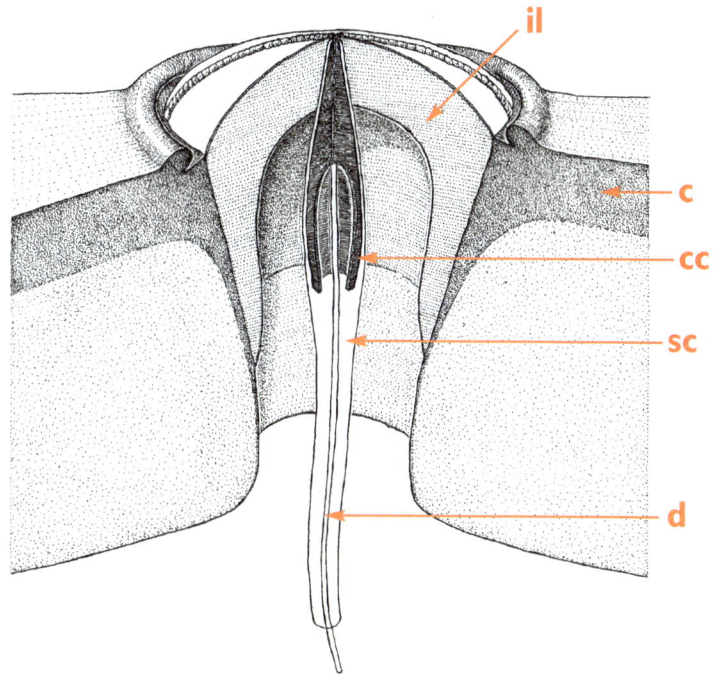

FIG. 7.23 *Campaniform sensilla shown in part sectional view. Cuticle (c), inner lamella of dome (il), cuticular connection to scolopale cell (cc), scolopale cell (sc), dendrite (d).*

worker bee would have to be able to com-pensate for the effects of cross-winds dur-ing the flight to and from the food source and hive. He and Lindauer[67] postulated that the bee computed the angle between (a) its goal and the sun and (b) the angle between its long body axis (heading) and the direction of ground movement beneath it (track) to compensate for drift (angle between heading and track, the actual path over the ground taken by the bee). This would require the assessment of wind speed and direction and the calcula-tion of heading and airspeed. Recent work by Riley and co-workers using radar to monitor bee flights has suggested that a simpler approach is in fact employed by the bee, which very probably compares the image movement directly with the solar angle[68]. By rotating its long axis it aligns the perceived direction of ground move-ment to the angle relative to the sun's azimuth that fits its intended track. Riley points out that von Frisch observed that the sun's azimuth angle signalled by the worker bee in the waggle dance following foraging flights involving cross-winds was the true sun compass direction to the source and not that of the heading that it must have flown.

Height, speed and energy costs

It is a long-held observation that honey bees fly lower on slow upwind flights than when moving fast downwind, and it is assumed that they try to maintain a constant preferred optical flow rate of images moving from front to back beneath them during flight[69]. It has been calculated that the preferred optical flow rate is about 3.5 rad/s for the honey bee[36], and it was exactly the same for bumble bees[68]. This has important implications for flight times and total energy expenditure during a foraging flight. For bumble bees it has been calculated that about 5 m/s is the most appropriate airspeed for minimum energy cost, but the average speed observed in still

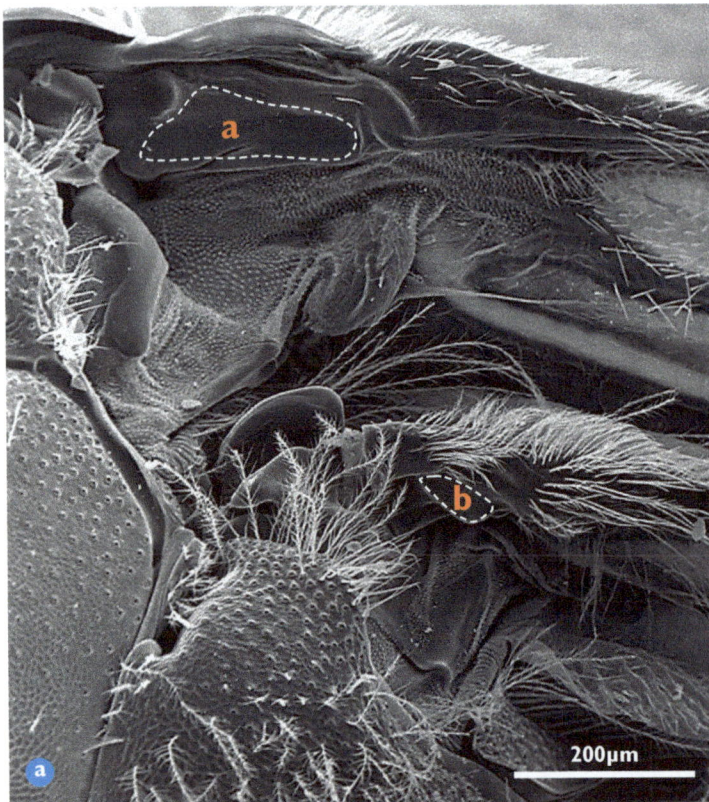

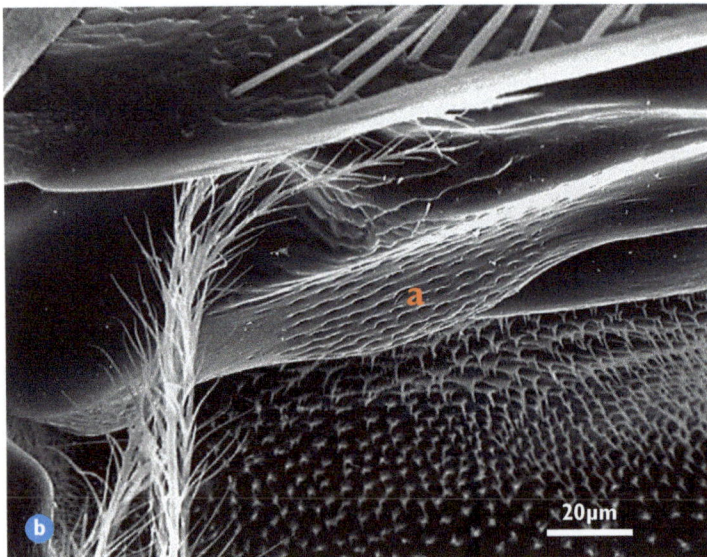

FIG. 7.24 *Micrographs showing campaniform sensillae.* **a** *View showing ventral surface of wing bases (wings extend to the right from the right edge of the micrograph) and body of worker honey bee, i.e. an 'armpit' view; a and b groups of campaniform sensillae.* **b** *A view of group a field of the campaniform sensillae.* **c** *A view of the group b field of campaniform sensillae.* **d** *Close-up view of the domes of the campaniform sensillae.*

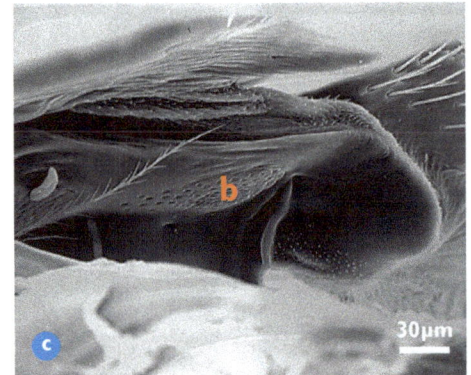

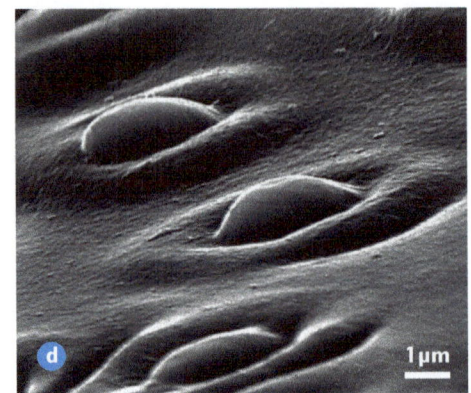

air was 7.1 m/s, implying that flight time is more important than the minimization of energy costs to the bumble bee[68]. Reducing the amount of time in flight might serve to maximize time available for pollen and nectar collection, thus overall increasing the energy intake per flight. Adjusting height of flight appropriately in headwinds or tailwinds minimizes both energy costs and travelling time[68].

Learning or orientation flights

Recent studies using both visual observation of marked bees and radar tracking of bees carrying transponders have revealed that from four to fourteen days of age bees leave the hive to make exploratory, orientation flights prior to foraging[70]. An orientation flight starts with a period of hovering facing the entrance to the hive, followed by a flight away from the hive. As bees gain experience they fly faster and further from the hive, thus increasing their appreciation of the position of the hive in relation to surrounding features. Once foraging starts the flights become still faster and the distances flown increase further; the bees also fly straighter, more direct paths. It was also found that the majority of orientation flights are confined to single quadrants around the hive.

Swarm flight

How does an entire swarm, which can be anything from 10 000 to 20 000 bees, maintain itself during the swarm flight? A detailed record of a swarm flight has been documented[71]. The swarm travelled 580 m from its first resting place to its new nest site. About 5% of the swarm, all older bees with knowledge of the area, had been observed searching around the hive for suitable nest sites during the few days before the swarm took to the air. It is assumed that when the swarm flies it is these scout bees that guide it.

Once the scout bees have selected the new nest site, they initiate swarm take-off by performing 'buzzing runs'. These are excited, zigzag runs, punctuated by bursts of wing buzzing in which they butt into groups and push through hanging nets of bees. Eventually, once most of the scout bees are engaged in this activity, the swarm beings to lift away. Immediately, a few scout bees leave the swarm and fly directly to the new nest site, where they start releasing assembly pheromones from their Nasonov glands (see chapter 8). Initially the swarm is roughly spherical in shape, and its flight begins very slowly, about 1 km/h. During this phase, the bees are checking for the airborne presence of their queen. After moving about 30 m, the swarm speeds up, reaching a maximum speed of 11 km/h at a height of about 3 m above the ground. There is no central cluster; the bees are dispersed evenly throughout the swarm cloud, and the overall shape becomes flattened vertically, into an ovoid.

It seems that throughout the flight the scout bees actively direct all the other bees by streaking through the swarm in the direction of the new site. On approaching the chosen site, the chemical cues are picked up by the bees within the swarm, and at about 80 m before it reaches its destination it begins to slow. This method of reproduction through developing new colonies is only possible on this scale because the workers assist the queen in preparation of the new nesting site and in the construction of the new nest. The worker bees also transport food reserves to the new site during the swarm flight, for each bee in the swarm carries an average of 36 mg of honey in its honey sac which amounts to approximately 40% of its weight[72].

How far do bees fly?

Bees in a colony regularly cover a 6 km radius around the hive in their foraging activity. In experiments, when flown to absolute exhaustion, they flew for 7.4 km without stopping[73], and foraging flights of up to almost 13.5 km have been recorded[74]. This may not seem far, but a 6-km flight is approximately 400 000 bee body lengths, and an equivalent distance for a human would be 600 km, more than a third of the length of the British Isles! Other insects such as butterflies and grasshoppers, which migrate across continents, can cover much more impressive distances of 250 miles in a single day, and the albatross can fly 5000 miles from North America to Brazil to find food for its young. Research has shown that migrating insects use favourable winds to increase their flight distance and, because of their small size, flying insects are operating in a very different medium from birds; to insects the air is more like syrup than the thin gas we and birds experience.

Mating flights

Drones and virgin queens mate in flight at some distance away from their birth colonies in **congregation areas**[75]. The drones assemble in congregation areas to which the queens fly and are then pursued by large numbers of drones which form what are described as 'drone comets': groups of drones trailing behind the flying queen. Queen capture, mating, ejaculation and separation take only a few seconds. The leading drone approaches the queen from below with his hindlegs trailing out beneath him and as the under-surface of his thorax touches the dorsal surface of the queen's abdomen the first two pairs of legs straddle her. Then, grasping the queen with all six legs, he inserts his partially everted endophallus into her open sting chamber and flips himself over backwards with such force that the pressure of body fluid in his abdomen causes the explosive ejaculation of semen into the queen's sting chamber, rupturing the endophallus, separating the briefly paired bees and mortally wounding the drone[76].

Energy sources for flight activity

For their size, insects are very efficient fliers. The primary fuel for flight is pure sugars, although longer-distance fliers like locusts will mobilize fat during extended flights. The ratio of power input to power output lies between 4% and 15% for insects in flight. The same ratio for birds like starlings and pigeons is around 10%, but physiologically the efficiency of muscle increases with increasing body size for all types of locomotion. Based on this relationship, we would expect insects to fly at lower efficiencies, between 1% and 2%, because their smaller muscles have to move faster and so would have higher costs associated with the chemistry of muscle contraction. As we have seen, various special features of the muscle physiology and aerodynamic qualities of flying insects allow them to be much better fliers than we might have predicted from our observations on other animals.

The honey bee's flight fuel is the sugar, sucrose, which it obtains as nectar from flowers. Sucrose is converted to glucose in the gut, or stored in the honey sac. It is then converted to a carbohydrate called trehalose, which passes into the haemolymph where it forms an important carbohydrate reserve.

In the early stages of a flight, glucose stored in the flight muscles is utilized, but this does not last long. Once it is exhausted, more glucose is released in the haemolymph from trehalose by activity of the enzyme trehalase, and transported to the flight muscles.

Radioactive tracer methods have been used to show how, during tethered flight at 1.1 m/s, 139–146 µg of glucose per minute is transported from the midgut to the haemolymph. Since the glucose level in the haemolymph remains constant during this time, and only a negligible amount is detectable at the rectum, we can assume that it is all directly metabolized. Many common flowers contain in their nectar at least 50 µg of sugar per flower, so this amount of fuel could easily be obtained from a few flower visits[77].

The direct use of fuel from the gut during flight can also be shown by experimentally flying bees to complete exhaustion after feeding them a known quantity of sugar solution. This has been done by tethering bees to a flight mill apparatus and measuring the duration and speed of their flight[73]. It was found that flight performance was directly related to the volume of sugar solution given. Bees that had drunk 5 µl (probably 5–10-flowers' worth of nectar, depending on weather conditions and time of day) flew for just over nine minutes, covering 1.9 km before becoming exhausted. Bees that drank 20 µl flew for almost half-an-hour and covered 7.4 km on average. The speed of flight, the frequency of wingbeats and aerodynamic power they generated were less dependent on the amount of fuel consumed, indicating that bees with more energy will use it to fly further at the same speed, rather than covering the same distance more quickly.

The energetics of flight

Flying is energetically very costly, and the metabolism of winged insects represents an extreme of physiological design among animals. The thoracic muscles of insects in flight exhibit the highest known mass-specific rates of oxygen consumption for any muscle tissue, exceeding even the metabolic rate of hovering hummingbirds by a factor of 2–3[8]. Exactly how insects manage to achieve these exceedingly high metabolic rates is still not fully understood. Densities of energy-generating organelles, the mitochondria, in insect flight muscle cells are high (up to 45% of their volume), but no higher than in humming birds. It is thought that differences in mitochondrial structure, or insect-specific respiratory enzymes, may be responsible for the increased metabolic rates achieved by insect flight muscle.

The metabolic rate of a bee during flight can be 50–100 times higher than the resting rate and must be maintained for long periods during foraging or mating flights. In order to maintain it, the bee's body must provide the flight muscles with three basic requirements: a plentiful supply of oxygen, an abundant supply of fuel and an appropriate temperature.

Oxygen supply

Flight muscle activity is obligately aerobic, meaning that it requires a constant supply of oxygen. Oxygen is supplied by the tracheal respiratory system (see chapter 6). The rate of oxygen consumption is reflected in the amount of tracheal tissue present in the muscles, and indirect, asynchronous flight muscles of bees have much higher densities of tracheae than do the direct flight muscles and the synchronous flight muscles of other insects. The supply is enhanced by the enlarged thoracic air sacs, which are almost isolated from the air supply to the rest of the body, thus ensuring that the high demand for oxygen in the thorax during flight does not result in oxygen starvation in other tissues.

The cuticular deformation of the thorax that contributes to the storage of energy during wing flapping also promotes the movement of air in the thoracic tracheal system, resulting from the rhythmic compression and expansion of the thorax that helps to ventilate the tracheae, thus facilitating rapid exchange of respiratory gases.

Efficiency of fuel use

While a bee is in flight, 80% of the energy consumed by the flight muscles is lost as heat, and up to half of the remainder is spent in accelerating and decelerating the wings. This means that only 5–10% of energy consumed by a flying bee actually goes into generating the aerodynamic forces necessary for flight.

Calculating the efficiency of flight involves comparing the amount of energy input with the amount of energy that is actually translated into aerodynamic work, or the mechanical power output. There are various ways to do this, all of which involve some theoretical calculations. To estimate power input, a gas analyser can be used to measure the amount of oxygen that has been consumed, or the amount of heat being generated can be measured, to estimate metabolic rate. Another method involves feeding known quantities of sugar to bees that have previously been completely exhausted. The amount of fuel used during a flight can be estimated by mass loss, which is reliable if the concentration of fuel that has been consumed is known, and it is assumed that any water loss is balanced by metabolic water production. To calculate mechanical power output, flight performance is monitored. The metabolic efficiency of honey bee flight has been estimated at between six and 13%[78].

The actual amount of power required for flight varies with flight speed. A basic aerodynamic prediction, resulting from the speed dependency of the aerodynamic components like lift and drag, is that a U-shaped curve results if power required is plotted against speed (fig. 7.25). This means that there in an optimum flight speed, somewhere between hovering and high-speed flight, which requires the lowest amount of power to sustain it. Although there is some disagreement among insect physiologists about the exact shape of the power–speed curve in insects[8], as mentioned earlier, recent direct measurements of flight speed of bumble bees showed that they flew in still air at about 2 m/s above the optimum[68].

Temperature control

Insects are said to be 'cold-blooded' animals, but that does not necessarily mean that they do not have the capacity to control their body temperature. In fact, many

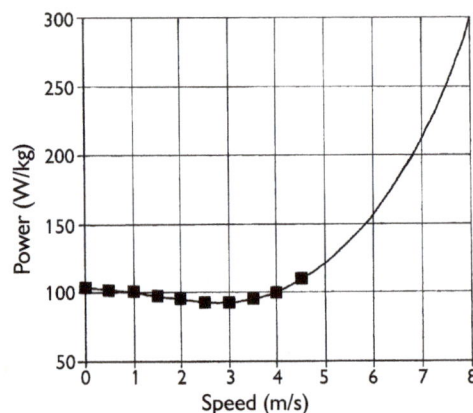

FIG. 7.25 *Required power as a function of airspeed.* Redrawn from Riley and Osborne[68].

insects, especially the larger ones, have very precise temperature control mechanisms. Active bees are normally endothermic, maintaining a body temperature higher than that of the environment. Heat is generated by metabolic reactions fuelled by sugars, and a constant supply is vital especially at low ambient temperatures. Honey bees will remain endothermic as long as they have access to a source of sugar or food in their honey sacs. Unlike solitary bees, they soon deplete their tissue reserves once this fuel is exhausted; they are then only able to crawl and soon die if they fail to locate a source of food.

It may be that social bees have evolved a requirement for a high body temperature because they spend the majority of their time in the warmth of the hive, surrounded by hive mates. Clustered bees, even small groups, keep their temperature as close as possible to 35°C[79], and clustering behaviour increases as ambient temperature decreases. Within the hive, overheating may be a problem at certain times, and individual bees move to the edges to cool down. The highest temperature is found in the area in and around the central brood cells.

This ability to generate and maintain a high body temperature is vital for flight. In order to fly, a honey bee must raise the temperature of its thorax to above 28°C, whatever the ambient temperature. Such a high temperature is necessary to provide an optimum temperature for the enzymes controlling the chemical reactions in flight muscles, and to allow a wingbeat frequency high enough to create enough lift.

Once flight is in progress, a bee must keep the temperature of its thorax fairly constant, to maintain a constant power output from the muscles. Since the rapid action of the muscles during flight generates a lot of heat, thermoregulation can be as much a matter of keeping cool as keeping warm, and the exact balance required will depend largely on the ambient temperature.

Honey bees are able to fly over a wide range of ambient temperature. Under natural conditions a minimum air temperature of 8.4°C has been reported for flight by honey bees leaving the hive and foraging[80] under artificial conditions; bees have been recorded flying at temperatures as low as 4.8°C[81]. At the other end of the scale, honey bees can sustain flight at very high ambient temperatures, up to 47°C, which enables them to live in hot, dry environments.

Warming up

When ambient temperatures are below 28°C honey bees warm up by 'shivering'. This involves functionally decoupling the wing muscles from the wings and rapidly contracting and relaxing them to generate metabolic heat. During warm-up, the muscles behave like synchronous flight muscle, with one muscle contraction per action potential, compared with the more normal 10 per action potential during the myogenic behaviour associated with flight. Shivering bees may appear to be resting as there are no visible vibrations of the wings and thorax because the muscle responses are isometric, thus causing no change in the length of the muscle fibres and therefore no deformation of the thorax.

Honey bees can also suffer from heat loss when flying, especially in cold weather or early in the day. If the thoracic temperature drops below the temperature necessary for flight, the bee must land and warm up again by shivering. Bees take more of these rest stops when the ambient temperature is low, or when they are carrying a heavy load of nectar and pollen.

Bees shiver before they leave the hive to forage. The temperature of the thorax during this warm-up has been measured, using tiny thermocouples placed on the cuticle[82]. Eighty per cent of the bees measured began flying once the thoracic temperature reached 34.1°C, which was around seven degrees higher than the ambient temperature. The temperature inside the thorax may be 1–1.5 degrees higher still[83]. The rate of warming was 3.4 degrees per minute, so in this example the bees had to spend a couple of minutes shivering to reach the appropriate temperature for flight. The other 20% of bees in these trials took off when their thoracic temperatures were only 1.6 degrees above ambient (27°C), and it continued to increase during the first third of their flight. These bees were demonstrating their capacity for emergency take-off, which is an important consideration when, for example, guard bees suddenly have to fly off to repel invaders.

The short pile of hairs covering the thorax of the bee provides some insulation by trapping a layer of warmer air next to the cuticle, but its length is limited by its effect in increasing drag during flight. As a result, most small flying insects are poorly insulated, or not insulated at all. Bumble bees can warm themselves up five times as fast, because their thorax is better insulated and is typically heavier than a honey bee's. It also has a lower surface area to volume relationship and thus loses heat more slowly. Young honey bees are unable to leave the warmth of the hive during the first few days of their lives because they do not immediately have the capacity to warm up by shivering.

Keeping cool

The thoracic temperature of honey bees has been monitored during continuous flights of up to 45 minutes duration whilst the bees were tethered to the arm of a roundabout[82]. Once flight had started, there was a pronounced cooling of 4–5 degrees during the first two minutes of flight, probably due to convective cooling of the thoracic surface by airflow. Then the thoracic temperature remained constant at between three degrees above ambient temperature (approximately 27°C) for the duration of the flight[83].

During flight the flight muscles produce large quantities of heat, just as they do during shivering, so why does the bee's body not continue to warm up? While flying, bees are unable to regulate flight muscle activity to alter body temperature, because the flight muscles have to be active at a constant rate to maintain flight. However, bees have a number of other ways of keeping cool, both physiological and behavioural. The effectiveness of these cooling mechanisms is demonstrated by flight at temperatures of up to 47°C when the thorax of the bee is maintained at around 45°C[84]. An internal temperature of 46°–48°C would be lethal.

The physiological cooling mechanisms involve shunting warm haemolymph to cooler parts of the body. In most large insects, bumble bees for example, heat can be transferred from the thorax to the abdomen via haemolymph that has passed between the flight muscles. The abdomen then acts as a heat radiator. However, honey bees possess a circulatory adaptation that prevents heat loss to the abdomen. Instead, they dissipate excess heat by evaporative cooling from the head. The temperature of the honey bee's head closely tracks the thoracic temperature, primarily because of passive conductive heat flow and heat transfer that can be promoted by rapidly pulsing haemolymph from the thorax into the head.

Honey bees have a special mechanism to minimize thoracic heat loss via the abdomen. Even with high thoracic temperatures during flight, a honey bee's abdominal temperature remains very close to ambient temperature. The differential is maintained partly by the presence of an insulating air sac in the abdomen (see chapter 6), and partly by a circulatory heat exchange mechanism that prevents heat passing from the thorax to the abdomen via the haemolymph.

As haemolymph flows from the heart, which lies just under the dorsal surface of the abdomen, to the thorax, it passes close to the abdominal air sacs, and is cooled. On its way through the narrow petiole to the thorax, it goes through a series of nine tight loops (fig. 7.26). In these convolutions (aortic loops), the cooled abdominal haemolymph is forced to flow very close to the warm haemolymph leaving the thorax. Heat is transferred from the haemolymph leaving the thorax to that returning from the abdomen. Heat exchange is encouraged by the slow speed of flow through the aortic loops. This mechanism, known as a counter-current heat exchanger, is a common method of thermoregulation in animals, for example, seals use it to reduce heat loss through their flippers when they are swimming in icy water, and penguins have a similar system in their legs to reduce body heat loss to the ice on which they stand. The aortic loops in bees also prevent large amounts of warm haemolymph from passing quickly into the abdomen. This mechanism is particularly important in bumble bees, which incubate their brood using heat from their abdomens and are able to maintain abdominal temperatures of 35°C[85]. Honey bees do not incubate their brood in this way and, being smaller insects, probably suffer less than bumble bees from excess heat production in the thorax.

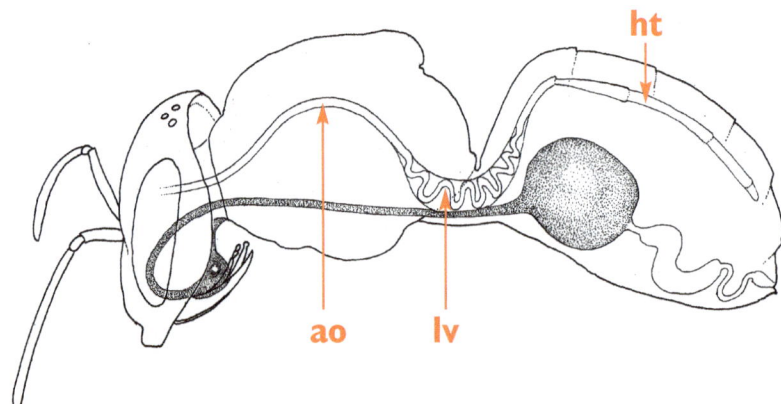

FIG. 7.26 *Diagram showing position of looped vessels or aortic loops (lv) between the thorax and abdomen in the honey bee. Aorta (ao), heart (ht).*

By retaining heat in the thorax using these mechanisms, honey bees are able to forage at times of day and year when temperatures are too low for other insects to be very active. Overheating problems do not end when a bee lands. When flight stops, thoracic temperature increases by up to 18% in the first 30–60 seconds. Body temperatures of 35°–41°C have been recorded in bees landing at feeding stations, and at rest their rate of cooling is four times as long as their warm-up rate during shivering[82].

At air temperatures of 30°–35°C, additional behavioural responses that facilitate cooling are employed. During flight, bees can reduce their thoracic temperature by one degree by dangling their legs rather than holding them tucked well in against the thorax and abdomen, thus exposing the ventral surface of the body to airflow.

Some grounded honey bees will regurgitate a drop of liquid, and hold it between proboscis and thorax, or spread it over the ventral surfaces with their legs. They may then suck it in again and repeat the process several times. This behaviour is known as **gobbetting** and leads to evaporative cooling of the head and thorax. The effectiveness of gobbetting is unclear and may be limited to excessively high temperature. There is no evidence that it actually cools down the thorax during flight at air temperatures of 30°–35°C[83], but it has been shown that if the thorax is artificially heated to 47°C, gobbetting can lower thorax temperature by 3–4 degrees[84].

Thermoregulation in drones

Drone honey bees are not quite as good at thermoregulation as workers. In flight, thoracic temperatures may be several degrees higher because drones are larger with a much higher body mass so their flight muscles have to do more work. This also means they cool down less quickly. Because they do not forage, they cannot employ the water droplet evaporative cooling mechanism and they are less willing to expend energy shivering to maintain high thoracic temperature. These factors limit drone flight to times when temperatures are more favourable.

Load carrying

When a bee is foraging successfully it collects nectar and/or pollen from a number of flowers and transports them back to the hive. Carrying a load requires an increase in power and in energy consumption. The metabolic rate of a honey bee in slow forward flight increases linearly with increased body mass due to the amount of nectar it is carrying[86]. As a result, thoracic temperature is higher when a bee is carrying a load, because more metabolic heat is generated.

How much can a bee actually carry? Nectar loading in bumble bees can effectively double their body mass[87], while wasps commonly return to the nest with nectar or prey loads weighing 50–70% of their body weight[7]. The maximum load lifting capacity for a variety of flying animals, from tiny bugs to bats and birds, has been measured by attaching progressively larger weights to their bodies and encouraging them to take-off[88]. It was found that all the animals could lift somewhere between 10% and 300% of their own body weight, with birds and bats at the lower end of the range and insects like butterflies and damselflies, known to use the clap and fling mechanism of lift enhancement, at the higher end of the range. Insects are better at lifting than birds or bats because a higher proportion of their body mass is flight muscle (20–50% of mass, compared with less than 20% for vertebrates). So far, it seems that no one has directly measured the load-lifting capacity of the honey bee, although measurements on other Hymenoptera revealed a capacity to lift a load between 50% and 100% of body mass[81], and it has been shown experimentally that the smaller euglossine bees (orchid bees) can carry an average of 95% of their own weight in plastic beads[89].

8. Glands:
chemical communication and wax production

Insects, like other animals, have a variety of glands whose secretions act as chemical messengers and play a crucial role in regulating bodily functions; for example, the products of different endocrine and neuro-endocrine glands regulate metabolism and homeostasis (the maintenance of a constant internal environment), reproduction, growth and development. The remarkable variety of physiological processes influenced by the hormones and neuropeptides produced by these glands include adult haemolymph sugar levels; fat metabolism and haemolymph lipid levels; water balance; gut contractility; cardiac activity; reproductive behaviour; egg production and the rate of sperm maturation; and the whole complex process of growth, moulting and metamorphosis found in insects. The hormones are released from the endocrine glands directly into the haemolymph and are carried to their target sites in this fluid medium, as are the neuropeptide hormones, released from the neurosecretory cells or nerve axons of the neuroendocrine system. Although the hormones come into contact with most tissues as they are transported, in many cases only cells that have the specific receptor for a particular hormone are affected. The amount of hormone circulating is, in general, very small, while the receptors on target cells are extraordinarily sensitive, with some target cells responding to hormone levels as low as 10^{-12} molar[1]. Insect hormones, like those in vertebrates, can be divided into two groups: those that can penetrate the surface membrane of all cells and act within the cell; and (the majority) those that are unable to penetrate the lipid phase of the target cell membrane[1]. The latter group produce quicker, short-lived responses, generally concerned with the regulation of metabolic events. When they bind with the receptor protein molecules on the cell surface membrane, they activate an intracellular messenger molecule that produces a cascade of enzyme reactions within the cell, serving to amplify the

effect of the hormone. In this way, a relatively low concentration of hormone molecules circulating in the haemolymph can influence many thousands of molecular reactions in the cells.

Chemical messengers are not only used to regulate the internal environment of an animal; when secreted to the exterior, they are used to communicate with other individuals. Known as **semiochemicals** when secreted externally, they may be used to communicate between individuals of the same species, in which case they are known as **pheromones**. They may be used in interactions between individuals belonging to different species, when they are classified as **allomones** of various types, depending upon the nature of the interaction. In recent years it has become clear that such intraspecific and interspecific chemical communication plays a very important part in the lives of animals. Pheromones appear to be of particular importance in insects: they are secreted by one individual and are received by a second individual of that species. They act either as a releaser, rapidly triggering a specific behaviour in that insect, or as a primer, producing a long-term change in that insect's behaviour. In insects, the majority of responses to pheromones are of the rapid, releaser type used in tasks such as alarm behaviour, orientation, nest recognition, trail marking and mate recognition[2]. Primer responses are perhaps more characteristic of mammalian species, although the queen mandibular pheromone of the honey bee, which exercises a fundamental level of control over worker and colony reproduction, provides an example of an insect primer pheromone. Primer pheromones can be both inhibitory and stimulatory in action, as demonstrated by the queen pheromone that both inhibits ovarian development in worker bees and attracts flying workers to a swarm cluster.

Chemical signals, particularly pheromones, have played a major role in the develop-

FIG. 8.1 *Worker honey bee exposing its Nasonov gland. The external openings of the 500–600 duct cells are present on the intersegmental membrane between terga VI and VII on the dorsal surface of the bee's abdomen.*

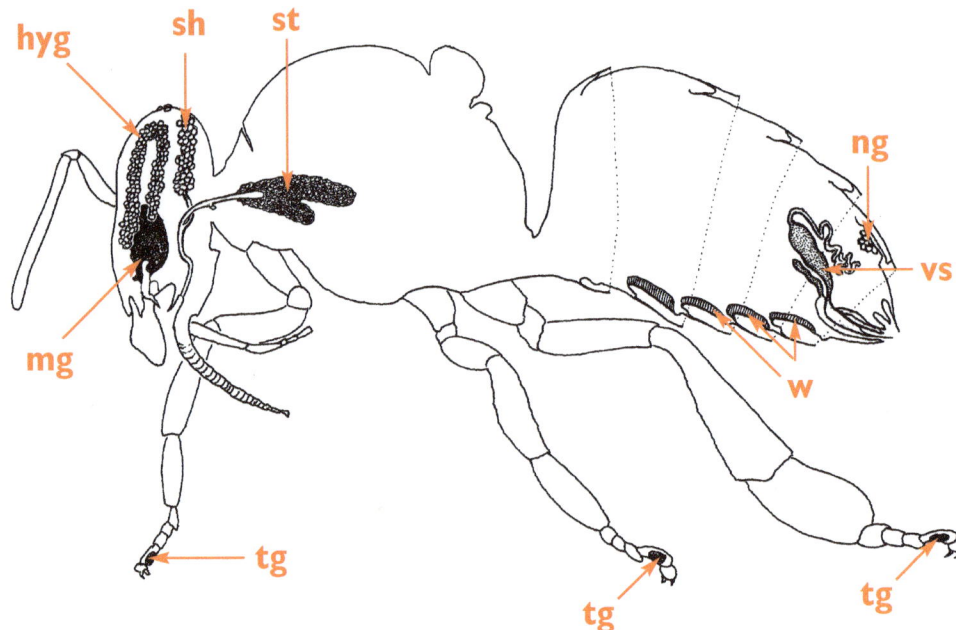

FIG. 8.2 *Worker honey bee showing the position of some of the major glands. Hypopharyngeal gland (hyg), wax glands (w), Nasonov gland (ng), mandibular gland (mg), venom sac (vs), head (sh) and thoracic (st) salivary glands, and tarsal glands (tg).*

ment of complex societies of highly social insects such as bees, ants and termites. Even among the social insects, honey bees stand out for their intricate chemical communication systems that establish and regulate the life of the entire colony. They are estimated to produce at least 36 pheromones[3]. Bees are richly endowed with pheromone glands whose compounds have to be secreted to the exterior; thus pheromone glands are exocrine (surface-opening) glands located over a range of sites on the body, for example, on intersegmental membranes beneath abdominal sclerites; around mouthparts, anal and genital regions; and on legs (fig. 8.2). Somehow, the product of these exocrine glands has to cross the impervious insect exoskeleton. There appear to be two main ways in which this is achieved. One way of conveying the secretions to the cuticular surface is for the cuticle to be invaginated, with narrow, cuticle-lined ducts extending down into the gland cells that normally lie just beneath the epidermis. The second method, which is considerably less well understood, is for the glands to be located near to cuticular areas that are riddled with tiny perforations, so that secretions can percolate from the secretory cells to the outer surface of the cuticle[4]. Once on

the body surface, most pheromones are volatile and are borne on air currents to the receptor sites. These consist primarily of olfactory receptors on the recipient's antennae, although olfactory receptors also may be located on other areas of the body. Some pheromones are not very volatile, for example, the queen pheromone, and they are believed to exert their effects through contact chemoreceptors on the antennae (see chapter 1) and body surface. Transfer of queen pheromone throughout the colony is described below (p. 170).

Most insect pheromones are aliphatic compounds. These are built up of straight or branched chains of carbon atoms derived from fatty acids or terpenes, many of which have been identified in plants. Often, pre-existing metabolites such as the insects' cuticular waxes are co-opted in the biosynthesis of pheromones, or naturally occurring plant compounds, such as those responsible for host-plant odours, are sequestered. The pleasant fragrance of the bee's Nasonov pheromone, discussed below, is dominated by floral odours. One of the components of the pheromone, the alcohol geraniol which derives its name from geranium, has a sweet rose odour and is the active constituent of oil of roses[6]. Another alcohol present in the pheromone (E,E)-farnesol, is also used in the perfume industry to emphasize the odour of sweet floral perfumes. Many of the pheromone compounds associated with the sting apparatus manifest strong floral odours such as lavender and jasmine.

In general, the composition and volatility of a particular pheromone reflect the nature of the behaviour that it evokes in the recipients. For example, pheromones used as a lure to attract male insects to females over a considerable distance, as in the case of many moths, need to have a high potency and a narrow specificity. The potency of these pheromones may be illustrated by the fact that the relatively small glands of the female insects produce a restricted amount of secretion that can nevertheless attract a male over distances as great as two kilometres in some instances. The sex pheromone-detecting receptor neurons of the males are characterized by a very high sensitivity to the molecule to which they are tuned. A few picograms of the stimulating pheromone component in the air is sufficient to elicit an increase in the firing rate of the male receptor neurons[7]. Detailed studies of the sex pheromone receptor neurons on the antennae of the male silk moth, Bombyx mori, show that about 100 pheromone molecule strikes per exposed receptor surface per second evoke a significant increase in the rate of impulse firing. However, male silk moths give a behavioural response to the pheromone, known as bombykol, if the airstream contains only 200 bombykol molecules per cm^3 of air. In these conditions, 40 out of the 40 000 receptor cells present on the antennae that are tuned to detect the pheromone will each receive one bombykol strike per second[8]. Since specificity is important in sex pheromones, the molecular weights of the compounds involved tend to be among the largest for pheromones. Compounds having few carbon atoms, for example less than five, cannot be assembled in enough different ways to provide a distinctive molecule for each

of a large number of species. On the other hand, the molecules cannot be made too large and complex or they will become too difficult and energetically expensive for the insect to manufacture, and they will not have a high volatility. In the Lepidoptera (butterflies and moths), in which the pheromones of a large number of species have been examined, those identified to date contain between 10 and 17 carbon atoms[9].

The chemical requirements for pheromones involved in sexual attraction, especially over a long distance, are very different from those that must elicit rapid recruitment to a nearby site of disturbance. The alarm pheromones of insects are of this type, generally warning against local and usually short-lived disturbances. Here, the need is for speed of reaction, with rapid movement of the stimulated insects to the site of pheromone release, then maintenance of those individuals at the site until the problem is dealt with, followed by rapid closure of the response. Most social insects' alarm pheromones are not species-specific, and it appears to be of advantage to the colony to be able to 'read' alarm signals coming from other species, particularly those that are dangerous to them. There is no need for large, complex molecules, hence most alarm pheromones are small, relatively simple molecules. Alarm pheromones tend to be highly volatile, diffusing to below threshold concentration within 2–3 minutes so that they are quite sharply demarcated in time. In the case of social insects, disturbances within the colonies are likely to be frequent, and are normally dealt with by the workers within that particular area. If the pheromone were to be too persistent, then gradually more and more workers would be drawn into each circle of disturbance and large areas of the colony would remain in a continual state of alarm[10].

Social insects, whose lives are regulated in so many ways through chemical communication, rely heavily on the presence of a large number of chemical messengers and an ability to convey subtly different signals with these pheromones. As more and more pheromones are identified, their components analysed, and the behaviours elicited by individual components described, it has become apparent that social insects are able to use single pheromones to subserve multiple functions. Widespread pheromonal parsimony has been found among the Hymenoptera, and the development of this communicative versatility is regarded as one of the major developments that enabled them to expand their sociality[11]. The queen pheromone of the honey bee offers a prime example of a multifunctional pheromone used in different social contexts. This pheromone serves as a sex attractant and an inhibitor of ovarian development and queen cell-building activity in workers. It also has a tranquillizing effect within the hive, as its continuous emission identifies the presence of the queen within the colony. Furthermore, the queen pheromone, in the presence of the Nasonov pheromone secreted by the workers, exhibits short-range, attractive properties to a swarm, an example of two pheromones originating from two different castes acting synergistically to release a certain behaviour[12]. Pheromones that are multifunctional normally work by combining several to many components, and by requiring a specific ratio of those components to elicit a particular stage of a behavioural sequence. Some multi-component pheromones release a series of steps in a particular behaviour. An example of this is shown by the African weaver ant, *Oecophylla longinoda*, where the multiple components of the mandibular gland secretion trigger a stepwise escalation of responses as ants approach an enemy[13,14]. The four chemical substances present have different volatilities so that when they are deposited by alarmed ants they produce different active spaces. Worker ants encountering the outermost space are alerted by hexanal and express alarm behaviour that causes them to engage in increased random motor activity. This is likely to carry them into the next active space where they encounter 1-hexanol that attracts them, causing them to move up the chemical gra-

dient to where they are stimulated by 3-undecanone, stimulating them to attack. Finally, they encounter 2-butyl-2-octenal that evokes a biting response. Examples of other multi-component pheromones can be found in this chapter and in chapter 9.

Insects also possess non-pheromone-secreting exocrine glands, producing other compounds that have to be secreted to the insect's exterior. Some compounds produced in exocrine glands are used in interspecific communication, for example, the contents of the venom glands of the worker honey bee. Other exocrine gland secretions include those of: the hypopharyngeal or maxillary glands that are used to provide larval food; the head and thoracic salivary glands; the postgenal or hypostomal glands; the wax glands. In this chapter we consider in detail examples of three exocrine glands: the Nasonov gland; the tarsal gland; and one non-pheromone-secreting gland essential in colony life, the wax gland. Some of the functions of the mandibular gland are described below (pp. 169–171), and those of the hypopharyngeal and salivary glands in chapter 5 (pp. 75–78).

The Nasonov gland

The Nasonov pheromone, produced by worker bees, is involved in several aspects of behaviour. It is used to attract the swarm to a suitable nest site and to ensure that the queen and accompanying workers enter and remain at the new site. It is also released when young flying workers are disorientated or when bees are feeding at low-odorant sugar feeders, and it can be used to mark non-odorant water sources[15].

Structure of the gland

The **Nasonov scent gland** lies beneath a thin, pale band of cuticle that forms the anterior part of the last abdominal segment (tergum VII)[16] (fig. 8.1) . The cuticle is not sclerotized, as the remainder of the tergum is. The Nasonov gland cells are situated below the anterior portion of this band of cuticle with their ducts opening

FIG. 8.3 *Openings of the Nasonov gland cells between terga VI and VII of the worker honey bees' abdomen.* **a** *View of abdomen in the region of the junction between terga VI and VII. The intersegmental cuticle has been exposed to reveal the band or canal onto which the Nasonov gland pheromone is secreted via the duct cell pores. Abdominal intersegmental membrane (am), canal region (cr).* **b** *As above, but closer in to show the canal region (cr) where the pores emerge. The pores can be seen as minute surface depressions in this region.* **c** *Area of cuticle in the canal region showing opening pores of the duct cells (pr).*

into a deep canal that traverses the band (fig. 8.3). Normally, this area is not visible since it is overlapped by the preceding segment (tergum VI), but when the bee flexes the tip of its abdomen the gland is exposed, together with the intersegmental membrane between tergum VI and tergum VII (figs 8.1 and 8.3).

The gland (fig. 8.4a) is derived from the outer epidermis and comprises some 500–600 separate gland cells, each with its own duct cell, together with a number of fat body and oenocyte cells[17,18]. Fat body cells are intimately associated with metabolic activities in insects and are frequently distended by stores of lipid, glycogen and, more rarely, protein. Oenocytes are very large cells, generally concerned with the secretion of lipoproteins. The roles of these two types of cell in Nasonov gland activity have not yet been determined. Several types of gland cell are found in exocrine glands, classified according to the cells' relationship to the outer cuticle and the means of egress of the glandular secretion[4]. The Nasonov gland cells are defined as class 3 cells since they have a ductule or canal cell lying between the gland cell and the surface[4]. This thin cell (fig. 8.4b, c) contains a cuticular duct lined with epicuticle, surrounded by a thin tube of cytoplasm. The duct opens onto the surface of the cuticle (fig. 8.4d) and penetrates the gland cell at its other end (fig. 8.4e, f). The ducts enter the cells at a region known as the end organ or ampulla. The cuticular lining of the duct is thrown into a series of

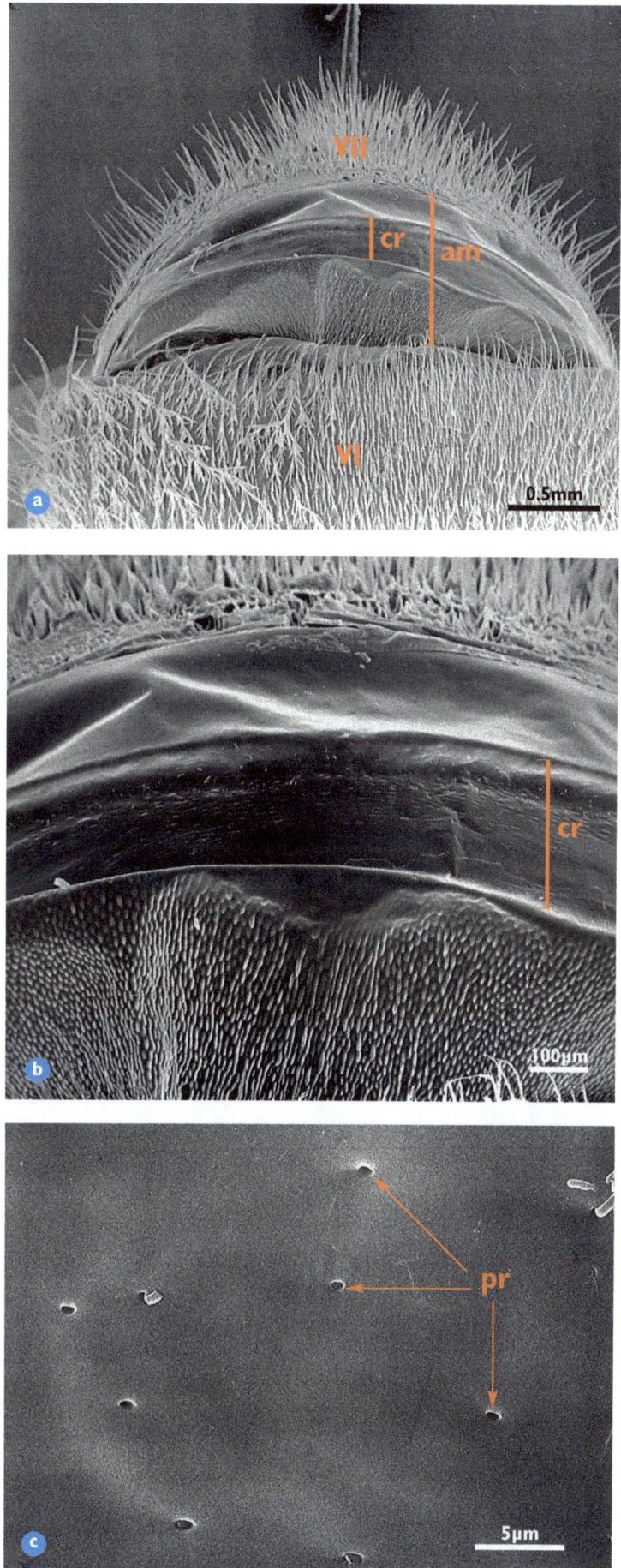

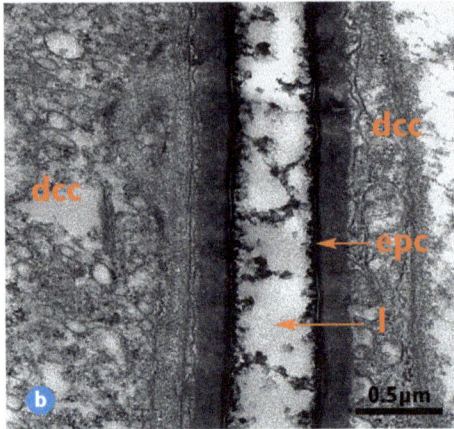

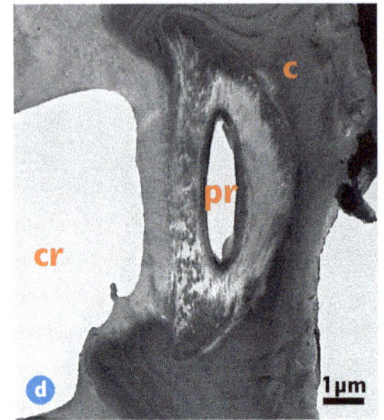

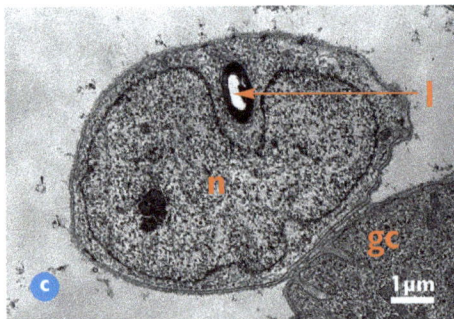

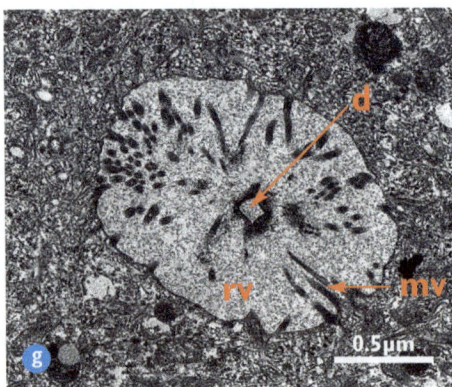

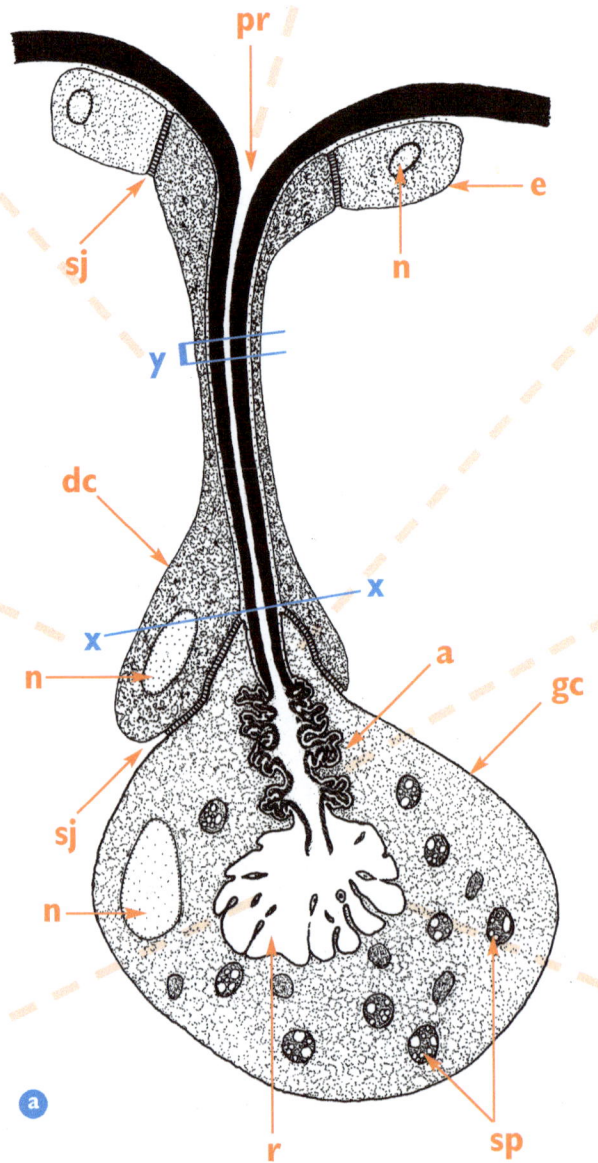

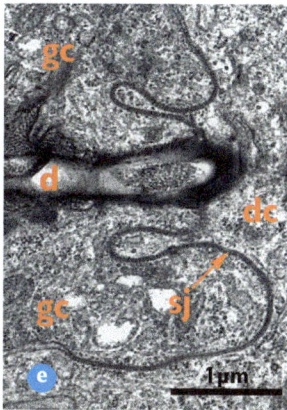

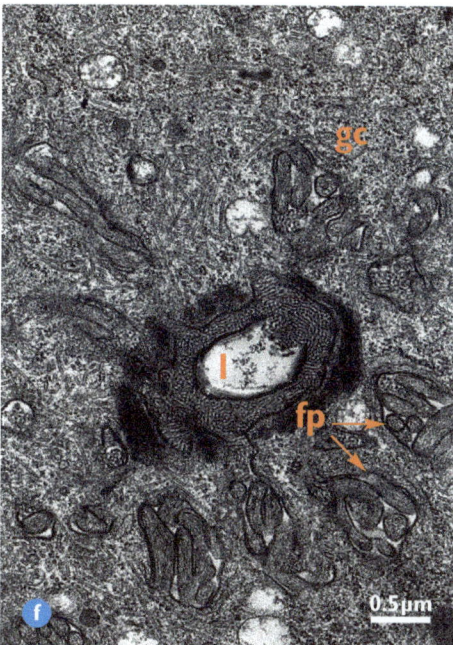

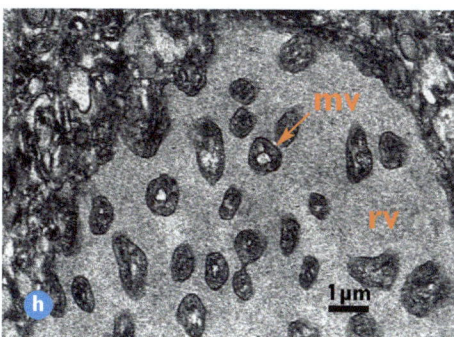

FIG. 8.4 *The Nasonov gland.* **a** *Diagrammatic section through one duct cell (gland cell component of the Nasonov gland). The duct cell (dc) is shown linking the gland cell to the canal region of the abdominal intersegmental cuticle. The proximal end of the duct terminates within the gland cell in a series of finger-like processes, the ampulla which itself terminates in the sac-like reservoir within the cytoplasm of the gland cell. Ampulla (a), reservoir (r), gland cell (gc), cell nucleus (n), epidermal cell (e), septate junction (sj), opening pore of duct cell (pr), secretory product (sp).* **b** *Longitudinal section of duct cell at level y on diagram* a *showing the thin epicuticular lining of the duct (epc) and duct cell cytoplasm (dcc). Lumen of the duct (l).* **c** *Transverse section (x—x) of duct cell in the region of its nucleus (n). Lumen of the duct (l) surrounded by epicuticle. Part of gland cell (gc).* **d** *Section through the cuticle (c) of the abdominal membrane in the region of the canal (cr) showing a pore (pr) lined with epicuticle close to its opening.* **e** *Contact region between a duct cell and its gland cell. The junction between the two cells is septate (sj) with the cross-bands (desmosomes) clearly visible. Duct (d) penetrating the gland cell at the level of the ampulla. Duct cell (dc).* **f** *Transverse section through the mid-region of an ampulla showing finger-like processes (fp) cut transversely containing bundles of tubules or microvilli within the cytoplasm of the gland cell (gc). Lumen of duct (l).* **g** *Transverse section through the sac-like reservoir vacuole (rv) showing microvilli (mv) draining from the gland cell cytoplasm into it. End of duct (d) penetrating the reservoir.* **h** *Enlargement of microvilli (mv) like those shown in* g *cut transversely. Reservoir vacuole (rv). Diagram courtesy of K Pell*

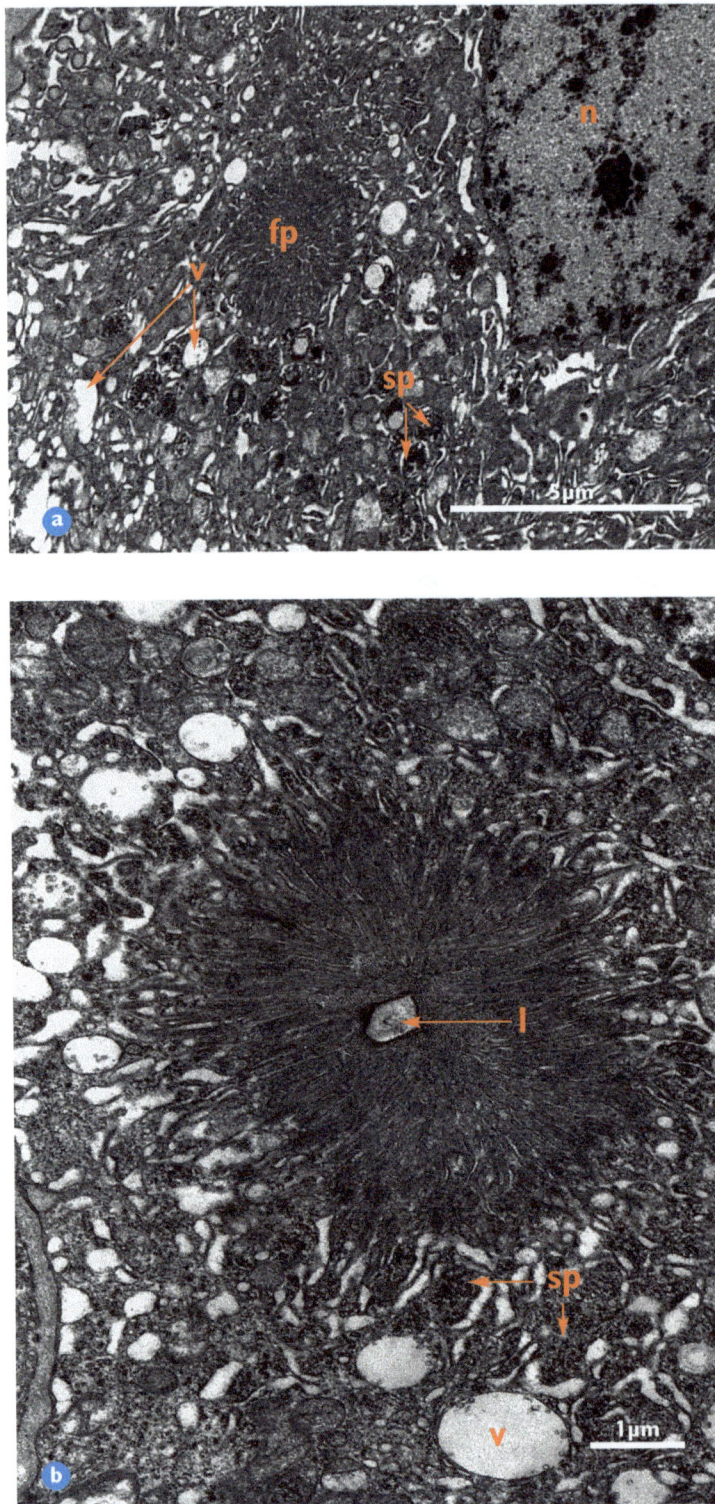

FIG. 8.5 *Active secretory cells in the Nasonov gland.* **a** *Active cell showing nucleus (n), small vacuoles (v), and accumulation of secretory products (sp) as dark granules. Bundles of microvilli entering finger-like process (fp).* **b** *Section through part of the duct region of the ampulla, with secretory products (sp) in the surrounding cytoplasm. Lumen of duct (l) within the ampulla region. Small vacuole (v).*

finger-like processes or chambers as it continues on into the cell for a short distance. The secretions of the gland cell appear to pass through microvilli into a sac-like reservoir region surrounding the end of the ampulla and from there pass through the ampulla and into the duct itself (fig. 8.4g, h). Both apparently active (fig. 8.5) and inactive (fig. 8.6) cells are normally present in the same tissue specimen.

In three-day-old pupae, the gland cells are not fully developed and lack their ducts (fig. 8.7a), the glandular region consisting largely of columnar epithelial cells, underlain by fat body cells. However, by the fifth pupal day, the gland cells are developing and making contact with the canal cells containing the ducts (fig. 8.7b). By seven days, fully developed gland cells are connected to the exterior by ducts (fig. 8.7c).

Composition of Nasonov pheromone

The composition and biochemistry of the Nasonov gland secretion has been investigated in some detail. Like many pheromones, it contains a number of components. Seven components have been identified by analysis of the secretion from single insects by capillary column gas chromatography, which resolves the components of the mixture into separate peaks[19]. These comprise the biosynthetically-related terpenoids geraniol, geranic acid, nerolic acid, nerol, (*E*)-citral, (*Z*)-citral and (*E,E*)-farnesol. This type of analysis also provides information about the ratio of the different components in the mixture[19]. The amount of the individual components present in foragers varies considerably between individual bees, but the relative proportions of the different components remains the same.

Composition of the secretion varies with age. Newly-emerged bees were found to produce little or no geraniol and no (*E,E*)-farnesol, and the amounts of both components increased with age, for example, around 0.3 µg of geraniol per bee was found in guard bees (normally 10–17 days old)[20,21]. The amount of geraniol present peaks at around 28 days in active foragers in spring and summer, falling thereafter, probably due to physiological ageing. The quantities produced also vary with the season: in winter the levels of geraniol and (E,E)-farnesol are low, while in spring, when foraging starts, the levels of these components increase.

When the response of foraging bees towards a synthetic mixture of the pheromone components was tested by bioassay, it was found that a mixture containing all of the components in their natural ratio was as attractive as the natural pheromone wiped from the canal into which the duct cells open[22]. Use of this synthetic mixture has permitted comparison of the attractiveness of the synthetic mixture with that of the mixture minus one component. These studies have shown that each component contributes to the attractiveness of the Nasonov pheromone. In tests of the attractiveness of the individual components, each showed some attractiveness for the bees, although some of the components were more attractive than others[22]. Bioassays conducted by several researchers have, however, yielded conflicting results for the ranking order of the attractiveness of the components. These variations may be due to the different methods employed and the fact that different behaviours were examined. Nevertheless, the majority of investigators agree that (E)-citral, geraniol and nerolic acid are the most important components for inducing clustering, pheromone release at the hive entrance, and in attracting foragers.

(E)-citral is the most active compound, and it is also one of the most volatile components, and the proportion present would be expected to decrease after release because of its volatility. However, the proportion of (E)-citral to geraniol present in the pheromone after release is the same as that found within glands that have not recently secreted the pheromone. A mechanism exists in the gland to maintain a constant proportion of (E)-citral during pheromone release. When Nasonov glands were stored in the inert compound

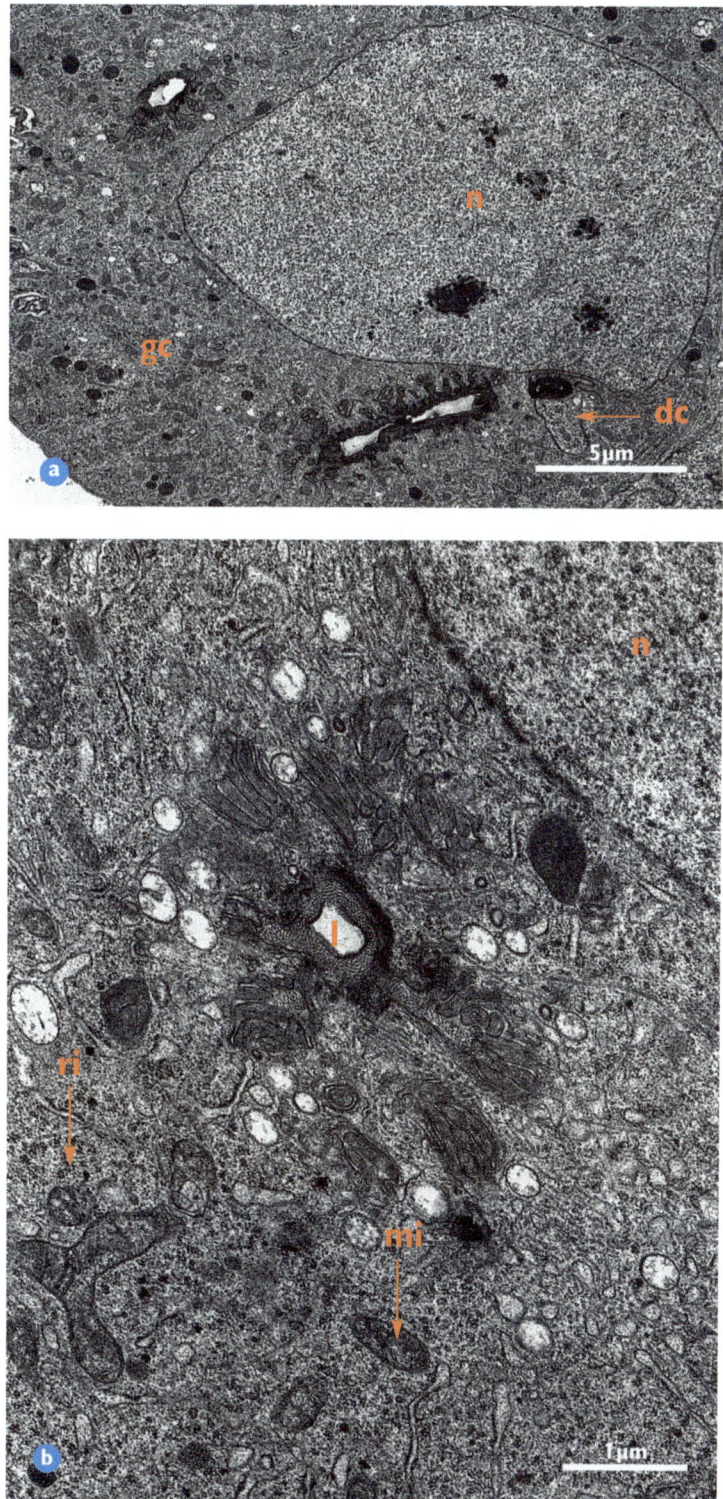

FIG. 8.6 *Inactive secretory cells in the Nasonov gland.* **a** *Inactive gland cell (gc) showing nucleus (n) and a duct with part of its associated duct cell (dc).* **b** *Similar position to fig. 8.5b showing much reduced complexity around the duct. Lumen of duct (l), nucleus (n), mitochondrion (mi), ribosomes (ri).*

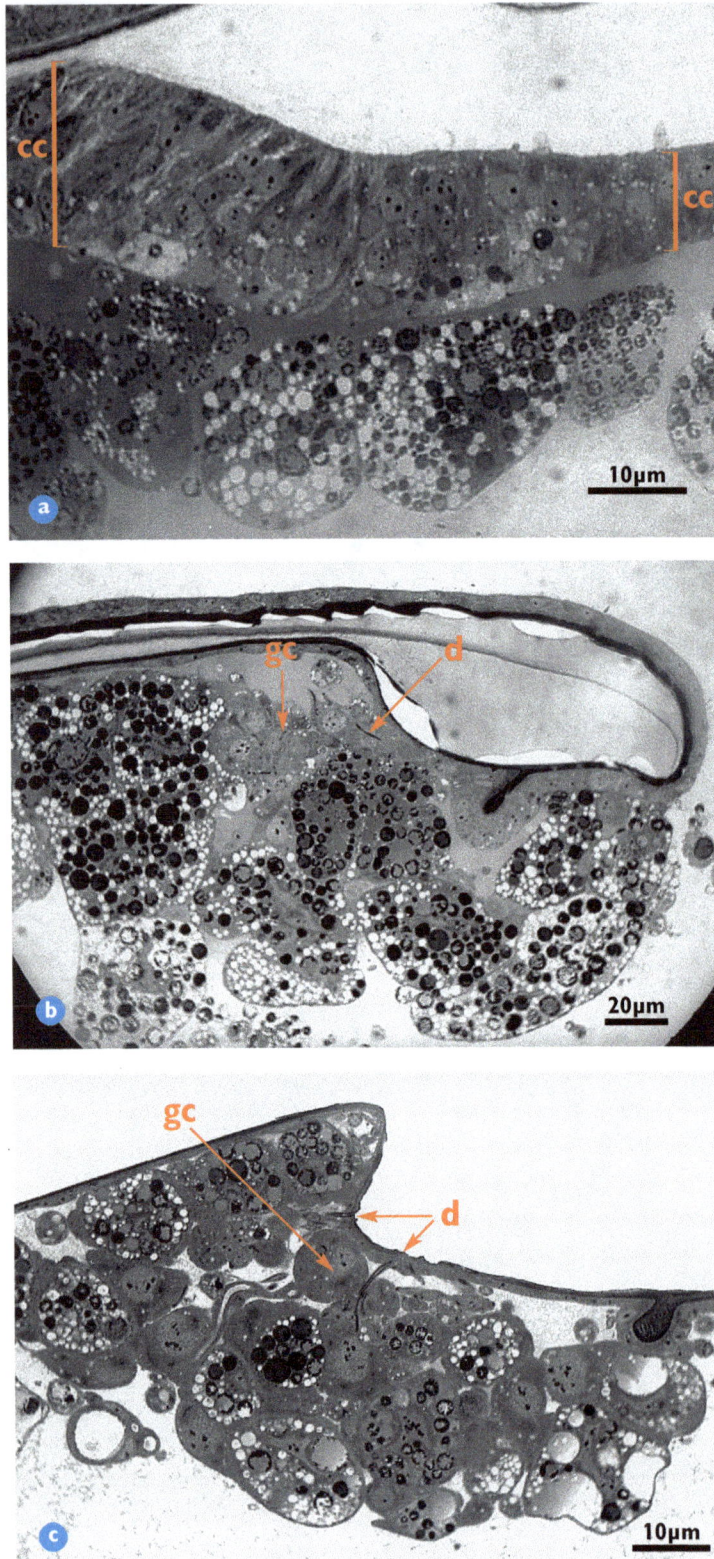

FIG. 8.7 *Nasonov gland during pupal development.* **a** *Three-day-old pupal stage showing columnar cells (cc) from which gland and duct cells develop.* **b** *Five-day-old pupal stage. Ducts (d) can now be seen among the gland cells (gc).* **c** *Seven-day-old pupal stage. Ducts (d) now connected to the outside through the cuticle. Gland cell (gc).*

hexane, it was found that the (*E*)-citral level rose to a maximum after 10 hours, while the geraniol level fell, suggesting that endogenous geraniol is converted into (*E*)-citral by the gland[20]. Further studies indicated that a specific enzymic process is involved in the conversion of geraniol to (*E*)-citral and that the enzymes responsible for geraniol metabolism are believed to be bound to the membranes within the gland cell[20]. It is clear from figure 8.5a, b that there is an abundance of membrane surfaces within the active gland cell. Further reactions occur as (*E*)-citral is converted to geranic acid.

Role of Nasonov pheromone in behaviour

Nasonov pheromone is employed in a number of behavioural contexts, all of which are concerned in some way with the orientation of the insect. The role of the Nasonov pheromone in swarming is perhaps most important. As the colony grows, and becomes overcrowded, it produces a new queen (see below, p. 176) and then divides, with the old queen leaving the nest with about half of the workers. The swarm clusters on the branch of a nearby tree or some other support. The first bees to arrive expose their Nasonov glands and fan their wings to disperse the pheromone (fig. 8.1), attracting other bees that are still airborne. If the queen is present, then the workers will settle, forming a cluster. Nasonov pheromone continues to be produced and dispersed to maintain the cluster. Scout bees then search for a new nest site, but sometimes they may have already begun searching three or four days before the old queen leaves the nest. Searching for a new nest site involves an extensive examination of the inside of a new cavity and the exploration of its surroundings. The potential nest site may be marked by Nasonov pheromone. When scouts have found a suitable site, they return to the cluster and dance on its outer surface to convey the distance and direction of the site. Normally, scouts investigate different sites, and a number of dances are then performed simultaneously on the surface of the cluster. The vigour of the dance signals the quality of the site, and the scouts performing the less vigorous dances may leave and explore the sites of the more vigorous dancers, later returning to perform a dance indicating one of these sites. As a result, some more workers from the cluster are recruited to visit the sites[23]. Eventually, the scouts agree on one site and mark its entrance with Nasonov pheromone[24].

The scouts then perform a 'buzzing' run on the cluster, thus breaking it up and launching the bees into the air whereupon they move off towards the new site. The scout bees fly through the swarm in the direction of the site, emitting Nasonov pheromone and returning to fly through the swarm again so that they almost appear to be leading the swarm along an aerial trail of pheromone. The swarm lands at the Nasonov pheromone-marked entrance forming a cluster, and the bees, including the queen, rapidly enter the cavity. Most workers emit Nasonov pheromone as they land at the entrance and while they are walking into the nest. In bioassay studies, the components most effective in inducing clustering were nerolic acid, (E)-citral and geraniol[24]. Nerol or (E,E)-farnesol alone showed little ability to induce cluster formation.

Nasonov pheromone still acts as an orientation aid even when the bees are well established in the hive or nest. Young bees making their first orientation flights frequently expose their Nasonov glands and fan on or near the nest entrance. Some older bees, including returning foragers, may show this behaviour if they have been disturbed in some way, or if there has been some disturbance around the nest or hive. They walk a short distance towards the hive and then stop and release their Nasonov gland secretion while fanning. This behaviour, which is repeated until they reach the entrance and disappear inside[15], is believed to help to guide other workers who are disorientated back to the hive. The bees face the hive during fanning

so hive odours from the wax comb, food stores, propolis, adult workers, drones and the queen are dispersed along with the pheromone; this may be important in helping disorientated bees to find their own hive among others[24], since the pheromone itself is neither race- nor colony-specific[15].

Bees also scent-mark a source of odourless water that they are collecting by exposing the Nasonov gland but not usually fanning[25]. They do not scent until they have made several visits, presumably as they need to establish that they have found a lasting source of water. The bees recruit other workers to that site by dancing on the comb and, since there is little or no odour associated with the water, recruits need more help in finding it than when they are recruited to flowers with a natural odour.

Bees rarely scent-mark flowers, but they mark dishes of odourless sugar water; the more concentrated the syrup, the longer the time spent scenting. In addition, bees walking on a substrate, at the hive or on a flower, may leave behind an oily substance known as the footprint pheromone[26], which can have an additive effect with Nasonov pheromone, as discussed below (pp. 175–176).

Queen pheromone

The secretion of the **mandibular gland** (fig. 8.2) of the queen comprises a blend of 24 identified compounds together with many as yet unidentified compounds[23]; it is called **queen pheromone** (also sometimes referred to as queen substance). The most abundant component is 9-oxy-(E)-2-decenoic acid (9-ODA), closely followed by the two optical isomers of 9-hydroxy-(E)-2-decenoic acid (–9-HDA and +9-HDA), together with methyl-p-hydroxybenzoate (HOB) and 4-hydroxy-3-methoxyphenylethanol (HVA)[3]. Queen pheromone plays an important role in short-range orientation of the workers during swarming, and also inhibits queen rearing, but it is not particularly volatile and cannot spread throughout a large colony by diffusion alone, so the workers assist in its dissemination by contact chemoreception[5]. The queen distributes the pheromone all over her body by grooming. Dissemination then starts with the worker bees of the queen's retinue: the queen is always attended by a circle of bees, who face her, either touching her with their tongue, other mouthparts and forelegs if they are 'lickers', or palpating her with their antennae, brushing them rapidly over her body if they are 'antennators'. The retinue bees pick up the queen's pheromonal secretion during their contact with her. The lickers, comprise around 10% of her retinue and pick up over half the secretion. The retinue bees groom their antennae, mouthparts and forelegs, spreading the queen pheromone

over these areas, and then act as messengers, moving for about 30 minutes through the colony making reciprocal antennal contacts with other workers. Contact chemoreceptors sensitive to queen pheromone are stimulated as the secretion is picked up and spread through the colony by these messenger bees. Meanwhile, another retinue takes their place around the queen and the process is repeated.

Synergistic interaction of separate pheromones is found to occur in a number of instances in insects. An example in honey bees, the interaction of Nasonov and footprint pheromone in food-source marking, is mentioned below. Another is the interaction of queen pheromone and Nasonov pheromone in swarming behaviour. When

FIG. 8.8 *Foot of a honey bee showing the arolium (ar) closed on a petal of a plant. Claw (cl) or ungues, fifth tarsomere (V).*

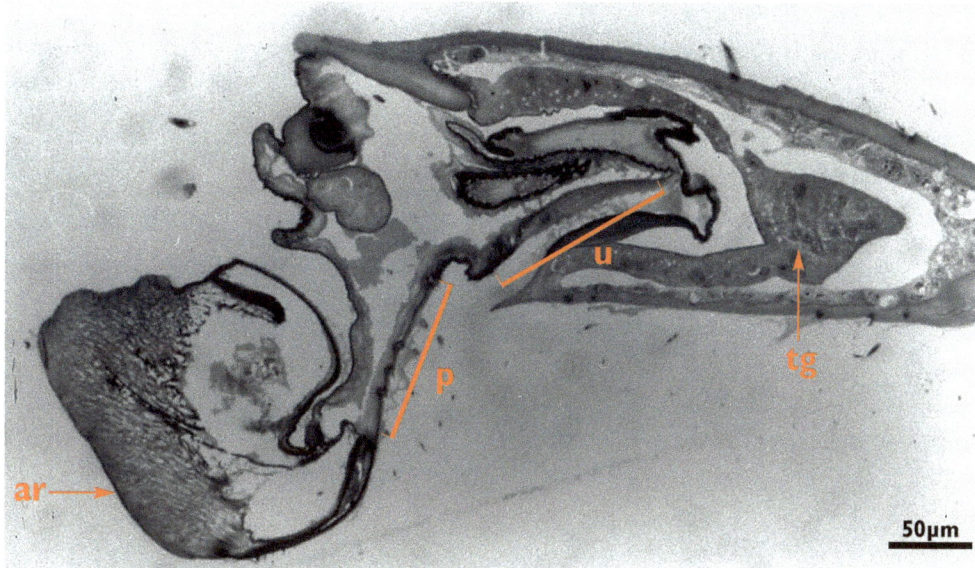

FIG. 8.9 *Light microscope section through the fifth tarsomere and pretarsus showing the position of the tarsal gland (tg). The arolium (ar) is expanded and the unguitractor plate (u) is hidden within the end of the fifth tarsomere. The planta (p) joins the arolium to the unguitractor plate. The claws are not visible in this section.*

the queen leaves the hive and settles on a nearby support she emits mandibular gland secretion. Workers in the proximity are attracted to her, settle on or around her and emit their Nasonov pheromone. This pheromone operates over a greater range and attracts the rest of the workers who left the hive to form a cluster around the queen. If the queen becomes lost during the flight to a new site, she lands and emits queen pheromone, which again attracts any nearby workers and they, in turn, attract more distant workers with their Nasonov secretion. Upon her arrival at the new site, the queen emits her pheromone to attract those bees still airborne nearby. The presence of queen pheromone indicates that she has entered the new nest and causes the remaining workers to enter the nest cavity rapidly. In experiments, swarms were offered sites that were unmarked, or were marked with queen pheromone or Nasonov pheromone alone, or with both pheromones; results showed that the queen pheromone was effective only over a short range, less than 10 metres[24]. This work also demonstrated that the queen pheromone alone was not attractive to scouts or to moving swarms but that it was attractive in the presence of the Nasonov pheromone.

The tarsal glands

When honey bees are allowed to walk over a clean glass plate, they sometimes deposit from their feet (fig. 8.8) an oily, colourless secretion that has a low volatility. This secretion, by virtue of the fact that it has been shown to affect the behaviour of other workers when deposited at the hive entrance or on flowers[15,27,28], is regarded as a pheromone, and has been called the **footprint**[24] or trail pheromone[14].

Structure of the gland

An examination of the production and role of this pheromone illustrates the fact that, despite being one of the most extensively researched insects, many features of the honey bee are still not well understood. The secretion is believed to originate from the tarsal glands, located in the fifth tarsomere of each leg in the worker, queen and drone. The **tarsal gland** (fig. 8.9) was first described by Arnhart in 1923 and is sometimes referred to as **Arnhart's gland**[29]. It consists of a unicellular layer of epithelial cells, each containing an abundance of cellular organelles consistent with a secretory activity (fig. 8.10). This glandular epithelium lies within a large cuticular sac that occupies much of the fifth tarsomere and is said to form the reservoir of the gland[29,30,31].

How does the pheromone reach the exterior of the cuticle?

The cells of the tarsal gland differ from those of the Nasonov gland in that they have no ducts to transport the secretory material either to the reservoir or to the exterior. The gland cells are bounded by cuticle and to enter the reservoir the secretion must cross this barrier. This type of gland cell is classified as class 1[4]. It is not clear in this and other insect class 1 gland cells how the secretory material crosses the cuticle — does it diffuse through the cuticle or does it make use of the pore canal system? The deeper layers of cuticle are normally crossed by pore canals that communicate near the outer epicuticular layers with narrower canals, such as the wax or epicuticular canals; however, these canals normally stop short of the surface. Nevertheless, in some glands composed of class 1 cells, the epicuticular canals do continue to the surface, and the canal system appears to be used for the transport of secretory material[4]. As this does not

FIG. 8.10 *Tarsal gland. a The tarsal gland (tg) contained within the fifth tarsomere (V). Pretarsus lies to the left of the photograph. b More detailed view of the junction of the pretarsus (to the right) with the fifth tarsomere (V) (to the left). Tarsal gland cells (tgc).* Star *indicates possible area where the tarsal gland opens to the outside where the unguitractor plate (u) folds back inside the end of the fifth tarsomere.*

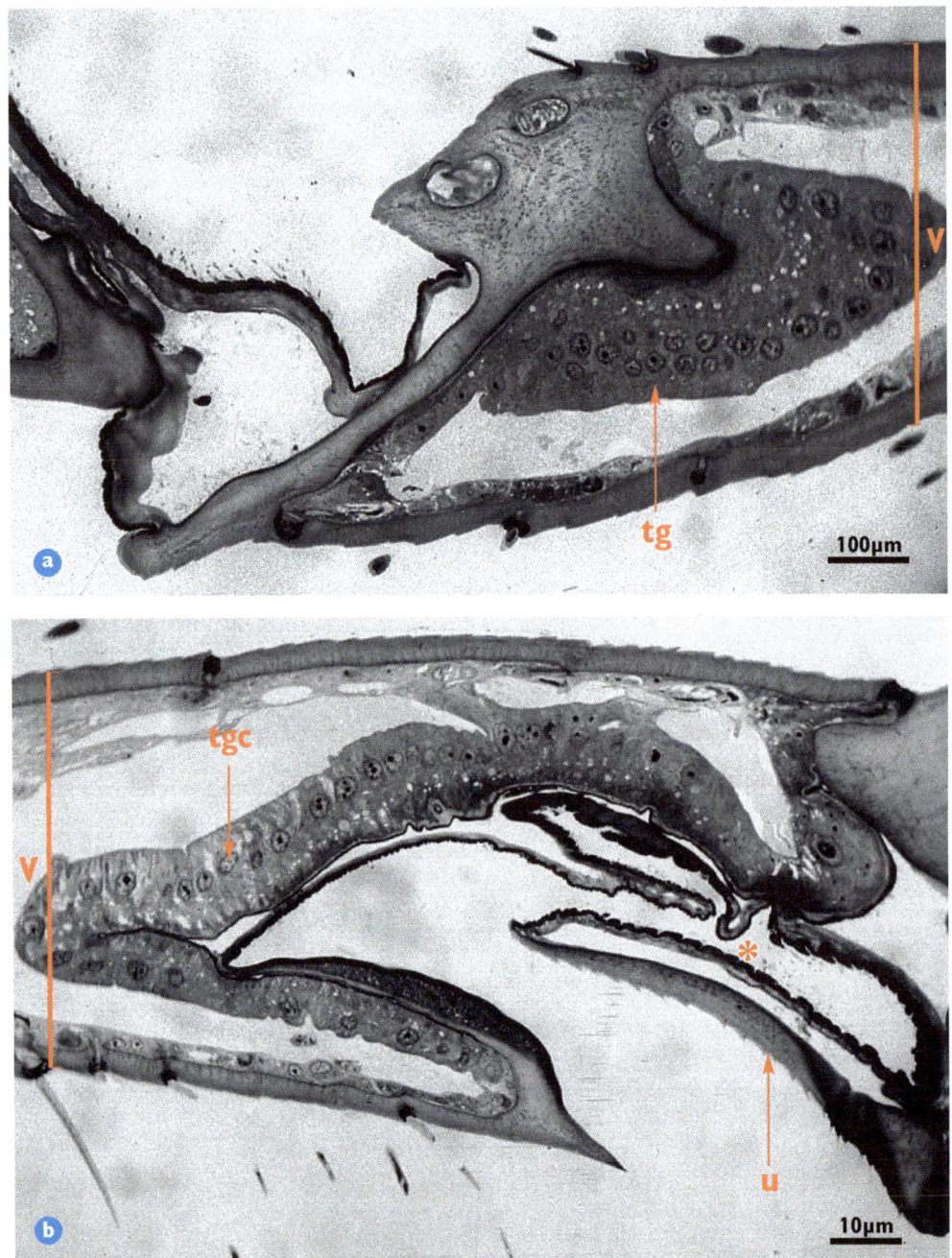

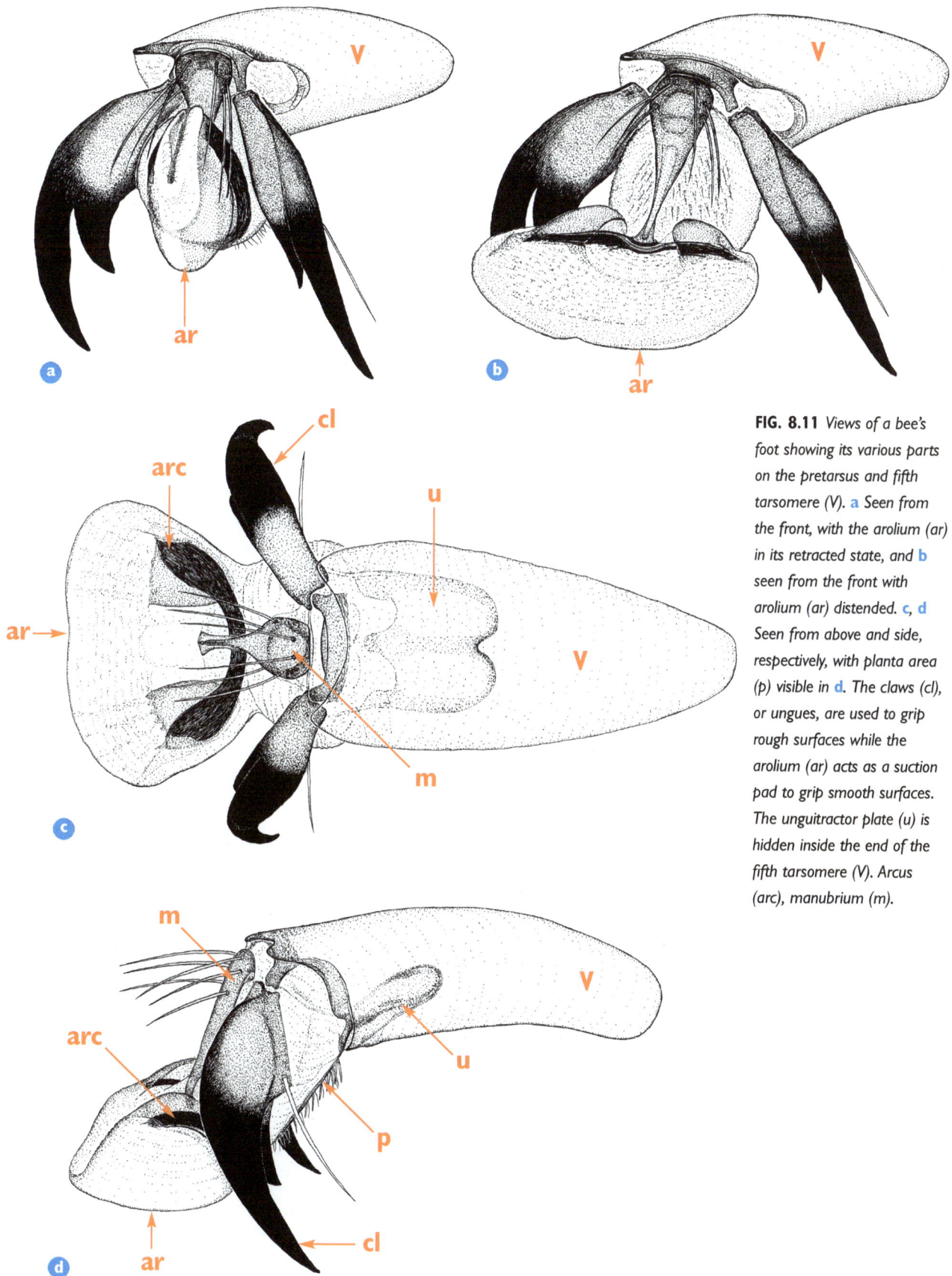

FIG. 8.11 *Views of a bee's foot showing its various parts on the pretarsus and fifth tarsomere (V).* **a** *Seen from the front, with the arolium (ar) in its retracted state, and* **b** *seen from the front with arolium (ar) distended.* **c, d** *Seen from above and side, respectively, with planta area (p) visible in* **d**. *The claws (cl), or ungues, are used to grip rough surfaces while the arolium (ar) acts as a suction pad to grip smooth surfaces. The unguitractor plate (u) is hidden inside the end of the fifth tarsomere (V). Arcus (arc), manubrium (m).*

appear to be the case in the bee, the means of egress of the secretion from the tarsal gland cells to the reservoir remains to be established; it may well be in the folds in the thin, pliable cuticle where the unguitractor plate of the pretarsus merges with the cuticle of the fifth tarsomere[28,30] (fig. 8.10b).

Since it is left behind when the bee has walked over a surface, the secretion must originate from some part of the ventral surface of the pretarsus or foot (fig. 8.11) and it has been assumed that the secretion flows out onto the inflatable arolium, or

footpad (fig. 8.11a, b). It was first suggested that the secretion passed out through holes in the tips of the large hairs or spines on the planta, a ventral plate that lies just behind the arolium; or through the squamous surface of the unguitractor, a second ventral plate that lies between the planta and the fifth tarsomere (fig. 8.12)[29,31]. However, examination by scanning electron microscope has shown that there are no apertures in the hairs or on the unguitractor plate (fig. 8.12b). A more recent study suggests that the proximal end of the cuticular sac or reservoir communicates with the exterior via a large transverse slit

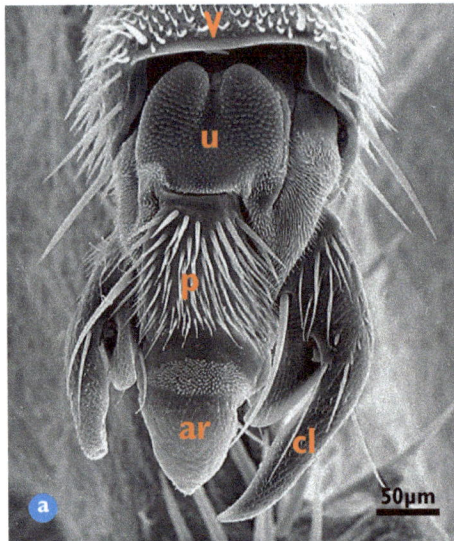

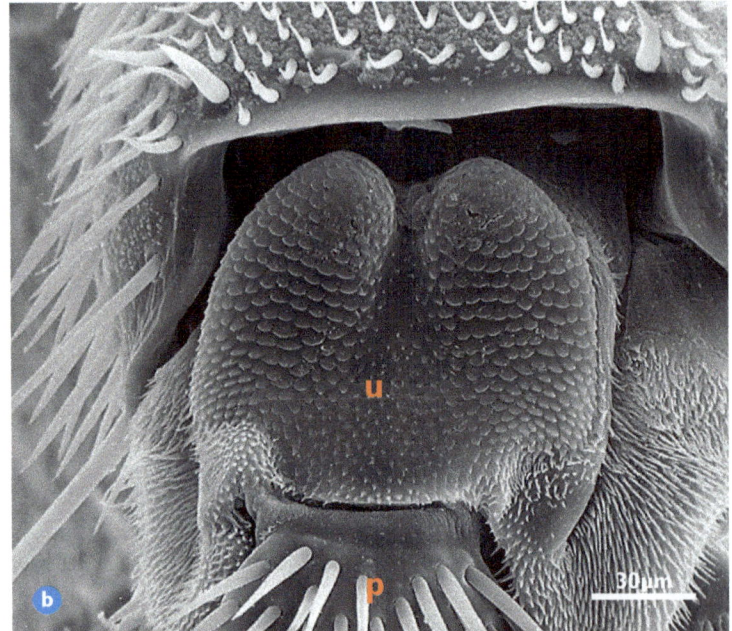

FIG. 8.12 *Ventral surface of the bee's foot.*
a Ventral view of the pretarsus of a worker bee. Arolium (ar), planta (p), unguitractor (u), claw (cl).
b Unguitractor plate (u) showing scaly surface.
c Arrow indicates region between unguitractor plate (u) and fifth tarsomere (V) from where tarsal gland secretion may emanate[30].

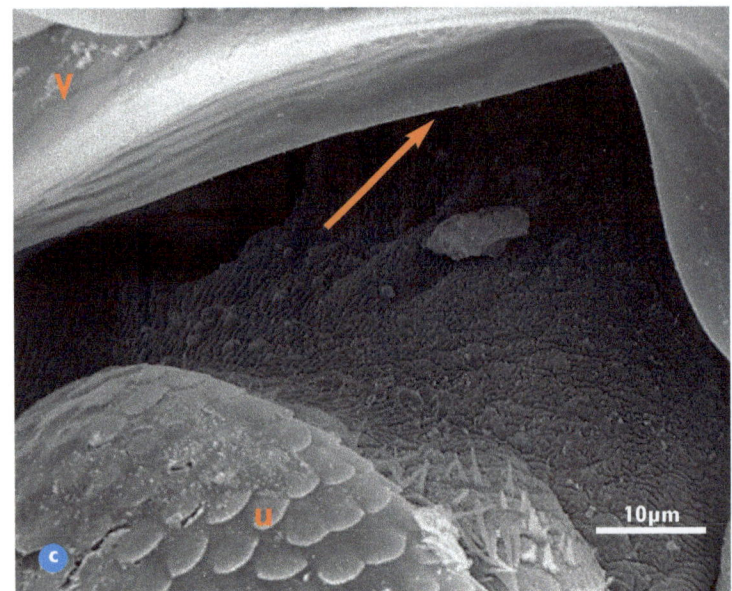

on the ventral surface of the foot in the region connecting the arolium with the fifth tarsomere (fig. 8.12c)[30]. It is not known if or how the bee controls the egress of the secretion from the gland.

Not all researchers agree that the tarsal glands are the site of production of the footprint pheromone[14,23]. Behavioural studies have shown that chemical extracts of various regions of the body, particularly the dorsal region of the thorax, are equally, if not more attractive, to the bees than extracts of the legs. For this reason they have proposed that the footprint pheromone is produced elsewhere on the body of the bee, possibly in the dermal glands[15]. If so, since only the legs normally touch the substrate when the forager is walking into the hive, the secretion must be spread to the feet somehow, perhaps by grooming. However, the ultrastructural studies identifying active glandular cells in the fifth tarsomere indicate that a secretion is indeed produced at this site[30].

The individual components of the pheromone have not yet been identified but in a preliminary analysis, compounds specific to each caste and sex were characterized: 12 for the queen, 11 for the worker and one for the drone[32,13,34]. A further difference lies in their rate of secretion, the tarsal gland of the queen secreting at a much higher rate than that of the worker and the drone. Nevertheless, there appear to be no gross structural differences between the glands of the sexes and the castes[33]. However, the secretion serves different purposes in workers and in the queen.

Role of footprint pheromone in the worker honey bee

In the worker, the pheromone appears to belong to those chemicals that assist the bee in orientation. Workers landing at the nest or hive deposit the footprint secretion as they walk inside. If a glass entrance tunnel to the hive is provided for the bees, the oily, persistent secretion accumulates on the glass as more and more bees enter.

Bees presented with a choice of this tunnel or a clean one much preferred the tunnel marked with the secretion. Furthermore, the attractiveness of the marked tunnel increased with the number of bees that had walked through it, up to a limit of about 400 workers[27]. A tunnel marked with the secretion of bees from another hive also proved attractive to the test bees, although slightly less so than one marked by bees from their own colony. It has also been noted that the odour of this pheromone will induce disorientated bees to expose their Nasonov glands at the hive entrance. Thus, it appears that the footprint pheromone aids in orientating workers that have become temporarily disorientated in the vicinity of the hive entrance. It can work in concert with the Nasonov pheromone, a highly volatile, airborne pheromone that presumably exerts its influence over a greater distance than the low-volatility footprint pheromone deposited on the substrate.

Experiments in which bees are trained to forage for sugar water on a glass dish have shown that such dishes are more attractive to other potential foragers than sugar water presented in a clean glass dish[24]; they were not being influenced by the Nasonov pheromone of previous foragers as those bees did not expose their Nasonov glands. Bees mark the foraging site with the footprint pheromone when they alight, and they appear to do this irrespective of whether they have foraged successfully at the site or not. It is only necessary for a bee to land briefly at a clean, empty site (without exposing its Nasonov gland) for it to prefer that site subsequently, and other bees also prefer such a site. It seems that workers mark a feeding site with the footprint pheromone, releasing alighting behaviour in bees recruited to a source of forage. In experiments, small blocks of plaster of Paris or squares of wire gauze were placed on the floor of a hive where returning bees walked over them and when the blocks were subsequently placed at foraging sites outside the hive, they proved to be attractive to foraging bees for many hours. This suggests that the secretions deposited at the hive entrance and on forage sites are the same pheromone[24]. It should be noted, however, that although most investigators agree on the results obtained when examining the tarsal secretion of workers, a few have found no attraction either to the hive entrance or forage sites[33].

The attractiveness of synthetic Nasonov pheromone on the recruitment of foragers to glass dishes of sugar water has been compared with that of the footprint pheromone[22]. Dishes marked with the footprint pheromone were visited more than clean dishes, while dishes marked with either footprint pheromone or synthetic Nasonov pheromone were equally attractive. However, dishes marked with both pheromones received a greater frequency of visits than any others. These experiments show that the footprint pheromone enhances the attractiveness of the Nasonov pheromone to foragers trained to visit a particular site for food.

Role of pheromones in inhibiting queen rearing and swarming

It has been suggested that the tarsal gland secretion of the queen plays a part in inhibiting the construction of queen cups and hence in inhibiting queen rearing and swarming. During the spring or summer, colonies multiply by swarming, the onset of which is marked by the construction of queen cups by the workers along the bottom edges of the comb. Swarming appears to be caused by overcrowding and congestion within the brood nest. Above a threshold of 2.3 worker bees per cm^3 of hive volume, the number of queen cups built is directly related to the colony density[28]. Colony overcrowding has the effect of limiting the queen's movements on the comb and she is said to be restricted from the path along the bottom edges of the comb in the brood nest of a very overcrowded colony. Such overcrowding could also restrict the circulation of messenger bees, thus resulting in overall reduction in the constant spread of queen pheromone

among the workers. This should free them to construct queen cups, but studies have shown that queen cup construction is not prevented by constantly supplying the overcrowded colony with an artificial source of pheromone, and the presence of some other inhibitory substance appears to be necessary. Since the queen deposits her tarsal gland secretion on the comb as she walks over it, tests were undertaken in which extracts of the tarsal gland were supplied, but queen cup construction was not inhibited. However, if a blend of both mandibular and tarsal extracts was supplied, construction was inhibited in an overcrowded colony in the swarming season[28]. It seems from these experiments that the footprint pheromone of the queen has some part to play in the control of swarming in overcrowded colonies. This provides a further example of behaviour regulated by two pheromones.

Beeswax and the construction of the comb

Comb construction is of the utmost importance to the colony. A swarm, on entering its new nest, cannot survive without cells in which to rear its brood and to store its honey and pollen (fig. 8.13). The priority given to this task by the workers is shown by the fact that in one study a feral colony had completed over 90% of its comb building within 45 days of finding and colonizing a new nest[23], despite the fact that beeswax is energetically very expensive to produce. It takes a kilogram of honey, plus an undetermined amount of pollen, to produce just 60 g of wax; the comb in an average hive 'costs' around 7 kg of honey[35].

Structure of the wax glands, and the production of wax

The four, paired wax glands are situated on the sternites or ventral plates of the fourth to the seventh abdominal segments of the worker (fig. 8.14a). The sternites overlap one another, each rear edge overlapping the front edge of the plate behind it (fig.

8.14b). Since the wax glands are located at the anterior of each sternite, they are normally not visible, although when active, the wax scale produced by the gland can be seen projecting just beyond the overlapping sternite. If the preceding sternite is removed, two large, oval, pale yellow areas of cuticle are revealed, which form the wax mirrors (fig. 8.14c). These polished, cuticular plates overlie the wax gland complex, which is comprised of three cell types: specialized glandular epithelial cells, fat body cells and oenocytes[36,37]. The wax is secreted through the cuticle of the wax mirrors. If the wax is removed from these plates, it can be seen to have a subregular, hexagonal pattern corresponding to that of the epithelial cells underlying the thin cuticle of the mirrors. Globular droplets of wax can be seen to exude through this surface, eventually coalescing to form a thin layer of wax. Subsequent droplets appear under this layer and become attached to it, building it up to form the several, irregular laminae of the scale covering the whole surface of the mirror. The scales usually range in thickness from 200–500 μm although scales up to 1 mm in thickness have been recorded[36].

Wax secretion in the glands, and transport across the cells

In spite of the great importance of wax to the bees and commercial interest to man, the full details of wax production are not fully understood. An examination of the fine structure of the gland allows some insight into the way in which the wax is produced, and highlights some of the problems. The wax mirror cuticle, secreted by the underlying epithelial cells, is thin, comprising a layer of orientated lipid, a layer of cuticulin and a homogeneous layer of procuticle, the whole being only just over 3 μm in thickness[37]. The outer cement layer, present in the majority of insect cuticles, is missing from the wax mirror cuticle. The cuticle possesses fine pore canals running from the epithelial cells into the epicuticle but not continuing through to the outer surface. These canals are thought to be involved in the transport of materials from the epidermal cells to the outer layers of the cuticle, where various processes take place as the laying down of cuticle, is completed. These include the secretion and repair of the fine wax layer that is present on the surface of most cuticle and the tanning or hardening of the outer cuticular layer. The pore canals may also function as a cytoskeleton to anchor the epithelial cells to the procuticle. Distally, the pore canals contain filaments, possibly of wax, which continue into very fine canals, known as wax canals, reaching to the cuticulin layer.

In the wax mirrors (fig. 8.15a, b), the thin procuticle is different in structure from the remainder of the bee cuticle and from other insect cuticles, in that the pore canals are packed with filaments around 10–30 nm in diameter. The apical membranes of the epithelial cells are thrown into folds where they approach the procuticle, leaving spaces between the cells and the cuticle. These intercellular spaces are

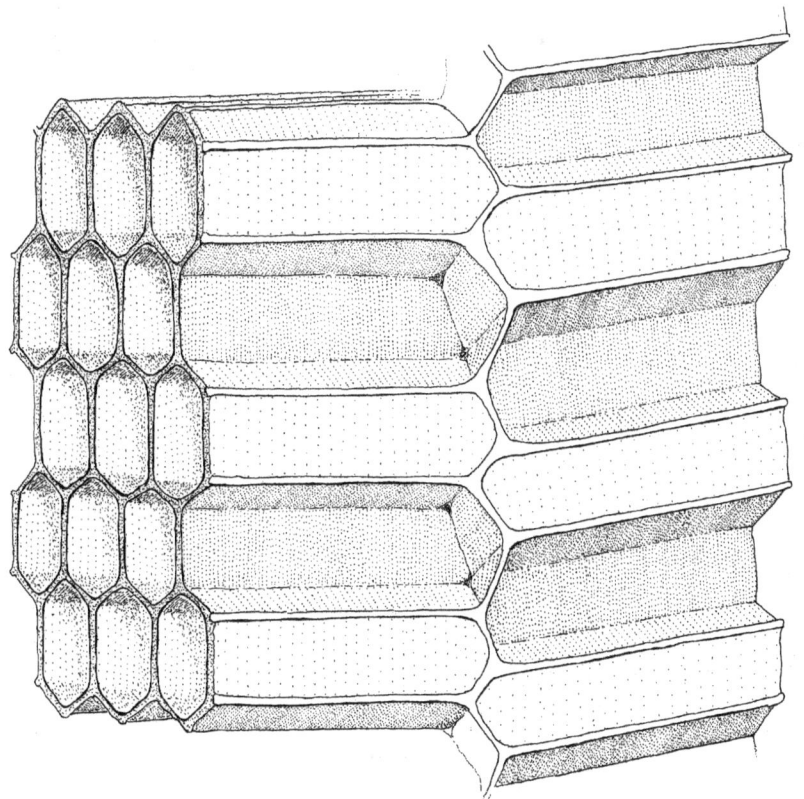

FIG. 8.13 Comb cells cut in section to show internal structure of the individual cells.

filled with twisted hanks of filaments within the pore canals that extend up into the procuticle to the cuticulin layer and down into the epithelial cells[37].

The epithelial cells elongate in the active gland to around 40 μm, forming a conspicuous palisade layer (fig. 8.15c). These cells have no canals running through the cuticle that are visible on the outer surface, nor do they make contact with the canal cells that drain to the exterior. Beneath the epithelial cells lie the fat body cells with the oenocytes among them, one for every two or three fat body cells, and the whole tissue is richly supplied with tracheae[37]. An association of these two types of cell with glandular epithelial cells is common

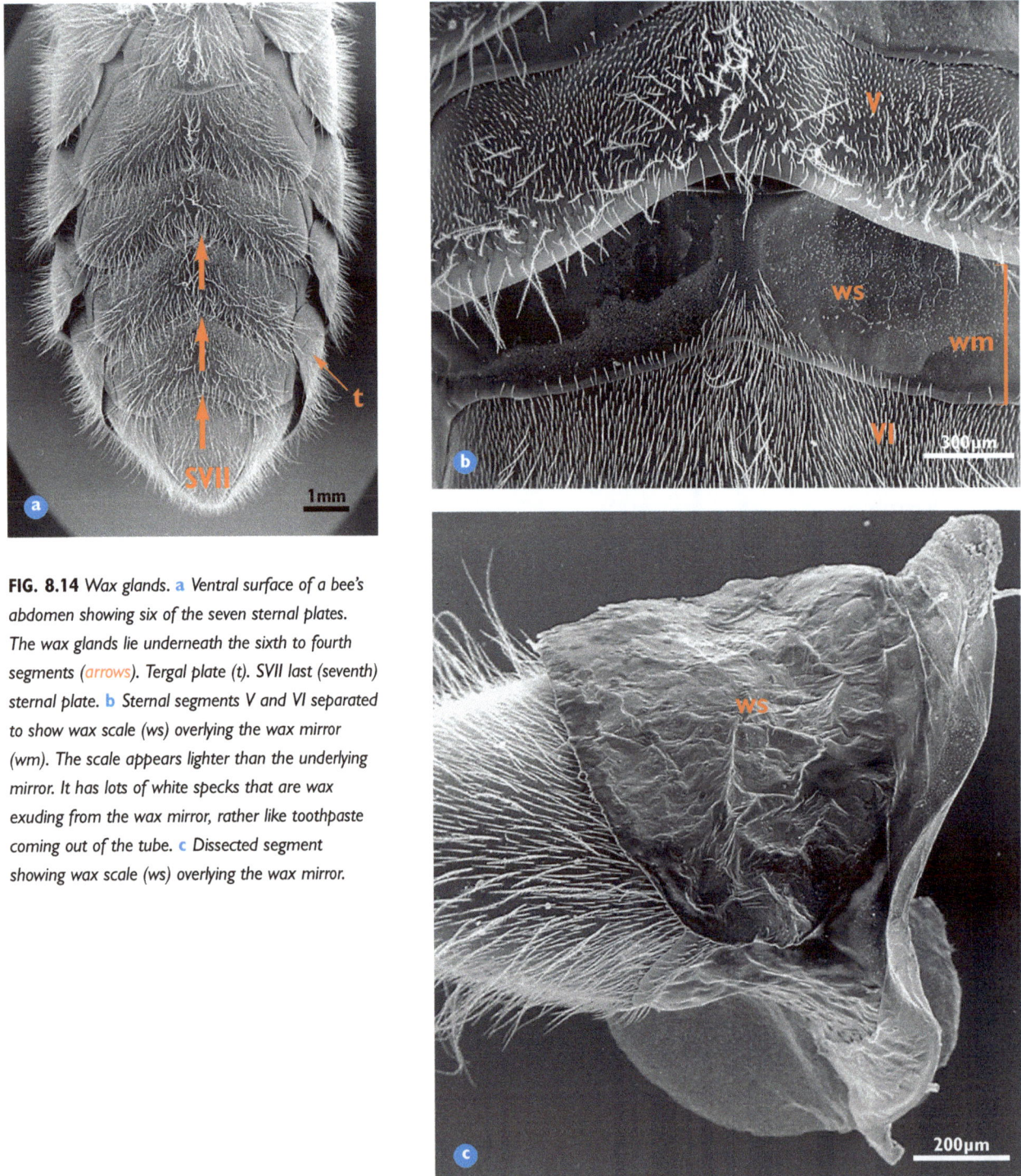

FIG. 8.14 Wax glands. **a** Ventral surface of a bee's abdomen showing six of the seven sternal plates. The wax glands lie underneath the sixth to fourth segments (arrows). Tergal plate (t). SVII last (seventh) sternal plate. **b** Sternal segments V and VI separated to show wax scale (ws) overlying the wax mirror (wm). The scale appears lighter than the underlying mirror. It has lots of white specks that are wax exuding from the wax mirror, rather like toothpaste coming out of the tube. **c** Dissected segment showing wax scale (ws) overlying the wax mirror.

among the exocrine glands of insects, including the Nasonov gland in the bee, and is said to indicate synergistic activity between these three cell types to produce the glandular secretion. The wax glands are not active throughout the worker's life; the cells tend to start increasing in size around days 2–3, maintaining maximum size between days 5–15 and thereafter slowly declining, gland activity being related to the periods spent capping cells or building and repairing the comb[23]. The development and resorption of the epithelial, oenocyte and fat body cells are thought to be closely linked to the duties workers perform within the hive. Histological studies have yielded evidence that during the secretory period material passes first from fat body cells and then from the oenocytes into the epithelial cells[37]. Radio-labelling of acetate precursors of fats suggests that different fractions are synthesized by the fat body cells and oenocytes; esters are produced by the former and hydrocarbons and free fatty acids by the latter. Both cell types then deliver their products to the epithelial cells[35]. The fat body and oenocytes appear to make the major contribution to wax secretion, but the role of the epithelial cells is not clear. Some authors report that they are unable to find within the epithelial cells several of the organelles commonly regarded as necessary for secretion[35] while others report their presence in cells of the active gland, together with increased nuclear size usually associated with a greater synthesis and secretory activity. These authors suggest that the cells are responsible for the synthesis of part of the protein fraction of the wax and possibly some other wax components[37]. The remaining fraction of the protein is variously reported as coming either from the haemolymph or the fat body.

How does the wax reach the surface of the wax mirrors?

Here again, the precise mechanism is unclear. The extensive system of pore canals filled with filaments, believed to consist of wax extending up into the wax

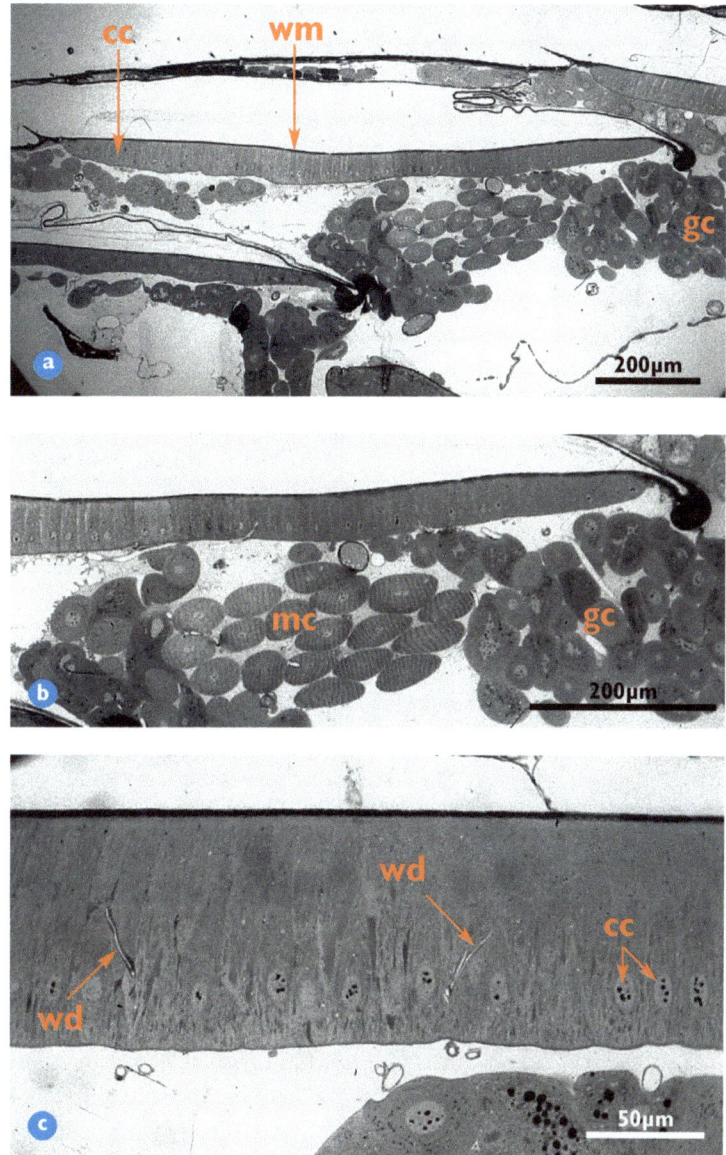

FIG. 8.15 Wax mirror. a Light microscope section of the sternum of a worker bee in the region of the wax mirror (wm). The columnar cells (cc) of the wax mirror are clearly visible beneath the cuticle that forms the mirror. The gland is associated with oenocyte and fat body cells (gc). b Higher power magnification of the area beneath the wax mirror. Muscle cells (mc), oenocyte and fat body cells (gc). c Columnar palisade cells (cc) beneath the wax mirror cuticle showing sections of wax ducts (wd).

canals, is thought to form part of the transport mechanism which brings wax or its precursors near to the surface of the wax mirrors. The localization of esterases (enzymes which hydrolyse the ester bonds within lipids) in the area of the pore canal has led to the suggestion that the final step in the synthesis of wax occurs in or near this area[38]. Even if this is the case, it still leaves the problem of the final stage in the transport across the outer layer of the cuticle, since no pores or canals have been found penetrating the surface[36,39]. Because the wax is hydrophobic, it is probably transported across the thin outer layer by lipophorins, a method of hydrocarbon transport by proteins known to occur in other insects. This would account for the presence of some of the proteins found in beeswax[40].

Composition of beeswax

The wax from *Apis mellifera* comb is a complex mixture of substances containing more than 300 individual components, although many of them are present in very small quantities[38]. It is this great diversity of composition that gives beeswax many of its unique properties. Beeswax differs quite markedly from plant waxes, having a relatively low melting point of 63°–64°C and being comparatively soft and plastic. Much attention has been paid to the composition of beeswax, in particular to that of *A. mellifera*, because of its commercial importance. The greater part of wax used commercially is purchased by the cosmetics industry (40%); pharmaceutical companies take another 25–30% and much of the remainder is used for church candles[23]. It is also used in polishes, engraving, dentistry and confectionery manufacture.

The major components of comb wax are hydrocarbons (14%); monoesters (35%); diesters (14%); hydroxypolyesters (8%) and free acids (12%)[38]. Each of these classes of lipid, together with the remaining five classes present, is a mixture of major components and a good number of minor components. While the composition of fresh *A. mellifera* comb wax differs very little, if at all, between samples, waxes from the Asian honey bees, *A. dorsata*, *A. cerana* and *A. florea* are different in composition both from *A. mellifera* and from each other. Comb wax differs somewhat in composition from freshly exuded scale wax. The latter is masticated with saliva to soften it, and the lipases present in the saliva hydrolyse the diacylglycerols (diesters) present in scale wax into the fatty acids and monoacylglycerols (monoesters) in comb wax.

Other material becomes incorporated into old comb wax, including the shed skins or exuviae of the larvae, as well as the silk threads spun by the latter.

How is comb-building activity regulated?

As mentioned above, comb construction (fig. 8.16) is energetically very expensive, and presumably for this reason comb building is quite tightly regulated in the nest or hive. When setting up a new colony, bees limit comb building to what is immediately necessary, building a small set of combs sufficient to establish brood and to provide storage space; they add additional combs only as required. Comb-building bees are normally those who have finished doing tasks in the central brood nest, such as feeding brood, capping brood cells and tending the queen, and who have changed to tasks associated with peripheral regions of the hive, such as receiving and storing nectar and pollen, repairing existing comb and building new comb[41]. The majority of these bees are aged between 10 and 20 days; their hypopharyngeal glands are no longer secreting brood food, but the enzymes needed for producing honey, and their wax glands reach a peak of activity. In established colonies, new comb building is triggered only when two conditions arise: when the bees are experiencing a very good nectar flow, coupled with shortage of comb space for nectar storage. Although neither of these variables by itself is an infallible indicator of the need for more storage space, having control of building tied to both the abundance of external nectar flow and the internal shortage of space serves to ensure that additional comb is constructed only when needed[41]. In autumn, as the queen shuts down her egg-laying in preparation for winter, workers start to use old brood nest cells for nectar storage if there is a good flow, instead of initiating new building. Building also ceases if the colony is preparing to swarm. While all these factors have been shown to affect the onset of new comb building, comb maintenance is

controlled solely by the continuation of an abundant nectar flow.

How do the individual bees sense the conditions under which comb needs to be built?
Two hypotheses have been proposed[41]: one suggested that, since the potential comb-builders are also the bees which receive and store the incoming nectar, if storage space is so short that they have to retain the nectar in their honey sac, distention of the sac will serve as a signal. The second hypothesis proposed that the increasing difficulty in finding a space to store the nectar, leading to an increase in the duration of honey sac distention, would act as a signal. As yet, it has not been possible to show which, if either, signal influences the bees' behaviour.

How workers build comb

The comb is a complex piece of architecture[41]. When construction begins at a new site, the bees remove any loose material from the roof surface of the site to provide a firm foundation for the comb. Then, nearly all of the bees present, except for the foragers, hang together in tight chains from this surface, forming a cluster that is maintained at a temperature of 35°C; this appears to be the best temperature for secretion and manipulation of wax[23]. Under these conditions, when the newly-swarmed colony has to build a comb, older bees, up to an age of around 30 days, participate in producing wax. Wax scales are formed on the wax mirrors of the bees after arrival at the nest site, and are then removed by the bee, using the spines that

FIG. 8.16 *Worker bee constructing a wax cell using its mandibles to manipulate the wax.*

form the pollen brush on the basitarsal joint of the hindleg (figs 5.36b and 5.37b). The wax scales are passed forward to the first pair of legs where they are held against the mandibles, mixed with saliva and chewed to increase the plasticity of the wax and to make it more malleable. As the scale is chewed, the wax is forced back into the cavity formed by the concave mandibles. Scales are not always successfully transferred and many fall to the bottom of the hive or nest. Although some are retrieved, they are more often abandoned and either found later and used by other workers during comb rebuilding operations or removed from the hive with waste material[35].

When comb construction begins in a natural cavity nest, individual bees laden with a number of scales climb up to the top of the cluster and deposit their masticated wax in little piles on the ceiling of the cavity. As more bees deposit their wax, the piles eventually merge to form a ridge several millimetres long and two to four millimetres deep[42]. At this stage, the bees begin constructing their regular, back-to-back array of hexagonal cells by first excavating a cavity the width of a worker cell in one side of the ridge and depositing the excavated wax around the edge of the cavity. Next, two cells are excavated on the other side of the ridge, with the wall separating the cells lying opposite the centre of the first cell. As wax is piled up on the rims of these new cells, their walls are gradually elongated; adjacent walls are laid out at an angle of 120 degrees. The bees then shave the wax down on both sides to form a smooth plane and deposit the excess wax on top of the walls so that the process can be repeated again and again until the cell is completed. While this is happening, the neighbouring cells are excavated and their walls built up in the same way. The hexagonal arrangement of the cells in the comb economizes on the use of wax, because the greatest number of cells per unit area can be obtained without any gaps between cells as there would be, for example, with circular cells; square cells would have a greater circumference and thus use more wax per unit volume. By staggering the two sides of the comb, cells can be built back-to-back, thus maximizing the number of cells in a given area[23]. When the adult bees emerge from the brood cells, the wax cappings are reused, and any vacated queen cells are torn down and the wax used elsewhere.

Three types of cell are built on the comb. There are two types of hexagonal cells: smaller ones for the workers, built at the top and in the middle of the comb; and larger cells for drones that are normally built along the lower edge. Both of these types of cell can be used to store honey and pollen, and even water briefly. Queen cells are large, irregular-shaped cones built hanging down from the bottom edge of the comb. These are constructed only when queen rearing is necessary and, even so, only 10–20 are built and subsequently removed when the first new queen has emerged. Cell size for each caste is generally fairly constant but can vary with the race of bee and with the age of the colony. Diameter of worker cells of *A. m. ligustica* and other European races, is 5.2–5.4 mm in diameter but it is 4.8–4.9 mm in the African race, *A. m. adansonii*. Newly-built comb shows remarkable precision: the thickness of the cell wall in *A. mellifera* is 0.072 ± 0.002 mm; the angle between adjacent cell walls is exactly 120 degrees; and each comb is constructed 0.95 cm from its neighbour[42]. There is a little more variation in size in older comb, where the cells are not quite so regular, probably having served for the development of several bees in succession or perhaps having been dragged out of shape by the weight of honey stored within[23,41,43]. The comb can support honey over 20 times its own weight: 1 kg of wax can support 22 kg of honey[23].

How do the workers manage to construct the comb with such precision? The hair plates in the neck are believed to permit the bees to determine the line of gravity (see chapter 4), so that they can build the combs hanging vertically and aligned parallel to each other at a regular distance. If the hair plates are disabled, the bees cannot build comb[44]. However, strangely, a group of workers subjected to zero gravity on a Challenger space shuttle mission was able to construct usable cells[45]. These conflicting results have so far defied explanation. Sense organs at the tip of the antennae (chapter 1) appear to be used to control the thickness and smoothness of the cell walls. As the bee makes planing movements over the cell wall with the curved edges of its mandibles, it simultaneously sweeps the tips of its antennae over the wall. This is thought to enable the bee to detect changing spatial patterns over the wall as it works on it. Removal of the tips of the antennae results in walls of varying thickness[46]. However, the antennae do not appear to be the structures that control the diameter or the angle between adjacent cell walls. Queens must be able to perceive the cell diameter in order to lay a suitable egg in the different sized drone and worker cells. Before laying an egg, the queen inserts her head and forelegs into a cell. After this inspection she is able to lay the appropriate fertilized (worker) or unfertilized (drone) egg. She is believed to use her forelegs as calipers to measure the diameter of the cell, since experimental amputation of the forelegs introduces inaccuracies in the type of egg laid. Removal of the tarsi has very little effect; removal of the tibiae and femora results in some mistakes, while the removal of the trochanter and coxal regions result in an almost total inability to measure the cell[35]. It seems probable that worker bees can use their forelegs in a similar manner to monitor the diameter of cells under construction, but this is impossible to check since they cannot manipulate wax and build cells without their forelegs.

Freshly prepared comb has a characteristic odour which appears to be due to the presence within the wax of a number of compounds produced by the bees, namely octanal, nonanal, decanal, furfural, benzaldehyde and 1-decanol. Empty comb and the volatiles from it appear to affect hoarding behaviour in foraging bees, and it has been suggested that these volatiles should be regarded as pheromones[6]. Evaluation of these volatiles has produced somewhat conflicting results: in some cases they increase hoarding behaviour, in others they decreased it, so their status as pheromones is still not established.

9. Defending the colony:
the sting

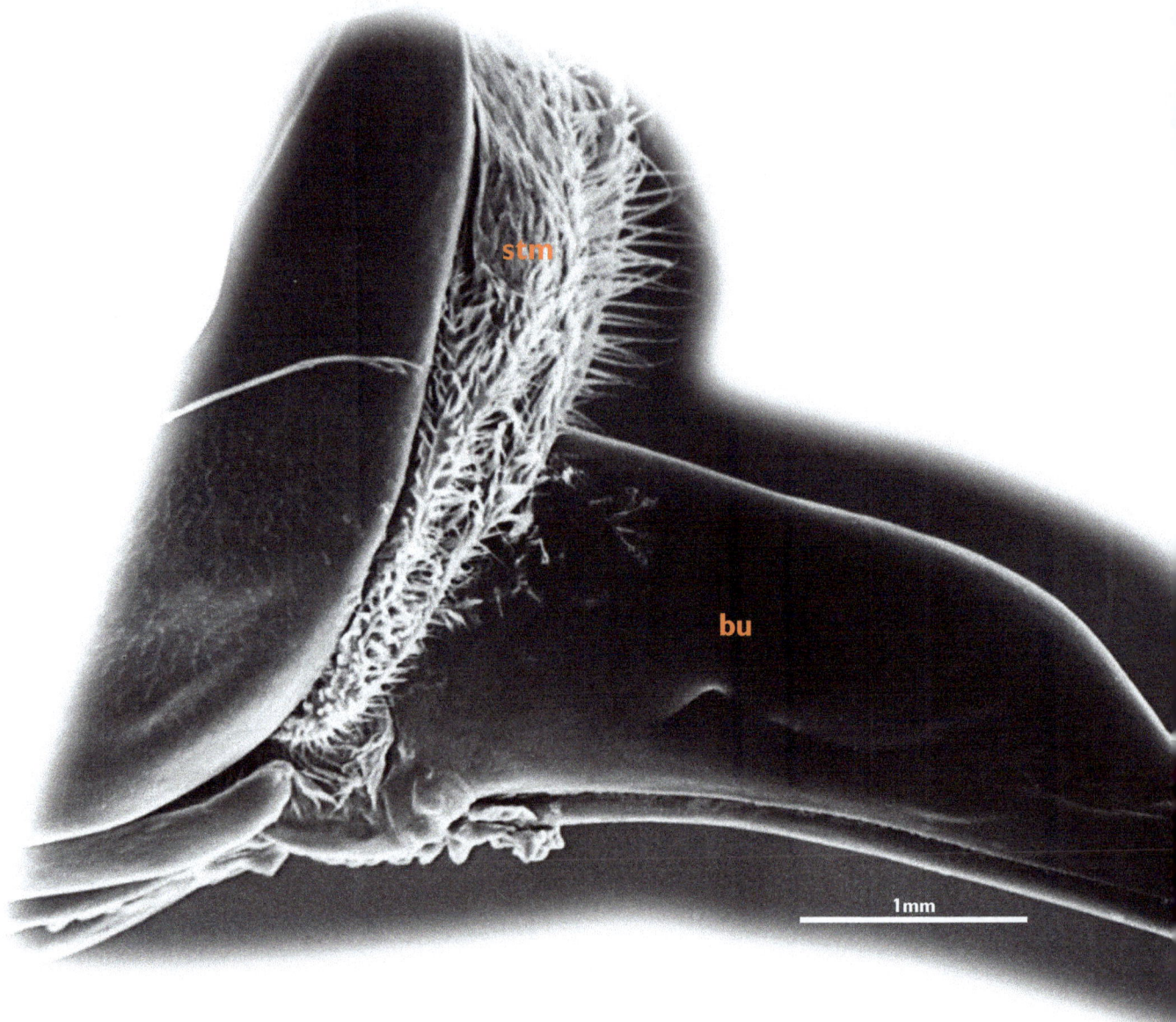

stm

bu

1mm

FIG. 9.1 *The shaft of the worker sting after removal from the sting chamber. The shaft consists of a dorsal stylet (sty) which expands into a bulb (bu) at its base, and two ventral lancets (la) bearing barbs (br) at the tip. Only one lancet is clearly visible in this micrograph. Overlying the bulb is the setaceous membrane (stm) which is covered in hairs. The membrane is exposed when the sting is extruded, allowing the volatile alarm pheromone held among the hairs to disperse into the surroundings.*

The nest or hive can provide other animals with a rich source of food. A wide variety of predators and parasites attack not only the honey stores, but also pollen and wax, and even the larval brood and the adult bees themselves. Man has long been one of the most serious predators but other vertebrates, including bears, hedgehogs, honey badgers, shrews, anteaters, toads and many birds, take honey and brood or catch adults at the nest entrance[1]. Among the invertebrates, ants and wasps do a great deal of damage. Moths are widespread pests, consuming honey, pollen and, in the case of wax moths, wax comb as well. Worker bees themselves may rob the honey stores of another colony.

Honey bees have developed defensive mechanisms in order to survive this predation. Individual bees exhibit a range of defensive behaviours which usually includes secreting alarm pheromones to

FIG. 9.2 *A bee showing the characteristic posture of an alerted guard bee with the forepart of the body raised, the mandibles open and the wings extended.*

alert and attract other bees, and which may culminate in stinging behaviour (fig. 9.1)[2]. If guard bees or other workers recognize a predator at the hive entrance, they adopt a characteristic posture, raising the body with the abdomen sloping upward; often the wings are extended and the mandibles held open (fig. 9.2). These bees may partially evert the sting and emit alarm pheromones, possibly secreting a pheromone from the mandibles as well, to alert and recruit their nestmates. Fanning of the wings may occur to disperse the pheromones. Some recruiting bees run into the hive with their stings everted and their wings spread wide[2]. Bees may initially be alerted by a variety of stimuli, including vibration of the substrate, mechanical stimulation of the bee and the presence of mammalian breath containing carbon dixoide.

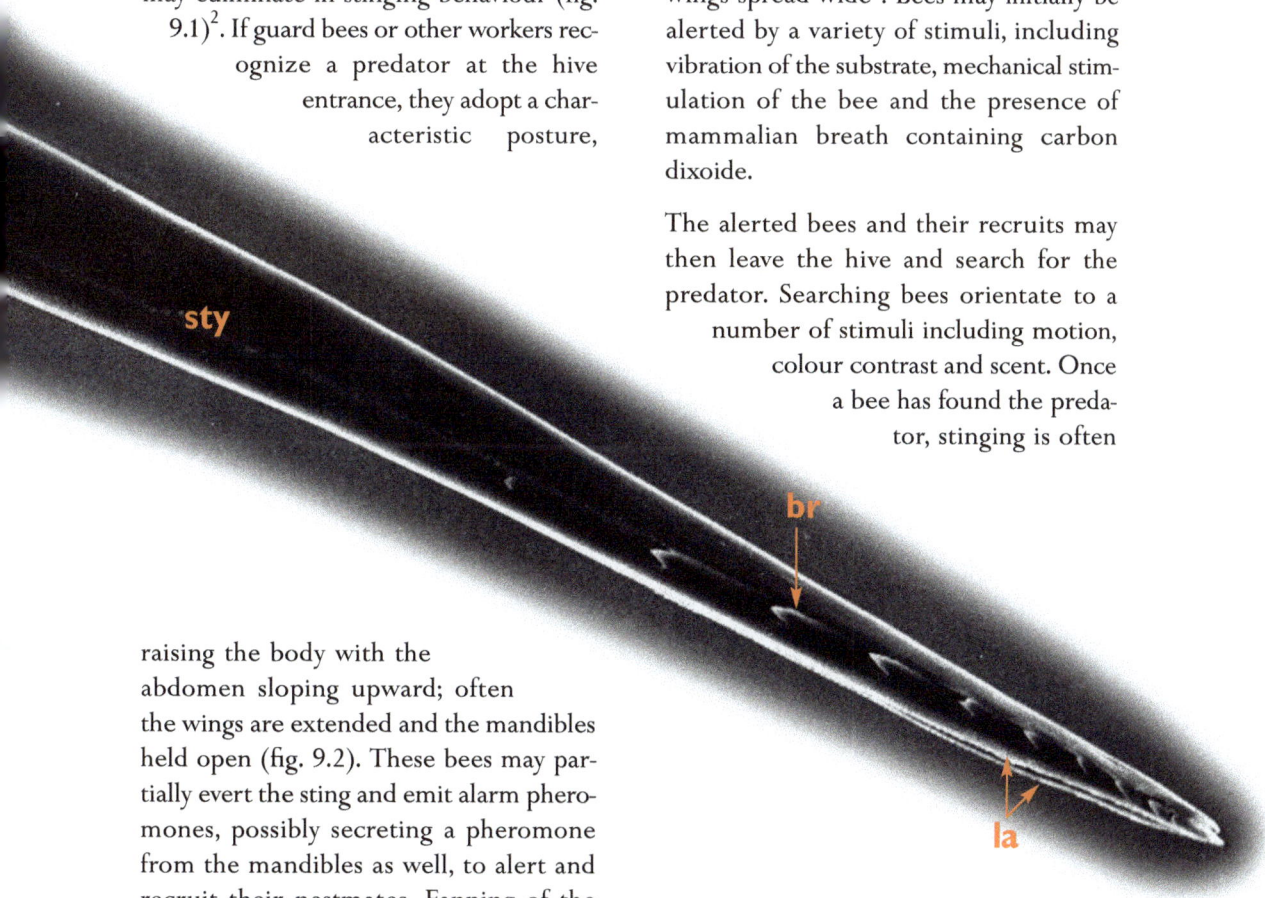

The alerted bees and their recruits may then leave the hive and search for the predator. Searching bees orientate to a number of stimuli including motion, colour contrast and scent. Once a bee has found the predator, stinging is often

sty

br

la

not its first resort, possibly because the sting is torn out as the bee leaves and it later dies. Pre-stinging behaviours can vary with the type of intruder, and can include buzzing the predator, burrowing into it, pulling hairs and biting. If stinging does occur, then alarm pheromones are emitted from the sting and other bees are attracted to the sting site. Stimuli likely to provoke stinging include rapidly moving objects, dark colours, rough textures and animal scents[2]. Stinging other insects is difficult except through the intersegmental membrane, so bees will display threat postures, run at other insects at the hive entrance or grapple with them to try to drive them off. Wasps often evoke 'shimmering' behaviour, where the bees shake from side to side. Robber bees from other colonies frequently show a characteristic flight pattern at the nest entrance, hovering and moving sideways. Guards recognizing this movement and perceiving the odour of the foreign bees attack them fiercely, holding onto them and attempting to sting[1].

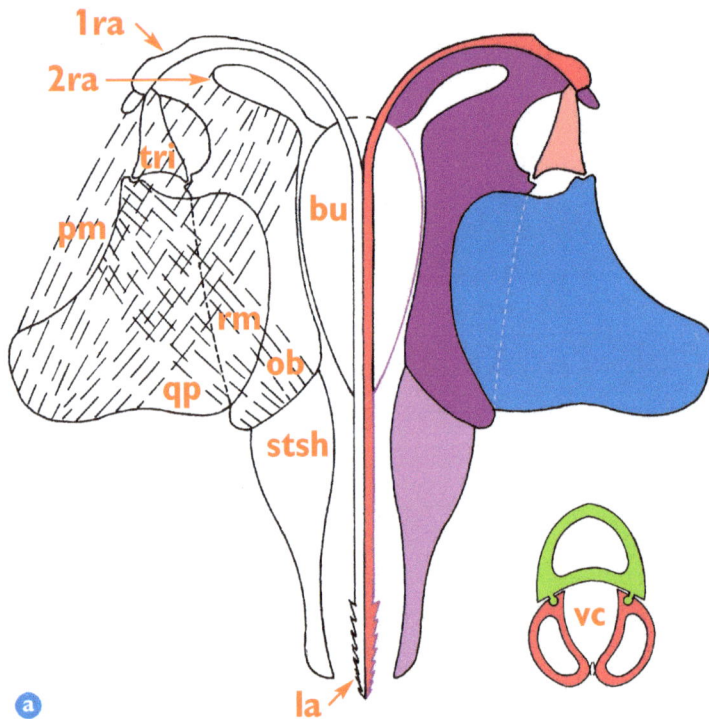

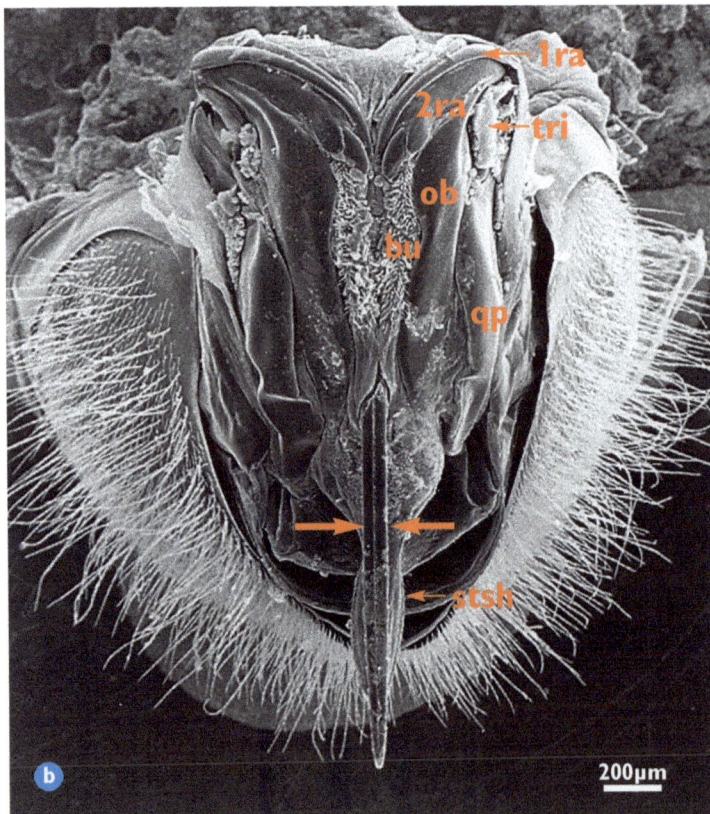

FIG. 9.3 a *The sting shaft and the elements of the basal motor apparatus viewed from the ventral surface. The lancets (la) are continuous with the first rami (1ra). The stylet and the bulb (bu) at its base, are continuous with the second rami (2ra). The three elements of the lever system by which the lancets are moved comprise the oblong plate (ob), the quadrate plate (qp) and the triangular plate (tri). The oblong plate is connected to the second ramus, and the triangular plate to the first ramus. The lower part of each oblong plate is extended into a long flap: together these form the sting sheath (stsh). The protractor (pm) and retractor (rm) muscles of the lancets are shown on the left. The actions of these two muscles are shown in fig. 9.4. Inset: transverse section through the distal region of the sting shaft showing the stylet and the lancets enclosing the venom canal (vc). Note the groove in each lancet that fits over a corresponding ridge on the stylet forming the track mechanism.* **b** *A ventral view of the sting lying in the opened sting chamber. The shaft of the worker is around 2.3 mm and that of the queen 3.0 mm in length[3]. Lancets (arrows); bulb (bu), partially covered by setaceous membrane; first ramus (1ra); second ramus (2ra); oblong plate (ob); extension of the oblong plate to form the sting sheath (stsh); triangular plate (tri); quadrate plate (qp).*

The sting apparatus

The sting (fig. 9.3) is modified from the female genitalia which form the ovipositor (egg-laying organ) in other insects and it is present in both the queen and the worker. It is normally retracted within a large chamber which is enclosed by the tergal and sternal plates of the seventh abdominal segment. The sting consists of a track-like mechanism that allows two barbed **lancets** to be moved back and forth relative to the unpaired **stylet** from which they are suspended, together with the motor apparatus that drives each lancet along the track. In addition, there are mechanisms for protracting the sting from its chamber and for ensuring that the movement of the lancets pumps venom along the **venom canal** between the stylet and lancets, into the integument of the target.

The track-like mechanism will be considered first. The long, tapering shaft of the sting consists of a dorsal stylet (fig. 9.1), barbed at the tip, bearing two ridges on its ventral surface, each of which fits into a groove on the dorsal surface of one of the lancets (fig. 9.3a, inset). Each lancet can slide back and forth on its rail. At the base of the shaft the stylet expands into a **bulb**,

the track for lancet movement continuing along its ventral surface. Beyond this, each track is continued by two chitinous arms which extend in an arch from the central shaft to the laterally situated levers of the **motor apparatus** (fig. 9.4). The arms consist of two elements, the first and second **rami**. On each side of the body, the first ramus is continuous with the lancet of that side and extends to the movable lever on the motor apparatus. The second ramus extends from the base of the stylet bulb to a chitinous plate of the motor apparatus. It also bears a ridge which fits into the groove on the first ramus, thus ensuring that the track on which the lancets move is continuous from the motor apparatus to the tip of the sting.

The basal motor apparatus which moves each lancet consists of a lever system of three chitinous plates, the **oblong**, **quadrate** and **triangular plates**, together with a pair of antagonistic muscles. Both muscles are attached between the quadrate and oblong plates (fig. 9.3). The alternate contraction of these muscles pulls the quadrate plate backwards and forwards relative to the oblong plate, which is attached firmly to the second ramus. The quadrate plate is connected to the triangular plate, and its

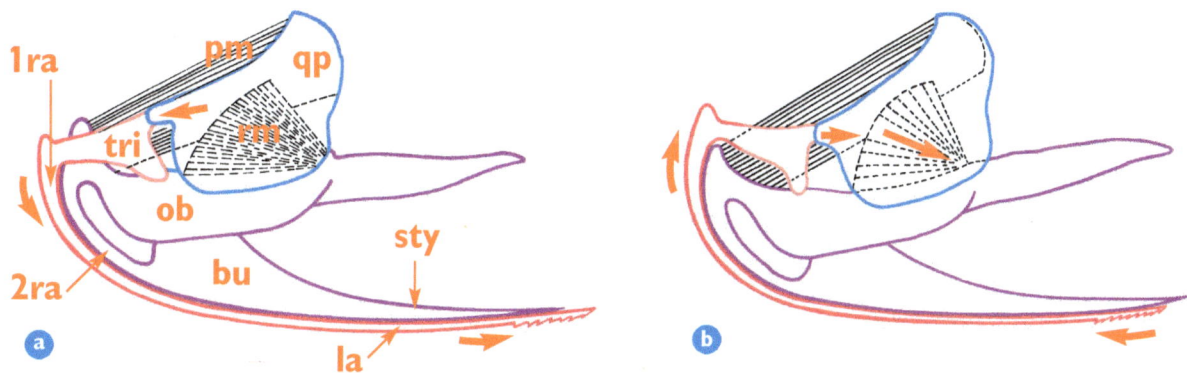

FIG. 9.4 *Side view of one half of the sting and basal motor apparatus to demonstrate the movement of the lancets (arrows). Each lancet (la) is continuous with the first ramus (1ra) which, in turn, connects to the triangular plate (tri). This plate is rocked up and down on its articulation with the oblong plate (ob) by the action of the protractor (pm) and retractor (rm) muscles. In* **a** *the quadrate plate (qp) is pulled anteriorly by the contraction of the protractor muscle. As the quadrate plate is connected to the triangular plate, its forward movement pushes the triangular plate downwards. This motion is transmitted to the first ramus and hence to the lancet so that they slide downwards and posteriorly along the track suspended from the second ramus (2ra), bulb (bu) and stylet (sty). The lancet is protracted beyond the stylet. In* **b** *the retractor muscle pulls the quadrate plate posteriorly, rotating the triangular plate upward, drawing the first ramus with it. This causes the lancet to slide anteriorly and hence to be retracted beneath the stylet. Redrawn from Snodgrass*[4].

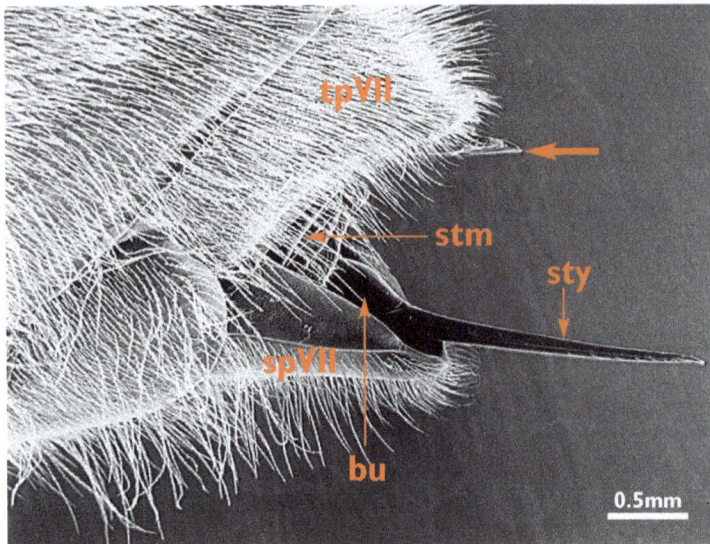

FIG. 9.5 *The emergence of the sting from the sting chamber. The bulb (bu) at the end of the stylet (sty) is visible and the setaceous membrane (stm), which runs above the bulb, can just be seen. The sting shaft emerges from the sting sheath (arrow) as it is turned downwards. The tip of the sheath can just be seen under the tergal plate of segment VII. Tergal plate of segment VII (tpVII); sternal plate of segment VII (spVII).*

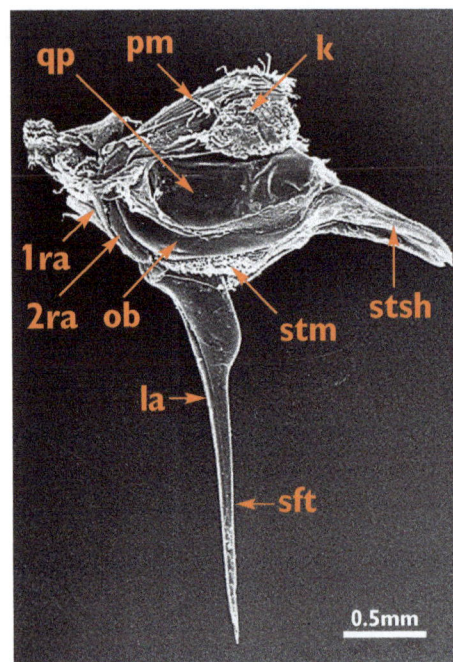

FIG. 9.6 *Left side of a protracted sting removed from the sting chamber showing how the sting shaft (sft) is turned downwards to facilitate entry into the target. The sting emerges from the sting sheath (stsh) formed from the flap-like extensions of the oblong plate (ob). First ramus (1ra); second ramus (2ra); Koschevnikov's gland (k); lancet (la); protractor muscle (pm); quadrate plate (qp); setaceous membrane (stm).*

movement rocks the triangular plate back and forth on its articulation with the oblong plate. Since the triangular plate is also connected to the first ramus, its movement is transmitted to the ramus and hence to the lancet. Figure 9.4 illustrates how the action of the muscles results in the movement of the lancets. When the protractor muscles contract, the quadrate plate is pulled forward. This movement depresses the triangular plate and pushes the ramus downward and posteriorly, so that the lancet continuous with it is protracted beyond the stylet. The retractor muscle pulls the quadrate plate back which then rotates the triangular plate upward, thus pulling the first ramus upward and retracting the lancet. The muscle systems on the two sides of the sting operate alternately so that one lancet is being protracted while the other is being retracted[4].

The stinging response

The stinging response of the bee involves several movements: flexure of the abdomen; protraction of the shaft of the sting from the **sting chamber** (fig. 9.5), accompanied by rotation of the basal motor apparatus; insertion of the **sting shaft**; and then deeper penetration and injection of venom.

The posterior part of the abdomen is flexed so that when the shaft is protracted it is facing downward toward the victim's integument. The abdomen is curved downward when the protractor muscles of the dorsal tergal plates and the retractor muscles of the ventral sternal plates contract (see fig. 6.12). When this happens, the basal motor apparatus of the sting turns posteriorly due to the flexion of the sternal plate of segment VII. When the posterior region of this sternal plate is depressed, the anterior part turns upward and backward into the abdomen against the base of the sting, forcing the basal apparatus to turn posteriorly and downward (fig. 9.6)[5]. Muscles running between the quadrate and triangular plates and a plate in the chamber roof are believed to

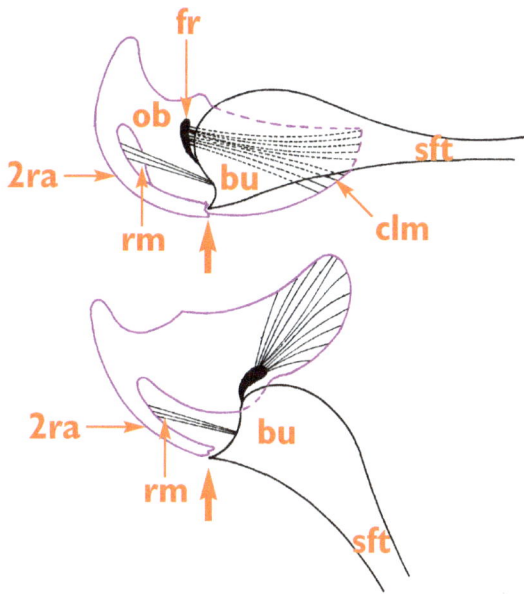

FIG. 9.7 *The shaft (sft) of the sting is deflected downwards by the action of a pair of muscles (clm) arising posteriorly on the inner surface of the oblong plates (ob) (only right oblong plate shown here), and extending forwards to the furcula (fr). The wishbone-shaped furcula sits astride the base of the bulb (bu) and when the deflector muscles on either side contract, the furcula is pulled back and the bulb and its attached shaft (sft) is tipped downwards (lower diagram). When retracted, the shaft is held in position mainly by the action of retractor muscles (rm) extending between the second rami (2ra) and the bulb. Hinge point of the bulb on the right-hand second ramus (arrow). Redrawn from Snodgrass[4].*

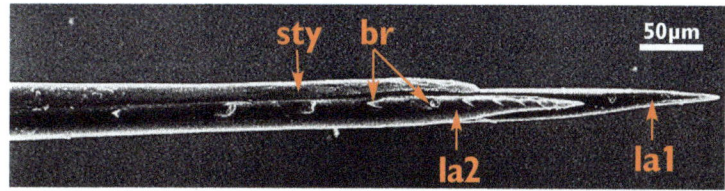

FIG. 9.8 *Tip of the sting shaft showing one lancet (la1) advanced in front of the other (la2). Stylet (sty); barbs on the lancet (br).*

the sting shaft is held in position by a pair of muscles running between the second rami and the base of the bulb (fig. 9.7), and it is enclosed in a sheath formed from two long tapering appendages extending from the oblong plates. These appendages do not turn downward during protraction, leaving the shaft exposed and almost perpendicular to the surface to which the bee is clinging.

The sting is inserted into the skin of the target by a quick, downward movement of the end of the abdomen, sometimes accompanied by a slight twist of the abdomen brought about by the torsion muscles of the anterior abdominal segments. Once inserted, deeper penetration is accomplished by the alternating movements of the lancets. After a lancet is thrust forward into the skin, it is anchored in place by the barbs on its distal end. The other lancet is then driven forward and holds the sting in the new position (fig. 9.8). The rapid alternating movement of the lancets drives the sting deeper into the skin. The protractor and retractor muscles of each lancet alternate as described earlier, but when the lancet is anchored in the target, the action of the retractor muscles cannot retract the lancet. Instead, the tension generated by the muscles depresses the anterior end of the oblong plate, thereby restoring the triangular plate to a position where the protractor muscles can again become effective. The same action allows the stylet to move into the wound with the lancets[4].

The venom

The venom pumped into the wound by the action of the lancets arises in the **venom gland**. The latter consists of a pair of long,

restore the basal apparatus to the retracted position[4].

Protraction of the sting shaft and bulb out of the sting chamber occurs simultaneously with rotation of the basal apparatus, but is brought about by a separate mechanism, namely contraction of the furcula muscles. The **furcula** is a wishbone-shaped plate that sits on top of the bulb at the base of the sting shaft. A muscle runs from the furcula across the base of the bulb to the oblong plate on either side of the body. Contraction of this pair of muscles exerts tension on the furcular plate and, in so doing, rotates the entire shaft of the sting downward (fig. 9.7). With this action, the sting shaft both emerges from the sting chamber and is turned downward to facilitate entry into the target. When retracted, convoluted tubules lying in the posterior part of the abdomen. Secretory cells occur along the length of the tubules, their small ducts opening into a common, chitin-lined duct. The duct opens into a **venom sac** or reservoir, and this, in turn, opens into the cavity of the bulb at the base of the sting (fig. 9.9). The cavity is open below and extends forward as a ventral groove that is continuous with the venom canal. Muscle bands are attached to the venom gland and are reported to move the secretion down into the venom sac[1]. The secretion of the venom glands accumulates in the venom sac and the venom is expelled by the action of valves attached to the lancets. Each lancet has an umbrella-like valve at its base, lying in the bulb (fig. 9.9). As the lancet is protracted, the valve opens, driving the fluid in front of it forward and

causing more venom from the venom sac to be sucked in behind it. At the end of the protraction phase, the valve collapses, allowing the venom to flow past it, and the cycle is repeated as long as the lancets are in action. The venom passes down the canal formed between the stylet and the two lancets and escapes into the wound via a small ventral opening between the two lancets near the tip of the sting.

Bee venom contains a mixture of proteins and peptides, the best known of which are the enzymes phospholipase A_2, hyaluronidase and acid phosphatase, together with the peptides melittin, the mast cell degranulating-peptide (MCD), allergen C, the neurotoxin apamin, and histamine[1,6,7]. This complex mixture of compounds, many of them proteins of high molecular weight, has the potential to elicit a pronounced allergic response. The main components contributing to this effect are: phospholipase A_2, hyaluronidase, acid phosphatase, allergen C and melittin[7]. In vertebrates, melittin is responsible for the rupture of red blood cells and mast cells, large cells containing granules that are found in connective tissue and along the course of some blood vessels. In the presence of melittin, and to an even greater extent MCD, the mast cells release the contents of their granules — notably histamine (much more than that contained in the original injection of venom), together

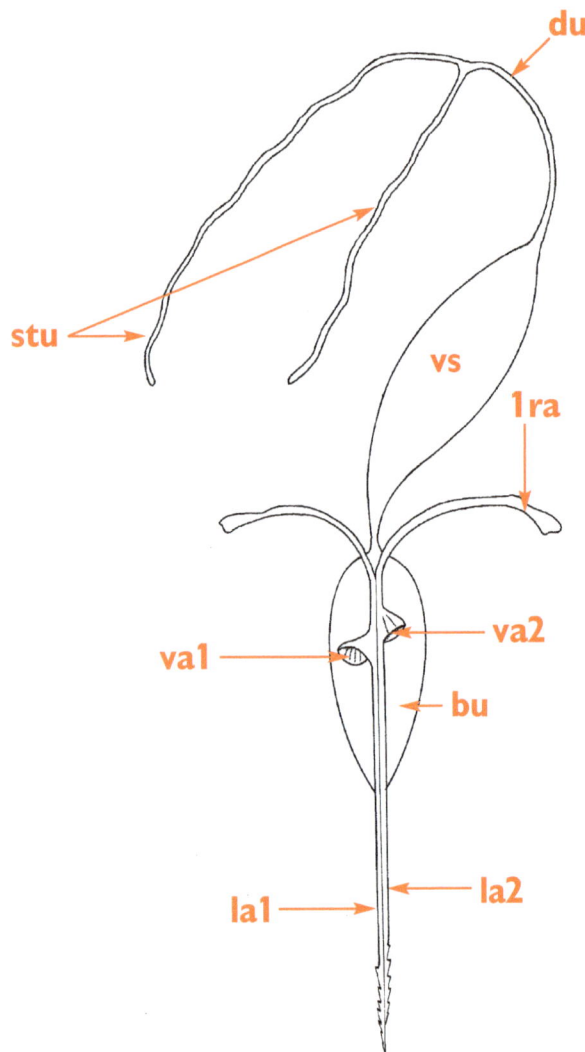

FIG. 9.9 *The venom gland consists of two long secretory tubules (stu) that unite to form a common duct (du) opening into the venom sac (vs). The venom sac opens into the cavity of the bulb (bu) and acts as a reservoir for the venom which is drawn through the bulb and into the venom canal by the action of the lancet valves (va). Lancets (la); first ramus (1ra). The valves arise from the dorsal surface of each lancet. The bulb cavity is open along its ventral surface permitting the valves to lie within the cavity and to be moved within it when the lancets move. As the lancet is moved posteriorly (la1) its valve within the bulb (va1) is extended, sweeping venom ahead of it and into the venom canal. As the lancet is retracted (la2), its valve (va2) collapses allowing fluid to move past it.*

with other chemicals that cause inflammation — into the extracellular fluid. These substances produce dilatation of blood vessels leading to swelling and redness, and serve to attract certain types of white blood cells (predominantly granulocytes) to the area. Histamine also leads to itching and pain. The hyaluronidase helps to digest the surrounding connective tissue, thus opening up passages for the diffusion of other venom components through the host tissue matrix. In humans, the effects of a sting which form an allergic reaction may take place on one or more of three levels: local, systemic (once the chemicals enter the bloodstream and circulate around the body), and anaphylactic[7]. Local reactions comprise an initial reaction followed three or four hours later by more extensive swelling, redness and itching, which may last for two or three days. A systemic reaction usually occurs within a few minutes of a sting and may involve a general rash, wheezing, nausea, abdominal pains and faintness. In extreme cases, where the victim is highly allergic to bee venom, anaphylaxis rapidly follows the onset of a sting and is potentially life-threatening. Wheezing, vomiting and confusion are followed by falling blood pressure and loss of consciousness. Death may result from circulatory collapse and respiratory obstruction, although fortunately such extreme reactions are rare. The fact that bee venom comprises so many compounds may reflect the wide divergence of pests and predators with which the worker bees have to cope. Histamine, while not toxic to vertebrates in the quantities present, forms a significant part of the venom's toxicity to other insects[1].

A second small gland, **Dufour's gland**, opens ventrally into the sting chamber. The function of this gland is unknown although it has been suggested that its secretion lubricates the sting, forms a waxy covering for eggs in queens, or functions to attach eggs to the bottom of cells[1].

The stinging apparatus of the worker has a membranous connection with the walls of the sting chamber. When the sting is embedded in thick skin, such as human skin, both the membranes and the muscles that connect the plates of the basal apparatus to the spiracular plate on each side are ruptured as the bee departs. The entire stinging apparatus including the sting shaft, the venom glands, the muscles and plates of the basal apparatus, the nerves innervating them and the entire terminal abdominal ganglion remain attached to the skin of the victim. The terminal parts of the alimentary canal also are removed. The isolated sting apparatus continues to function as the basal apparatus continues to drive the lancets further into the skin and venom is still secreted. The worker dies after this disruption to the abdomen but, presumably, the insertion of a sting that continues to pump venom into the target long after the bee itself might have been brushed off by the victim more than compensates for the loss of a worker. Obviously this would not be the case for the queen, so her sting apparatus is more firmly attached within the sting chamber and therefore is not lost when it withdraws from the target. The queen only uses her sting on rival queens, of which there may be several, and her venom glands and sac are larger than those of the worker.

Neural control of lancet movement

The protractor and retractor muscles producing the rhythmical, alternating movements of the lancets into the target are innervated from the **terminal abdominal ganglion** of the central nervous system. This ganglion also innervates the furcula muscles that rotate the shaft downward, together with the muscles that hold it in place in its retracted position. The ganglion lies in the anterior region of segment VII of the worker and queen, and is formed by the fused segmental ganglia of segments VIII, IX and X[4]. The ninth segmental nerve of this ganglion innervates each ipsilateral set of four muscles. The individual neurons innervating the protractors and retractors have been identified[8]. Five neurons supply the protractor muscles and six

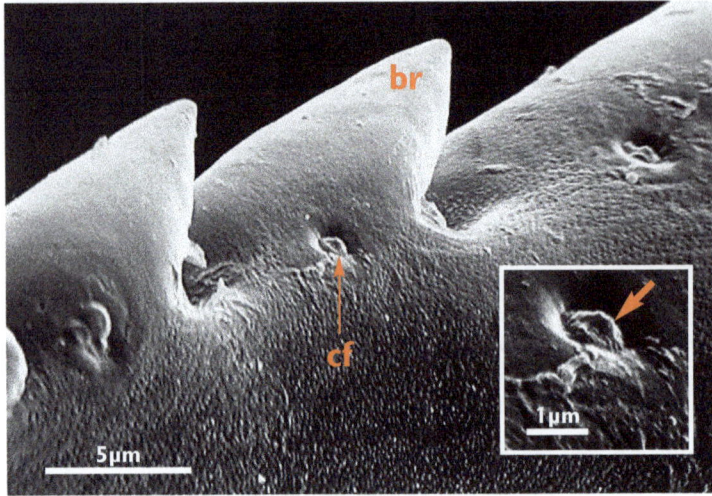

FIG. 9.10 *Campaniform sensillae (cf) situated at the base of each barb (br). These mechanoreceptors are stimulated by deformations of the cuticle. Inset: The campaniform sensilla consists of a dome-shaped area of thin cuticle (arrow) with the dendrite of a single sensory cell inserted into a thickened bar at the top of the dome. Stresses set up in the plane of the surface cuticle produce changes in the shape of the dome which stimulate the sensory cell (see chapter 7 for further details of the campaniform senillae).*

neurons the retractor muscles, their large cell bodies being located posterolaterally in the ganglion near the emergence of the ninth segmental nerve in which their axons run. Recordings of both muscle and nerve activity have shown that the furcula muscles on both sides of the sting chamber are continuously contracted during the stinging response, reflecting their action in protracting the sting. The initial contraction of these muscles is followed by the onset of alternating contractions of the protractor and retractor muscles of the basal motor apparatus. Each muscle is activated in antiphase to the homologous muscle on the other side, thus producing the alternating movements of the lancets[8]. After isolation from the body and rupture of the connectives linking the last abdominal ganglion to the rest of the central nervous system, the motor pattern recorded from the muscles and nerves remains similar to that found in the intact insect. This suggests that the rhythmic movement of the muscles is produced by a central pattern generator (CPG), located in the terminal abdominal ganglion. Central pattern generators are responsible for generating

rhythmic sequences of motor activity in both vertebrates and invertebrates, for example: swimming movements in fish; walking in cats; flight in locusts; and sound production in insects. In the spinal cord, or in the ganglia of insects, there are sets of premotor interneurons, appropriately wired up to each other and to the motor neurons, forming the CPG so that the correct sequence of motor activity is generated as long as the CPG is active.

The role of the sting sense organs

There are a number of sense organs associated with the movable parts of the sting in the bee and in other stinging Hymenoptera[3,9]. In the honey bee, campaniform sensillae are present on the lancets and the stylet in both the queen and the worker. The sensillae are associated with the base of the barbs (fig. 9.10) except at the very tip. These sensillae are mechanoreceptors, sensitive to deformations of the cuticle (for a detailed account of their structure and function, see chapter 7). In this location, they can detect distortion of the barbs as they are forced deeper into the skin and deformation of the entire shaft. Distortion of a barb has been shown to evoke a burst of nerve impulses in the eighth segmental nerve of the terminal abdominal ganglion[8].

Also present are three hair plates comprising 20–30 socketed hairs (fig. 9.11). One is situated on the anterolateral edge of the oblong plate beneath the triangular plate, where the sensillae can detect movement of the triangular plate relative to the oblong plate. A second hair plate is situated on the second ramus, with its sensillae engaging with the first ramus. The sensillae will be deflected by movement of the first ramus relative to the second ramus, hence this hair plate is in a position to detect the forward thrust of the lancet attached to the ramus. A third plate with rather smaller, peg-like sensillae is found on a hinge of the second ramus, near to the bulb. It has been suggested that the sensillae of this plate may also detect the

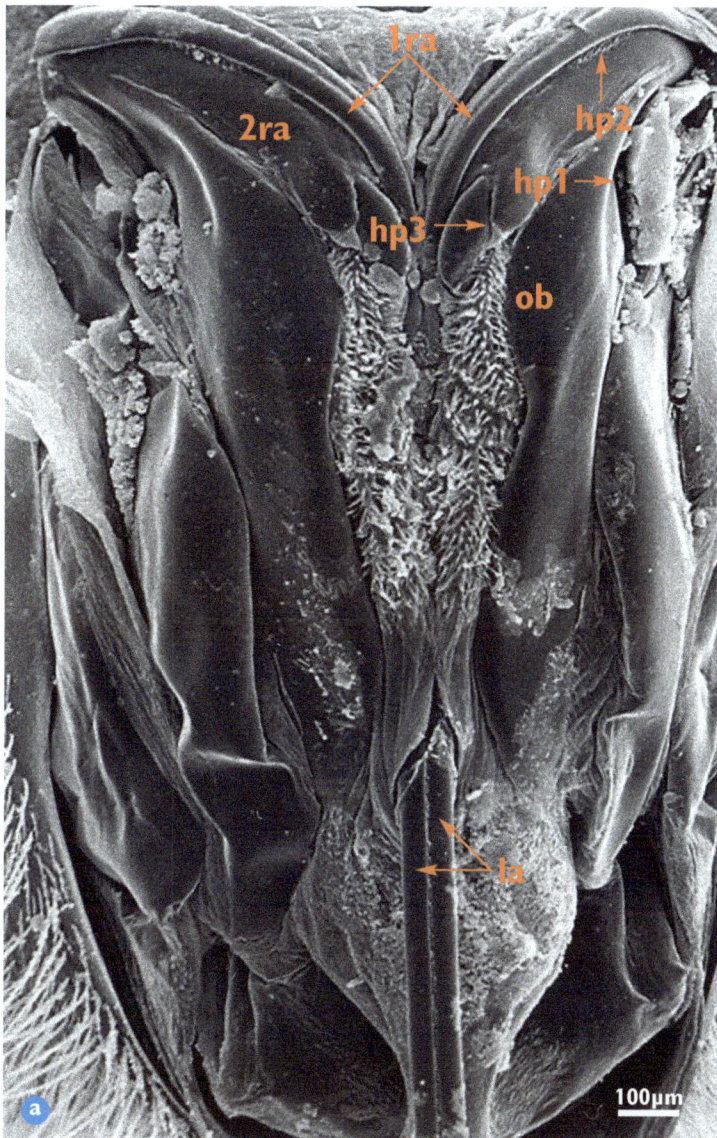

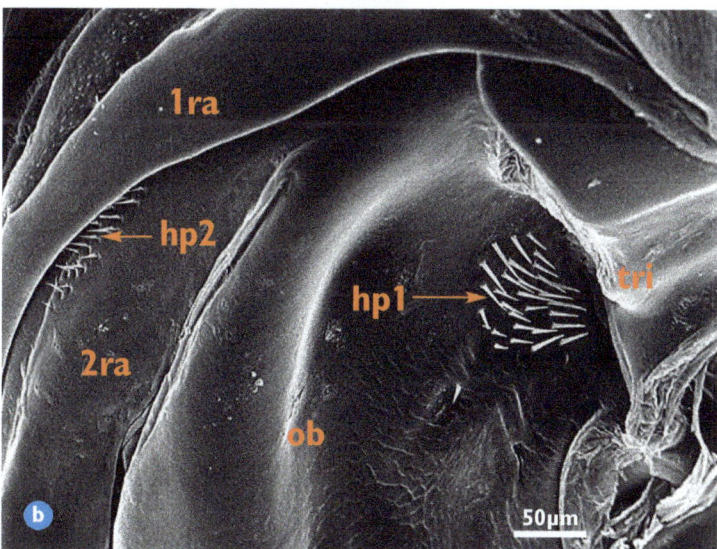

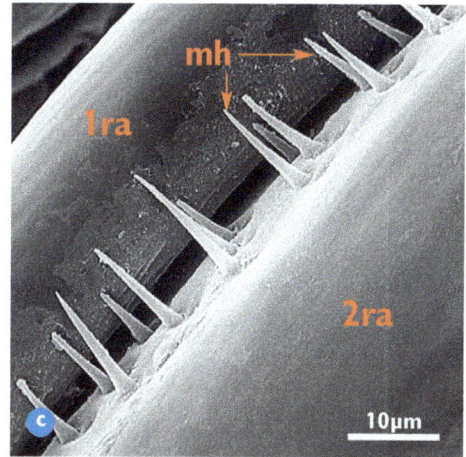

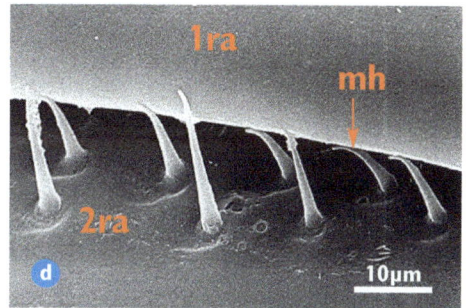

FIG. 9.11 a Three hair plates are associated with each basal apparatus to monitor the movements of the lancets (la). Hair plate 1 (hp1) is situated on the anterior edge of the oblong plate (ob) where the mechanoreceptive sensillae present are stimulated by movement of the triangular plate relative to the oblong plate. Hair plate 2 (hp2) is located on the second ramus (2ra) where its sensillae are stimulated by the movement of the first ramus (1ra). Hair plate 3 (hp3) is found on the second ramus, near to the point where it is hinged to the bulb (see also fig. 9.7 for location of the hinge). The sensillae of this plate may also detect the thrust of the lancets and possibly the extent to which the sting is protracted. b Part of the triangular plate (tri) has been removed to show hair plate 1 (hp1) on the oblong plate (ob). The intact triangular plate deflects the socketed hairs when it moves relative to the oblong plate. Hair plate 2 (hp2) can be seen on the second ramus (2ra), its sensillae lying under the first ramus (1ra). c Some of the mechanoreceptive hairs (mh) comprising hair plate 2 on the second ramus (2ra). d Shows how the sensillae (mh) are bent when the first ramus (1ra) moves relative to the second ramus (2ra).

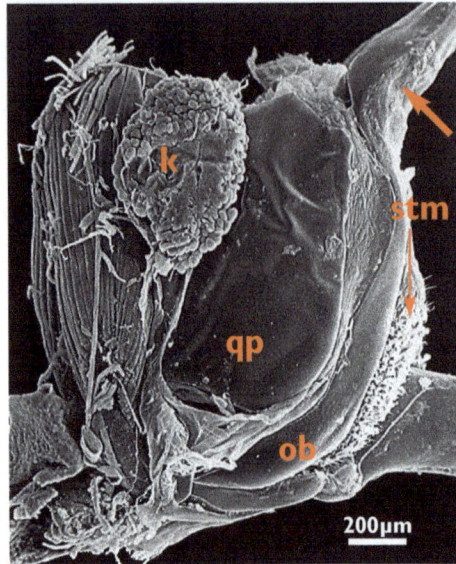

FIG. 9.12 *Outer surface of the left side of the sting apparatus. Koschevnikov's gland (k) lies on the upper part of the quadrate plate (qp) between this plate and the spiracular plate (not shown here). Its secretion flows across the quadrate and oblong (ob) plates and eventually accumulates among the hairs of the setaceous membrane (stm). The proximal region of the sting sheath (arrow) also secretes volatile compounds.*

thrust of the lancets and possibly the extent of protraction of the sting[3].

The relationship between cycle period, duration of individual muscle activity, and the interval between muscle activity in the lancet protractors and retractors on both sides of the body have been measured. It was found that, although each variable increased in proportion to the tension built up in the sting as it penetrated into an object, the linear relationship between the variables was preserved. The preservation of this relationship is believed to enable sufficient strength to be generated to pierce objects of variable resistance, while keeping the forward stroke length of the lancets constant[8]. If the sting apparatus is deprived of the sensory signals from the sting sensillae, the basic stinging response continues but the relationship between these variables is lost. Deprivation of signals from one side results in greater variation in muscle activity on that side. These results suggest that the rhythmic muscle

activities of the stinging response are generated by a CPG consisting of a pair of oscillators, one in each half of the terminal ganglion, which are loosely coupled to one another. However, the activity of each oscillator can apparently be modulated by input from the sensory receptors on its own side. Modulation by the sensory feedback is believed to be necessary to allow the stinging response to function under different conditions.

Alarm pheromones

Pheromones that alert and attract other bees, and which may provoke stinging if additional appropriate stimuli are present, are released when the worker sting is protracted and when the sting apparatus is left buried in an intruder. The mandibular gland also produces a pheromone which has been implicated in alarm behaviour. Most insect pheromones consist of a blend of volatile compounds, and the alarm pheromone is no exception. As many as 40 volatile compounds have been reported in extracts of honey bee stings[10] but it is unlikely that all of these are active components of the alarm pheromone. Some of the compounds reported may have other roles, while others may be precursors or breakdown products, or even contaminants. The components that elicit a behavioural response, the active components, are simple molecules of low molecular weight and high volatility. They disperse rapidly, thus eliciting a rapid response from members of the colony, but do not persist for long unless danger signals continue.

The first component of the alarm pheromone identified was isopentyl acetate[11] (called isoamyl acetate in earlier works), which recent work suggests is produced in **Koschevnikov's gland**[12]. This paired gland also produces other volatile compounds, namely esters and alcohols, which may contribute to the alarm pheromone. Koschevnikov's gland is located on the upper part of the quadrate plate, and consists of a mass of cells lying along an intersegmental membrane that runs between the quadrate plate (fig. 9.12) and a plate in

Table 9.1 Worker bee responses to pheromone compounds.

Alerts bees at the hive entrance	Releases stinging behaviour	Inhibits foraging activity in bees	Inhibits scenting
isopentyl acetate*	isopentyl acetate	isopentyl acetate*	isopentyl acetate*
n-butyl acetate	n-butyl acetate*	n-butyl acetate*	
n-hexyl acetate		n-hexyl acetate	
		n-octyl acetate*	n-octyl acetate*
2-nonyl acetate			2-nonyl acetate*
		n-decyl acetate	
eicosanol acetate			
octanoic acid		octanoic acid*	octanoic acid
1-butanol	1-butanol		1-butanol
1-pentanol	1-pentanol*	1-pentanol	1-pentanol*
		isopentyl alcohol*	
2-heptanol	2-heptanol		
1-octanol	1-octanol	1-octanol*	1-octanol
	2-nonanol		
	9-octadecen-1-ol		
	p-cresol		
			(Z)-11-eicosen-1-ol*
	2-heptanone†	2-heptanone†	2-heptanone*†

*Indicates very effective
† 2-heptanone is secreted by the mandibular glands
Adapted from Free[19]

the roof of the sting chamber, the spiracular plate. Each gland contains large, egg-shaped cells that exude their secretions into narrow ducts opening onto the intersegmental membrane, and from there the collective secretion flows into the sting chamber via the quadrate and oblong plates[12]. In the sting chamber, the secretion accumulates on the setaceous membrane, which connects the lower edges of the two oblong plates, curving dorsally over the base of the sting bulb and extending posteriorly in a hairy lobe (fig. 9.12). The secretion of Koschevnikov's gland is held among the hairs of the setaceous membrane, which is exposed if the sting is protracted and when the sting is torn out, allowing the volatile compounds present to evaporate into the surroundings. Koschevnikov's gland is also present in the queen,

and the secretion of the mated queen is reported to be highly attractive to workers, although the function and chemistry of the gland in the queen is not known[10]. It appears to degenerate in one-year-old mated queens.

A second glandular structure producing alarm pheromones has been reported in the sting. The proximal region of the sting sheath has been shown to possess a secretory activity (fig. 9.12)[13]. The secretions are elaborated by enlarged epithelial cells and reach the surface via pore canals present in the cuticle of the sting sheath. The volatile compounds produced by the sting sheath membranes have been found to elicit alarm responses, including alerting, attraction and stinging from guard bees; indeed, the setaceous membrane and the sting sheaths were shown to be the most

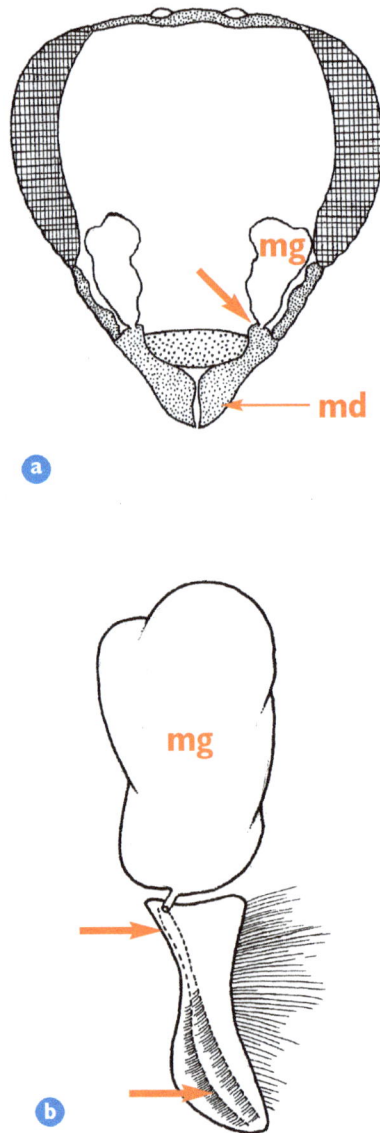

FIG. 9.13 a *Frontal view of the head of a worker bee showing the paired mandibular glands (mg) lying ventrolaterally within the head capsule, immediately above the mandibles (md). The short duct of the gland opens just above the inner base of the mandible (arrow).* **b** *The secretion of the mandibular gland (mg) flows along the groove (arrows) on the inner surface of the mandible (see also chapter 5.1, fig. 5.6, p. 74).*

active regions of the sting in eliciting this behaviour[13]. However, the identity of the volatile components of the sting sheath membranes has not been established.

Many of the volatiles identified in extracts of the sting have been tested for their effectiveness in eliciting various elements of defensive behaviour in bees. Release of alarm pheromone at the hive entrance alerts those bees present, particularly the guard bees. A few of them may recruit other bees, running inside the hive with the sting protracted to expose the setaceous membrane. Some bees are attracted by the alarm pheromone and move towards the hive entrance, joining in the defensive behaviour, although other bees may be repelled by the pheromone. It is not clear what determines these differences in response. In addition, the release of alarm pheromone at the hive entrance is found to decrease foraging activity and inhibit scenting behaviour. When swarming bees settle around their queen, they expose their Nasonov glands and fan their wings to disperse the pheromone emitted, but release of alarm pheromone in the vicinity inhibits scenting and no additional workers join the group. Some examples of the releasing effect of individual alarm components for different behaviours are shown in table 9.1. A single compound is normally found to be less effective than the exposed sting, i.e. than the complete blend of alarm pheromone components present in their normal proportions. Isopentyl acetate is generally accepted as the most effective volatile in eliciting the whole range of defensive behaviours and this, together with the compound (Z)-11-eicosen-1-ol, can replicate the activity of the natural sting pheromone when tested on a moving lure[14]. Although bees are alerted, attracted to and congregate round a stationary lure in which stings are embedded, they need the additional stimulus of movement in order to attack[15]. The main effect of (Z)-11-eicosen-1-ol appears to be to prolong the activity of the more volatile isopentyl acetate as it evaporates[14]. Not all of the compounds tested are equally effective, and not all of them elicit the same behaviours; however, care must be taken in interpreting the results of stimulation with single chemicals since, in the case of most insect pheromones, the blend and the proportions of the chemicals in the blend are the important factors in eliciting behaviour. Recordings from individual interneurons in the ventral nerve cord of the bee confirm the importance of the blend of compounds. No neurons tuned to respond to single compounds were encountered; instead individual neurons were tuned to a range of the volatile compounds found in the alarm pheromone[16].

Why does the bee have so many different components in its alarm pheromone?

It would appear that some of the components are specialized for different functions, for example, *n*-octyl acetate is consistently inhibitory or repellent in its effect, while eicosanol acetate only releases alerting behaviour[17]. The relative amounts of the different components may change with age: isopentyl acetate, for example, is absent in worker bees up to the age of three days. Only small quantities are present from day 4 to day 7; thereafter, the amount present rapidly increases to a maximum at 2–3 weeks of age, when bees are undertaking guard duty. This is followed by a gradual decline to half the maximum quantity as the bees become foragers[11]. This change in the proportion of the compound present also occurs in other components[18]. If some of the components are specialized for different functions, and the proportions of the components are changing even slightly, then the bee may be conveying subtly different messages as its role in the colony changes with age.

The mandibular glands of the worker produce 2-heptanone, a pheromone which is said to have a role in eliciting alarm behaviour, although not all authors agree. The glands lie just above the base of each mandible at the front of the head, and consist of an axial cavity into which the ducts of the surrounding gland cells empty their

secretions. Each gland opens at the inner side of the base of the jaw, adjacent to the inner surface of the mandible, and its secretion flows along the groove which runs across the mandible (fig. 9.13; see also chapter 5.1, fig. 5.6, p. 74). These glands produce 10-hydroxy-2-decanoic acid as well as 2-heptanone. In younger nurse bees, this secretion forms the main component of the food given to the larvae. The size of the gland and the amount of 2-heptanone produced increases with age, reaching a maximum in the foraging bee[18]. Although 2-heptanone will attract guards and other workers when presented at the hive entrance, it is 20–70 times less potent than extracts of the sting[10], and on its own will not promote aggressive behaviour. Some authors report a change from attraction of guards in summer, to repulsion of guards in winter[18], and very high concentrations of the pheromone will result in repulsion of workers at any time of the year[10]. Foraging bees are repelled in the short term from visiting flowers marked with 2-heptanone and it has been suggested that bees utilize this pheromone to mark flowers that do not provide nectar or pollen, or to mark a flower after visiting in order to warn other bees that the nectary has been emptied.

References

Chapter 1. The antennal sense organs

1. FRISCH, K VON (1967) *The dance language and orientation of bees* (translated by Leigh E Chadwick). The Belknap Press of Harvard University Press; Cambridge, MA, USA; 566 pp.

2. MEISAMI, E (1991) Chemoreception. *In* Ladd Prosser, C (ed.) *Neural and integrative animal physiology.* John Wiley and Sons; New York, USA; pp 335–434.

3. MENZEL, R; ERBER, J (1978) Learning and memory in bees. *Scientific American* 239(1): 102–108.

4. KEELE, C K; NEIL, E; JOELS, M (1982) *Sampson Wright's applied physiology.* Oxford Medical Publications; Oxford, UK; 613 pp (13th edition).

5. CARR, W E; GLEESON, R A; TROPIDO-ROTHENTHAL, H G (1990) The role of perireceptor events in chemosensory processes. *Trends in Neurosciences* 13: 212–215.

6. AMOORE, J E; JOHNSTON, J W; RUBIN, M (1964) Odor classification / stereochemical theory of odor. *Scientific American* 210: 44–49.

7. LAURENT, G (1996) Odor images and tunes. *Neuron* 18: 473–476.

8. KAUER, J S (1991) Contributions of topography and parallel processing to odor coding in the vertebrate olfactory pathway. *Trends in Neuroscience* 14: 79–85.

9. ZACHARUK, R Y (1985) Antennae and sensilla. *In* Kerkut, G A; Gilbert, L J (eds) *Comprehensive insect physiology, biochemistry and pharmacology.* Pergamon Press; Oxford, UK; pp 1–69.

10. MASSON, C; MUSTAPARTA, H (1990) Chemical information processing in the olfactory system of insects. *Physiological Reviews* 70: 199–215.

11. SLIFER, E H; SEKHON, S S (1961) Fine structure of the sense organs on the antennal flagellum of the honey bee, *Apis mellifera* Linnaeus. *Journal of Morphology* 109(3): 351–381.

12. ALTNER, H; PRILLINGER, L (1980) Ultrastructure of invertebrate chemo-, thermo- and hygroreceptors and its functional significance. *International Review of Cytology* 61: 69–139.

13. ESSLEN, J; KAISSLING, K E (1976) Zahl und Verteilung antennaler Sensillen bei der Honigbiene (*Apis mellifera* L.). *Zoomorphologie* 83(3): 227–251.

14. GETZ, W M; AKERS, R P (1993) Olfactory response characteristics and tuning structure of placodes in the honey bee, *Apis mellifera* L. *Apidologie* 24(3): 195–217.

15. VARESCHI, E (1971) Duftunterscheidung bei der Honigbiene — Einzelzell-Ableitungen und Verhaltensreaktionen. *Zeitschrift für Vergleichende Physiologie* 75: 143–173.

16. KAISSLING, K E; RENNER, M (1968) Antennale Rezeptoren für queen substance und Sterzelduft bei der Honigbiene. *Zeitschrift für Vergleichende Physiologie* 59: 357–361.

17. PHAM-DELÈGUE, M H; ETIEVANT, P; MASSON, C (1991) Allelochemicals mediating foraging behaviour: the bee–sunflower model. *In* Goodman, L J; Fisher, R C (eds) *The behaviour and physiology of bees.* CAB International on behalf of the Royal Entomological Society and International Bee Research Association; Wallingford, UK; pp 163–184.

18. ARNOLD, G; MASSON, C; BUD-HARUGSA, S (1985) Comparative study of the antennal lobes and their afferent pathway in the worker bee and the drone (*Apis mellifera*). *Cell and Tissue Research* 242: 593–605.

19. FREE, J B (1987) *Pheromones of social bees.* Chapman and Hall; London, UK; 218 pp.

20. YOKOHARI, F; TOMINAGA, Y; TATEDA, H (1982) Antennal hygroreceptors of the honeybee, *Apis mellifera* L. *Cell and Tissue Research* 226(1): 63–73.

21. LACHER, V (1964) Elektrophysiologische Untersuchungen an einzelnen Rezeptoren für Geruch, Kohlendioxyd, Luftfeuchtigkeit und Temperatur auf den Antennen der Arbeitsbiene und der Drohne (*Apis mellifica* L). *Zeitschrift für Vergleichende Physiologie* 48: 587–623.

22. ERBER, J; PRIBBENOW, B; BAUER, A; KLOPPENBURG, P (1993) Antennal reflexes in the honey bee: tools for studying the nervous system. *Apidologie* 24(3): 283–296.

23. THURM, U (1963) Die Beziehungen zwischen mechanischen Reizgrössen und stationären Erregungszuständen bei Borstenfeld Sensillen von Bienen. *Zeitschrift für Vergleichende Physiologie* 46: 351–382.

24. McIVER, S (1985) Mechanoreception. *In* Kerkut, G A; Gilbert, L J (eds) *Comprehensive insect physiology, biochemistry and pharmacology*. Pergamon Press; Oxford, UK; pp 71–132.

25. KEVAN, P G; LANE, M A (1985) Flower petal microtexture is a tactile cue for bees. *Proceedings of the National Academy of Sciences of the USA* 82(14): 4750–4752.

26. SNODGRASS, R E (1956) *Anatomy of the honey bee*. Comstock Publishing Associates and Cornell University Press; Ithaca, NY, USA; 334 pp.

27. MOHL, B (1987) Sense organs and the control of flight. *In* Goldsworthy, G J; Wheeler, C H (eds) *Insect flight*. CRC Press; Boca Raton, FL, USA; pp 76–97.

28. GULLAN, P J; CRANSTON, P S (1994) *The insects*. Chapman and Hall; London, UK; 491 pp.

29. KIRCHNER, W H (1993) Acoustical communication in honeybees. *Apidologie* 24(3): 297–307.

30. KIRCHNER, W H; TOWNE, W F (1994) The sensory basis of the honeybee's dance language. *Scientific American* 270(6): 74–80.

31. KIRCHNER, W H (1994) Hearing in honeybees: the mechanical response of the bee's antenna to near field sound. *Journal of Comparative Physiology, A* 175: 261–265.

32. DRELLER, C; KIRCHNER, W H (1995) The sense of hearing in honey bees. *Bee World* 76(1): 6–17.

33. MICHELSEN, A; KIRCHNER, W H; LINDAUER, M (1986) Sound and vibrational signals in the dance language of the honeybee, *Apis mellifera*. *Behavioral Ecology and Sociobiology* 18: 207–212.

34. BLIGHT, M (1997) *Personal communication*. Rothamsted Experimental Station; Harpenden, Herts, UK.

Chapter 2. Vision in the bee

1. SNODGRASS, R E (1956) *Anatomy of the honey bee*. Comstock Publishing Associates and Cornell University Press; Ithaca, NY, USA; 334 pp.

2. SEIDL, R (1982) *Die Sehfelder und Ommatidien Divergenzwinkel von Abeiterin, Konigen und Drohne der Honigbiene* (Apis mellifera). D thesis; Tech. Hochsch., Darmstadt, Germany.

3. WEHNER, R (1981) Spatial vision in arthropods. *In* Autrum, H (ed.) *Vision in invertebrates. Handbook of sensory physiology. VII/6C*. Springer-Verlag; Berlin, Germany; pp 288–616.

4. VARELA, F G; PORTER, K R (1969) Fine structure of the visual system of the honey bee (*Apis mellifera*). 1. The retina. *Journal of Ultrastructural Research* 29: 236–244.

5. LAND, M F (1981) Optics and vision in invertebrates. *In* Autrum, H (ed.) *Comparative physiology and evolution of vision in invertebrates. Handbook of sensory physiology. VII/6B*. Springer-Verlag; Berlin, Germany; pp 471–592.

6. LAND, M F (1989) Variations in the structure and design of compound eyes. *In* Stavenga, D G; Hardie, R C (eds) *Facets of vision*. Springer-Verlag; Berlin, Germany; pp 90–111.

7. MENZEL, R; WUNDERER, H; STAVENGA, D S (1991) Functional morphology of the divided compound eye of the honeybee drone (*Apis mellifera*). *Tissue and Cell Research* 23(4): 325–335.

8. AUTRUM, H; STOECKER, M (1950) Die Verschmelzungsfrequenzen des Bienen-auges. *Zeitschrift für Naturforschung* 56: 38–43.

9. CHITTKA, L; MENZEL, R (1992) The evolutionary adaptation of flower colours and the insect pollinator's colour vision. *Journal of Comparative Physiology, A* 171(2): 171–181.

10. FRISCH, K VON (1967) *The dance language and orientation of bees* (translated by Leigh E Chadwick). The Belknap Press of Harvard University Press; Cambridge, MA, USA; 566 pp.

11. GOLDSTEIN, E B (1989) *Sensation and perception.* Wadsworth Publishing Co; Belmont, CA, USA; 590 pp.

12. ZEKI, S (1993) *A vision of the brain.* Blackwell Scientific Publications; Oxford, UK; 366 pp.

13. WALD, G (1964) The receptors of human colour vision. *Science* 162: 230–239.

14. MENZEL, R; BLAKERS, M (1976) Colour receptors in the bee eye — morphology and spectral sensitivity. *Journal of Comparative Physiology, A* 108: 11–33.

15. BARTH, F G (1991) *Insects and flowers: the biology of a partnership.* Princeton University Press; Princeton, NJ, USA; 408 pp.

16. DAUMER, K (1956) Reizmetrische Untersuchungen des Farbensehens der Bienen. *Zeitschrift für Vergleichende Physiologie* 38: 413–478.

17. HELVERSEN, O VON (1972) The relationship between difference in stimuli and choice frequency in learning experiments with the honeybee. *In* Wehner, R (ed.) *Information processing in the visual systems of arthropods.* Springer-Verlag; Berlin, Germany; pp 323–345.

18. BACKHAUS, W (1991) Color opponent coding in the visual system of the honeybee. *Vision research* 31(7/8): 1381–1397.

19. CHITTKA, L; BEIER, W; HERTEL, H; STEINMANN, E; MENZEL, R (1992) Opponent colour coding is a universal strategy to evaluate the photoreceptor inputs in Hymenoptera. *Journal of Comparative Physiology, A* 170(5): 545–563.

20. SWIHART, S L; GORDON, W C (1971) Red photoreceptor in butterflies. *Nature, London* 231: 126–127.

21. GOULD, J L; GOULD, C G (1988) *The honey bee.* Scientific American Library; New York, USA; 239 pp.

22. WEHNER, R (1989) The hymenopteran skylight compass: matched filtering and parallel coding. *Journal of Experimental Biology* 146: 63–85.

23. FRISCH, K VON (1949) Die Polarisation das Himmelsichts als orientierender Faktor bei den Tanzen der Bienen. *Experentia* 5: 142–148.

24. SCHINZ, R H (1975) Structural specialization in the dorsal retina of the bee, *Apis mellifera. Cell and Tissue Research* 162(1): 23–34.

25. WEHNER, R (1989) Neurobiology of polarisation vision. *Trends in Neurosciences* 12: 353–359.

26. LABHART, T (1980) Specialized photoreceptors at the dorsal rim of the honeybee's compound eye: polarization and angular sensitivity. *Journal of Comparative Physiology, A* 141: 19–30.

27. ROSSEL, S; WEHNER, R (1984) How bees analyse the polarization pattern in the sky. Experiments and model. *Journal of Comparative Physiology, A* 154(5): 607–615.

28. LABHART, T (1988) Polarisation — opponent interneurons in the insect visual system. *Nature, London* 331: 435–437.

29. LINDAUER, M (1971) *Communication among social bees.* Harvard University Press; Cambridge, MA, USA; 161 pp (3rd edition).

30. DYER, F C (1987) Memory and sun compensation by honey bees. *Journal of Comparative Physiology, A* 160: 621–633.

31. CARTWRIGHT, B A; COLLETT, T S (1979) How honeybees know their distance from a near-by visual landmark. *Journal of Experimental Biology* 82: 367–372.

32. SRINIVASAN, M V; LEHRER, M; ZHANG, S W; HORRIDGE, G A (1989) How honeybees measure their distance from objects of unknown size. *Journal of Comparative Physiology, A* 165: 605–613.

33. LEHRER, M (1991) Locomotion does more than bring the bee to new places. *In* Goodman, L J; Fisher, R C J (eds) *The behaviour and physiology of bees*. CAB International on behalf of the Royal Entomological Society of London and the International Bee Research Assocation; Wallingford, UK; pp 185–202.

34. HORRIDGE, G A; ZHANG, S W; LEHRER, M (1992) Bees can combine range and visual angle to estimate absolute size. *Philosophical Transactions of the Royal Society, B* 337(1279): 49–57.

35. MENZEL, R; CHITTKA, L; EICH-MULLER, S; GEIGER, K; PEITSCH, D; KNOLL, P (1990) Dominance of celestial cues over landmarks disproves map-like orientation in honey bees. *Zeitschrift für Naturforschung, C* 45(6): 723–726.

36. KIRSCHFELD, K (1976) The resolution of lens and compound eyes. *In* Zettler, F; Weiler, R (eds) *Neural principles in vision*. Springer-Verlag; Berlin, Germany; pp 354–370.

37. DARTNALL, H J; BOWMAKER, J K; MOLLEN, J D (1983) Human visual pigment: microspectrophotometric measurements from the eyes of seven persons. *Proceedings of the Royal Society of London, B* 220: 115–130.

38. ROSSEL, S (1989) Polarization sensitivity in compound eyes. *In* Stavanga, D G; Hardie, R C (eds) *Facets of vision*. Springer-Verlag; Berlin, Germany; pp 298–316.

39. SOMMER, E (1979) *Untersuchungen zur topographischen Anatomie der Retina und zur Sehfeldtopologie im Auge der Honigbiene*, Apis mellifera *(Hymenoptera)*. Dissertation; University of Zurich, Switzerland.

40. SRINIVASAN, M V (1994) Pattern recognition in the honeybee: recent progress. *Journal of Insect Physiology* 40(3): 183–194.

41. HERTZ, M (1933) Über figurale Intensitäten und Qualitäten in der optischen Wahrnehmung der Biene. *Biologisches Zentralblatt* 53: 10–40.

42. GOULD, J L (1986) Pattern learning by honey bees. *Animal Behaviour* 34: 990–997.

43. ANDERSON, A M (1977) A model for landmark learning in the honey-bee. *Journal of Comparative Physiology, A* 114: 335–355.

44. COLLET, T S; CARTWRIGHT, B A (1983) Eidetic images in insects: their role in navigation. *Trends in Neurosciences* 6: 101–105.

45. HATEREN, J H VAN; SRINIVASAN, M V; WAIT, P B (1990) Pattern recognition in bees; orientation discrimination. *Journal of Comparative Physiology, A* 167(5): 649–654.

46. HUBEL, D H; WIESEL, T N (1968) Receptive fields and functional architecture of monkey striate cortex. *Journal of Physiology* 195: 215–243.

47. O'CARROLL, D (1993) Feature detecting neurons in dragonflies. *Nature, London* 362: 541–543.

Chapter 3. The dorsal ocelli

1. GOODMAN, L J (1981) Organisation and physiology of the insect dorsal ocellar system. *In* Autrum, H (ed.) *Handbook of sensory physiology. Comparative physiology and evolution of vision in invertebrates. Volume VII/6C.* Springer-Verlag; Berlin, Germany; pp 201–286.

2. SCHRICKER, B (1965) Die Orientierung der Honigbiene in der Dämmerung, zugleich ein Beitrag zur Frage der Ocellenfunktion bei Bienen. *Zeitschrift für Vergleichende Physiologie* 49: 420–458.

3. GOODMAN, L J (1970) The structure and function of the insect dorsal ocellus. *In* Treherne, J (ed.) *Advances in insect physiology* 7; pp 97–195.

4. KALMUS, H (1945) Correlations between flight and vision, and particularly between wings and ocelli in insects. *Proceedings of the Royal Entomological Society of London, A* 20: 84–96.

5. HOMANN, H (1924) Zum Problem der Ocellenfunktion bei den Insekten. *Zeitschrift für Vergleichende Physiologie* 1: 541–578.

6. WILSON, M (1978) The functional organisation of locust ocelli. *Journal of Comparative Physiology* 124: 297–316.

7. PAN, K C; GOODMAN, L J (1977) Ocellar projections within the central nervous system of the worker honey bee, *Apis mellifera*. *Cell and Tissue Research* 176(4): 505–527.

8. IBBOTSON, M R; GOODMAN, L J (1990) Response characteristics of four wide-field, motion-sensitive descending interneurons in *Apis mellifera*. *Journal of Experimental Biology* 148: 255–279.

9. BIDWELL, N J; GOODMAN, L J (1993) Possible functions of a population of descending neurons in the honeybee's visuo-motor pathway. *Apidologie* 24(3): 333–354.

10. ROWELL, C H F; REICHERT, H (1986) Three descending interneurons reporting deviation from course in the locust. II Physiology. *Journal of Comparative Physiology* 158: 775–794.

11. BIDWELL, N J (1992) *Response characteristics of motion-sensitive descending interneurons in the worker honey bee (*Apis mellifera*). PhD thesis; University of London, UK; 338 pp.

12. KASTBERGER, G (1990) The ocelli control the flight course in honeybees. *Physiological Entomology* 15(3): 337–346.

Chapter 4. The bee's response to gravity

1. MARLER, P R; HAMILTON, W J (1966) *Mechanisms of animal behaviour*. John Wiley and Sons; New York, USA; 771 pp.

2. ECKERT, R; RANDALL, D; AUGUSTINE, G (1988) *Animal physiology, mechanisms and adaptations*. W H Freeman and Co; New York, USA; 683 pp.

3. HORN, E (1985) Gravity. *In* Kerkut, G A; Gilbert, L J (eds) *Comparative insect physiology, biochemistry and pharmacology*. Pergamon Press; Oxford, UK; pp 557–576.

4. LINDAUER, M; NEDEL, J O (1959) Ein Schweresinnesorgan der Honigbiene. *Zeitschrift für Vergleichende Physiologie* 42(4): 334–364.

5. LINDAUER, M (1971) *Communication among social bees*. Harvard University Press; Cambridge, MA, USA; 161 pp (3rd edition).

6. THURM, U (1963) Die Beziehungen zwischen mechanischen Reizgrössen und stationären Erregungszuständen bei Borstenfeld-Sensillen von Bienen. *Zeitschrift für Vergleichende Physiologie* 46(4): 351–382.

7. MARKL, H (1962) Borstenfelder an den Gelenken als Schweresinnesorgane bei Ameisen und anderen Hymenopteran. *Zeitschrift für Vergleichende Physiologie* 45: 475–569.

8. HORN, E (1975) Mechanisms of gravity processing by leg and abdominal gravity receptors in bees. *Journal of Insect Physiology* 21(3): 673–679.

9. VOWLES, D M (1954) The orientation of ants. II. Orientation to light, gravity and polarised light. *Journal of Experimental Biology* 31: 356–375.

10. JANDER, R (1963) Insect orientation. *Annual Review of Entomology* 8: 95–114.

11. FRISCH, K VON (1967) *The dance language and orientation of bees* (translated by Leigh E Chadwick). The Belknap Press of Harvard University Press; Cambridge, MA, USA; 566 pp.

12. GOULD, J L; GOULD, C G (1988) *The honey bee*. Scientific American Library; W H Freeman and Co; New York, USA; 239 pp.

13. KELLY, J P (1981) Vestibular system. *In* Kandel, E R; Schwartz, J H (eds) *Principles of neural science*. Edward Arnold; London, UK; 731 pp.

Chapter 5. Feeding

1. SEELEY, T D (1985) *Honey bee ecology: a study of adaptation in social life.* Princeton University Press; Princeton, NJ, USA; 201 pp.

2. SNODGRASS, R E (1956) *Anatomy of the honey bee.* Comstock Publishing Associates and Cornell University Press; Ithaca, NY, USA; 334 pp.

3. HADLEY, N T (1986) The arthropod cuticle. *Scientific American* 265: 98–106.

4. WHITEHEAD, A T; LARSEN, J R (1976) Ultrastructure of the contact chemoreceptors of *Apis mellifera* L. (Hymenoptera: Apidae). *International Journal of Insect Morphology and Embryology* 5(4/5): 301–315.

5. CHAPMAN, R F (1995) Mechanisms of food handling by chewing insects. *In* Chapman, R F; Boer, G de (eds) *Regulatory mechanisms in insect feeding.* Chapman and Hall; New York, USA; pp 3–31.

6. MEYER, W (1956) Propolis bees and their activities. *Bee World* 37(2): 25–36.

7. WINSTON, M (1987) *The biology of the honey bee.* Harvard University Press; Cambridge, MA, USA; 281 pp.

8. SEELEY, T D (1995) *The wisdom of the hive: the social physiology of honey bee colonies.* Harvard University Press; Cambridge, MA, USA; 295 pp.

9. RIBEIRO, J M C (1995) Insect saliva: function, biochemistry and physiology. *In* Chapman, R F; Boer, G de (eds) *Regulatory mechanisms in insect feeding.* Chapman and Hall; New York, USA; pp 72–97.

10. RUTTNER, F; TASSENCOURT, L; LOUVEAUX, J (1978) Biometrical–statistical analysis of the geographic variability of *Apis mellifera* L. I. Material and methods. *Apidologie* 9(4): 363–381.

11. DADE, H A (1962) *Anatomy and dissection of the honeybee.* Bee Research Association; London, UK; 158 pp.

12. KINGSOLVER, J G; DANIEL, T L (1995) Mechanics of food handling by fluid-feeding insects. *In* Chapman, R F; Boer, G de (eds) *Regulatory mechanisms in insect feeding.* Chapman and Hall; New York, USA; pp 32–73.

13. MEISAMI, E (1991) Chemoreception. *In* Ladd Prosser, C (ed.) *Neural and integrative animal physiology.* John Wiley; New York, USA; pp 317–434.

14. BERNAYS, E A; CHAPMAN, R F (1994) *Host–plant selection by phytophagous insects.* Chapman and Hall; New York, USA; 312 pp.

15. FRINGS, H; FRINGS, N (1949) The loci of contact chemoreceptors in insects. A review with new evidence. *American Midland Naturalist* 41: 602–658.

16. DOSTAL, B (1958) Reichfähigkeit und Zahl der Reichsinneselemente bei der Honigbiene. *Zeitschrift für Vergleichende Physiologie* 41(2): 179–203.

17. SLIFER, E H; SEKHON, S S (1961) Fine structure of the sense organs on the antennal flagellum of the honey bee, *Apis mellifera* Linnaeus. *Journal of Morphology* 109(3): 351–381.

18. ERICKSON, E H; CARLSON, S D; GARMENT, M B (1986) *A scanning electron microscope atlas of the honey bee.* Iowa State University Press; Ames, IA, USA; 292 pp.

19. GALIC, M (1971) Die Sinnesorgane an der Glossa, dem Epipharynx und dem Hypopharynx der Arbeiterin von *Apis mellifera* L. (Insecta, Hymenoptera). *Zeitschrift für Morphologie der Tiere* No. 3: 201–228.

20. FRISCH, K VON (1967) *The dance language and orientation of bees* (translated by Leigh E Chadwick). The Belknap Press of Harvard University Press; Cambridge, MA, USA; 566 pp.

21. KUNZE, G (1933) Einige Versuche über den Antenngeschmackssinn der Honigbiene. *Zoologische Jahrbücher Abteilung für Allgemeine Zoologie und Physiologie der Tiere* 52: 465–512.

22. CRANE, E (ed.) (1975) *Honey: a comprehensive survey*. Heinemann in co-operation with Bee Research Association; London, UK; 608 pp.

23. WHITEHEAD, A T; LARSEN, J R (1976) Electrophysiological responses of galeal contact chemoreceptors of *Apis mellifera* to selected sugars and electrolytes. *Journal of Insect Physiology* 22(12): 1609–1616.

24. BEETSMA, J; SCHOONHOVEN, L M (1960) Some chemosensory aspects of the social relations between the queen and the worker in the honeybee (*Apis mellifera* L). *Proceedings Koninklijke Nederlanse Akademie·van Wetenschappen* 69: 645–647.

25. LINDAUER, M (1971) *Communication among social bees*. Harvard University Press; Cambridge, MA, USA; 161 pp (3rd edition).

26. STOFFOLANO, J G (1995) Regulation of a meal in Diptera, Lepidoptera and Hymenoptera. *In* Chapman, R F; Boer, G de (eds) *Regulatory mechanisms in insect feeding*. Chapman and Hall; New York, USA; pp 210–247.

27. CRANE, E (1990) *Bees and beekeeping: science practice and world resources*. Heinemann Newnes; Oxford, UK; 614 pp.

28. BARTH, F G (1991) *Insects and flowers. The biology of a relationship*. Princeton University Press; Princeton, NJ, USA; 408 pp.

Chapter 6. Respiration

1. SNODGRASS, R E (1956) *Anatomy of the honey bee*. Comstock Publishing Associates and Cornell University Press; Ithaca, NY, USA; 334 pp.

2. WIGGLESWORTH, V B; LEE, W M (1982) The supply of oxygen to the flight muscles of insects: a theory of tracheole physiology. *Tissue and Cell* 14: 501–518.

3. WEIS-FOGH, T (1964) Diffusion in insect wing muscle, the most active tissue known. *Journal of Experimental Biology* 41: 229–256.

4. CASEY, T M (1989) Oxygen consumption during flight. *In* Goldsworthy, G J; Wheeler, C H (eds) *Insect flight*. CRC Press; Boca Raton, FL, USA; pp 258–272.

5. KESTLER, P (1984) Respiration and respiratory water loss. *In* Hoffmann, K H (ed.) *Environmental physiology and biochemistry of insects*. Springer-Verlag; Berlin, Germany; pp 137–183.

6. MILLER, P L (1981) Ventilation in active and in inactive insects. *In* Herreid, C F; Fourtner, C R (eds) *Locomotion and energetics in arthropods*. Plenum Press; New York, USA; pp 367–390.

7. ALLEN, M D (1959) Respiration rates of worker honeybees of different ages and at different temperatures. *Journal of Experimental Biology* 36(1): 92–101.

8. ROTHE, U; NACHTIGALL, W (1989) Flight of the honeybee. IV. Respiratory quotients and metabolic rates during sitting, walking and flying. *Journal of Comparative Physiology, B* 158: 739–749.

9. LIGHTON, J R B; LOVEGROVE, B G (1990) A temperature-induced switch from diffusive to convective ventilation in the honey bee. *Journal of Experimental Biology* 154: 509–516.

10. BAILEY, L (1954) The respiratory currents in the tracheal system of the adult honey-bee. *Journal of Experimental Biology* 31(4): 589–593.

11. KAARS, C (1981) Insects — spiracle control. *In* Herreid, C F; Fourtner, R (eds) *Locomotion and energetics in arthropods*. Plenum Press; New York, USA; pp 337–366.

12. BAILEY, L; BALL, B V (1991) *Honey bee pathology*. Harcourt Brace Jovanovich; Sidcup, Kent, UK; 193 pp (2nd edition).

Chapter 7. Flight

1. KUKALOVÁ-PECK, J (1983) Origin of the insect wing and wing articulation from the arthropod leg. *Canadian Journal of Zoology* 61: 1618–1668.

2. DICKINSON, M H; HANNAFORD, S; PALKA, J (1997) The evolution of insect wings and their sensory apparatus. *Brain, Behaviour and Evolution* 50: 13–24.

3. BRODSKY, A K (1994) *The evolution of insect flight*. Oxford University Press; Oxford, UK.

4. KINGSOLVER, J G; KOEHL, M A R (1985) Aerodynamics, thermoregulation and the evolution of insect wings: differential scaling and evolutionary change. *Evolution* 39(3): 488–504.

5. WOOTTON, R J (1986) The origin of insect flight: where are we now? *Antenna* 10: 82–86.

6. RUFFIEUX, L; ELOUARD, J; SARTORI, M (1998) Flightlessness in mayflies and its relevance to hypotheses on the origin of flight. *Proceedings of the Royal Society, B* 265: 2135–2140.

7. COELHO, J R; HOAGLAND, J (1995) Load-lifting capacities of three species of yellowjackets (*Vespula*) foraging on honeybee corpses. *Functional Ecology* 9(2): 171–174.

8. DUDLEY, R (2000) *The biomechanics of insect flight: form, function and evolution*. Princeton University Press; Princeton NJ, USA.

9. BARNARD, R H; PHILPOTT, D R (1995) *Aircraft flight*. Longman; Harlow, UK.

10. ELLINGTON, C P (1999) The novel aerodynamics of insect flight: applications to micro-air vehicles. *Journal of Experimental Biology* 202: 3439–3448.

11. DUDLEY, R; ELLINGTON, C P (1990) Mechanics of forward flight in bumble bees. *Journal of Experimental Biology* 148: 19–88.

12. HEPBURN, H R; RADLOFF, S E; FUCHS, S (1999) Flight machinery dimensions of honeybees *Apis mellifera*. *Journal of Comparative Physiology, B* 169(2): 107–112.

13. DICKINSON, M H; LEHMANN, F; SANE, S P (1999) Wing rotation and the aerodynamic basis of insect flight. *Science* 284: 1954–1960.

14. BENNETT, L (1966) Insect aerodynamics: vertical sustaining force in near hovering flight. *Science* 152: 1263–1266.

15. WOOD, J (1970) A study of the instantaneous air velocities in a plane behind the wings of certain Diptera flying in a wind tunnel. *Journal of Experimental Biology* 52: 17–25.

16. WEIS-FOGH, T (1973) Quick estimates of flight fitness in hovering animals, including novel mechanisms for lift production. *Journal of Experimental Biology* 59: 169–230.

17. COOTER, R J; BAKER, P S (1977) The Weis-Fogh clap and fling mechanism in Locusta. *Nature, London* 269: 53–54.

18. ELLINGTON, C P (1984) The aerodynamics of insect flight I–VI. *Philosophical Transactions of the Royal Society of London, B* 305: 1–181.

19. ELLINGTON, C P; VAN DER BERG, C; WILLMOTT, A P; THOMAS, A L R (1996) Leading-edge vortices in insect flight. *Nature, London* 384: 626–630.

20. GRODNITSKY, D L; MOROZOV, P P (1993) Vortex formation during tethered flight of functionally and morphologically two-winged insects, including evolutionary considerations on insect flight. *Journal of Experimental Biology* 182: 11–40.

21. WOOTTON, R J (1992) Functional morphology of insect wings. *Annual Review of Entomology* 37: 113–140.

22. WOOTTON, R J (1981) Support and deformability in insect wings. *Journal of Zoology* 193: 447–468.

23. SNODGRASS, R E (1956) *Anatomy of the honey bee*. Comstock Publishing Associates and Cornell University Press; Ithaca, NY, USA; 334 pp.

24. NEVILLE, A C (1993) *Biology of fibrous composites: development beyond the cell membrane*. Cambridge University Press; Cambridge, UK.

25. PRINGLE, J W S (1957) *Insect flight*. Cambridge Monographs in Experimental Biology, No. 9. Cambridge University Press; Cambridge, UK; 133 pp.

26. DADE, H A (1962) *Anatomy and dissection of the honeybee*. Bee Research Association; London, UK; 158 pp.

27. DUDLEY, R (1999) Unsteady aerodynamics. *Science* 284: 1937–1939.

28. DANFORTH, B N (1989) The evolution of hymenopteran wings: the importance of size. *Journal of Zoology* 218: 247–276.

29. ROWELL, C H F; REICHERT, H (1986) Three descending interneurons reporting deviation from course in the locust. *Journal of Comparative Physiology, A* 158: 775–794.

30. EGELHAAF, M (1989) Visual afferences to flight steering muscles controlling optomotor response of the fly. *Journal of Comparative Physiology, A* 165: 719–730.

31. GOTZ, K G (1968) Flight control of *Drosophila* by visual perception of motion. *Kybernetik* 4: 199–208.

32. REICHARDT, W (1973) Musterinduzierte Flugorientierung. Verhaltens-Versuche an der Fliege *Musca domestica. Naturwissenschaften* 60: 122–138.

33. COLLETT, T S; KING, A J (1975) Vision during flight. *In* Horridge G A (ed.) *The compound eye and vision in insects.* Clarendon Press; Oxford, UK; pp 437–466.

34. PREISS, R; SPORK, P (1995) How locusts separate pattern flow into its rotatory and translatory components (Orthoptera: Acrididae). *Journal of Insect Behaviour* 8: 763–779.

35. FRIEDRICH, R W; SPATZ, H C; BAUSENWEIN, B (1994) Visual control of wing beat frequency in *Drosophila. Journal of Comparative Physiology, A* 175: 587–596.

36. ESCH, H E; BURNS, J E (1996) Distance estimation by foraging honeybees. *Journal of Experimental Biology* 199: 155–162.

37. SRINIVASAN, M V; ZHANG, S W; BIDWELL, J (1997) Visually mediated odometry in honeybees. *Journal of Experimental Biology* 200: 2513–2522.

38. SRINIVASAN, M V; ZHANG, S W; LEHRER, M; COLLETT, T S (1996) Honeybee navigation *en route* to the goal: visual flight control and odometry. *Journal of Experimental Biology* 199: 237–244.

39. DAVID, C T (1986) Mechanisms of directional flight in wind. *In* Payne, T E; Birch, M C; Kennedy, C E J (eds) *Mechanisms in insect olfaction.* Clarendon Press; Oxford, UK; pp 49–57.

40. FLETCHER, W A; GOODMAN, L J; GUY, R G; MOBBS, P G (1984) Horizontal and vertical motion detectors in the ventral nerve cord of the honeybee, *Apis mellifera. Journal of Physiology* 351: 16.

41. GOODMAN, L J; FLETCHER, W A; GUY, R G; MOBBS, P G; POMFRETT, C D J (1987) Motion sensitive descending interneurons, ocellar L_D neurons and neck motorneurons in the bee: a neural substrate for visual course control in *Apis mellifera. In* Menzel, R; Mercer, A (eds) *Neurobiology and behavior of honeybees.* Springer-Verlag; Berlin, Germany; pp 158–77.

42. GOODMAN, L J; IBBOTSON, M R; POMFRETT, C D J (1990) Directional tuning of motion-sensitive interneurons in the brain of insects. *In Higher order sensory processing.* Manchester University Press; Manchester, UK; pp 27–48.

43. IBBOTSON, M R; GOODMAN, L J (1990) Response characteristics of four wide-field motion-sensitive descending interneurons in *Apis mellifera. Journal of Experimental Biology* 148: 255–279.

44. IBBOTSON, M R (1991) Wide-field motion-sensitive neurons tuned to horizontal movement in the honeybee, *Apis mellifera. Journal of Comparative Physiology, A* 168: 91–102.

45. IBBOTSON, M R (1991) A motion-sensitive visual descending neurone in *Apis mellifera* monitoring translatory flow-fields in the horizontal plane. *Journal of Experimental Biology* 157: 573–577.

46. GOODMAN, L J; IBBOTSON, M R; BIDWELL, N J (1991) Spatial, temporal and directional properties of motion-sensitive visual neurons in the honeybee. *In* Goodman L J; Fisher, R C (eds) *The behaviour and physiology of bees.* CAB International on behalf of the Royal Entomological Society and the International Bee Research Association; Wallingford, Oxon, UK; pp 203–226.

47. GRIES, M; KOENIGER, N (1996) Straight forward to the queen: pursuing honeybee drones (*Apis mellifera* L.) adjust their body axis to the direction of the queen. *Journal of Comparative Physiology, A* 179: 539–544.

48. GOODMAN, L J (1965) The role of certain optomotor reactions in regulating stability in the rolling plane during flight in the desert locust, *Schistocerca gregaria. Journal of Experimental Biology* 42: 385–407.

49. ROWELL, C H F; PEARSON, K G (1983) Ocellar input to the flight motor system of the locust: structure and function. *Journal of Experimental Biology* 103: 265–288.

50. MIZUNAMI, M (1994) Information processing in the insect ocellar system: comparative approaches to the evolution of visual processing and neural circuits. *Advances in Insect Physiology* 25: 151–265.

51. GEWECKE, M; NIEHAUS, M (1981) Flight and flight control by the antennae in the Small Tortoiseshell (*Aglais urticae* L; Lepidoptera). I. Flight balance experiments. *Journal of Comparative Physiology, A* 145: 249–256.

52. GEWECKE, M; HEINZEL, H G (1980) Aerodynamic and mechanical properties of the antennae as air-current sense organs in *Locusta migratoria.* I. Static characteristics. *Journal of Comparative Physiology, A* 139: 357–366.

53. HEINZEL, H G; GEWECKE, M (1987) Aerodynamics and mechanical properties of antennae as air-current sense organs in *Locusta migratoria.* II. Dynamic characteristics. *Journal of Comparative Physiology, A* 161: 671–680.

54. CAMHI, J M (1969) Locust wind receptors: 1. Transducer mechanics and sensory response. *Journal of Experimental Biology* 50: 335–348.

55. BACON, J; TYRER, M (1979) The innervation of the wind-sensitive head hairs of the locust, *Schistocerca gregaria. Physiological Entomology* 4: 301–309.

56. GEWECKE, M; PHILIPPEN, J (1978) Control of the horizontal flight-course by air-current sense organs in *Locusta migratoria. Physiological Entomology* 3: 43–52.

57. GILBERT, C; GRONENBERG, W; STRAUSFELD, N J (1995) Occulomotor control in calliphorid flies: head movements during activation and inhibition of neck motor neurons corroborate neuroanatomical predictions. *Journal of Comparative Neurology* 361: 285–297.

58. ROBERT, D; ROWELL, C H F (1992) Locust flight steering. II. Acoustic avoidance manoeuvres and associated head movements. *Journal of Comparative Physiology, A* 171: 53–62.

59. ROBERT, D; ROWELL, C H F (1992) Locust flight steering. I. Head movements and the organisation of correctional manoeuvres. *Journal of Comparative Physiology, A* 171: 41–51.

60. EGGERS, F (1928) Die Stiftführenden Sinnesorgane. *Zool. Baust.* 2(1) (Quoted in Pringle, J W S (1957) *Insect flight.* Cambridge University Press; Cambridge, UK; 133 pp.).

61. ZAĆWILICHOWSKI, J (1933) Über die Innervierung und die Sinnesorgane der Flügel der Honigbiene (Apis mellifica L.). *Bulletin of the International Academy of Cracovie (Academy of Polish Science), B* II: 275–289. (Quoted in Pringle, J W S (1957) *Insect flight.* Cambridge University Press; Cambridge, UK; 133 pp.).

62. ALTMAN, J S; TYRER, M (1974) Insect flight as a system for the study of the development of neuronal connectives. *In* Browne L B (ed.), *Experimental analysis of insect behaviour.* Springer-Verlag; Heidelberg, Germany; pp 159–179.

63. DICKINSON, M H (1990) Linear and nonlinear encoding properties of an identified mechanoreceptor on the fly wing measured with mechanical noise stimuli. *Journal of Experimental Biology* 151: 219–244.

64. DICKINSON, M H (1990) Comparison of encoding properties of campaniform sensilla on the fly wing. *Journal of Experimental Biology* 151: 245–261.

65. DICKINSON, M H (1992) Directional sensitivity and mechanical coupling dynamics of campaniform sensilla during chordwise deformation of the fly wing. *Journal of Experimental Biology* 169: 221–233.

66. FRISCH, K VON (1967) *The dance language and orientation of bees* (translated by Leigh E Chadwick). The Belknap Press of Harvard University Press; Cambridge, MA, USA; 566 pp.

67. LINDAUER, M (1976) Foraging and homing flight of the honeybee: some general problems of orientation. *In* Rainey, R C (ed.) *Insect flight.* Symposia of the Royal Entomological Society, No. 7; Blackwell; Oxford, UK; pp 199–216.

68. RILEY, J R; OSBOURNE, J L (2001) Flight trajectories of foraging insects: observations using harmonic radar. *In* Woiwod, I P; Reynolds, D R; Thomas, C D (eds.) *Insect movement: mechanisms and consequences.* CAB International; Wallingford, UK; pp 129–157.

69. HERAN, H; LINDAUER, M (1963) Windkompensation und Seitenwindkorrektur der Bienen beim Flug über Wasser. *Zeitschrift für Vergleichende Physiologie* 47: 39–55.

70. CAPALDI, E A; SMITH, A D; OSBORNE, J L; FAHRBACH, S E; FARRIS, S M; REYNOLDS, D R; EDWARDS, A S; MARTIN, A; ROBINSON, G E; POPPY, G M; RILEY, J R (2000) Ontogeny of orientation flight in the honeybee revealed by harmonic radar. *Nature, London* 403: 537–540.

71. SEELEY, T D; MORSE, R A; VISCHER, P K (1979) The natural history of the flight of honey bee swarms. *Psyche* 86: 103–113.

72. COMBS, G F (1972) The engorgement of swarming worker honey bees. *Journal of Apicultural Research* 11(3): 121–128.

73. HANAUER-THIESER, U; NACHTI-GALL, W (1995) Flight of the honeybee VI: energetics of wind tunnel exhaustion flights at defined fuel content, speed adaptation and aerodynamics. *Journal of Comparative Physiology, B* 165: 471–483

74. SEELEY, T D (1995) *The wisdom of the hive: the social physiology of honey bee colonies.* Harvard University Press; Cambridge, MA, USA; 295 pp.

75. RUTTNER, F; RUTTNER, H (1966) Untersuchungen über die Flugaktivität und das Paarungsverhalten der Dronen. 3 Flugweite und Flugrichtung der Drohnen. *Zeitschrift für Bienenforschungen* 8(9): 332–354.

76. KOENINGER, G; KOENINGER, N; FABRITIUS, M (1979) Some detailed observations of mating in the honey bee. *Bee World* 60: 53–57.

77. COMBA, L; CORBET, S A; HUNT, L; WARREN, B (1999) Flowers, nectar and insect visits: evaluating British plant species for pollinator-friendly gardens. *Annals of Botany* 83: 369–383.

78. NACHTIGALL, W; HANAUER-THIESER, U; MÖRZ, M (1995) Flight of the honeybee VII: Metabolic power versus flight speed relation. *Journal of Comparative Physiology, B* 165: 484–489.

79. CAHILL, K; LUSTICK, S (1976) Oxygen consumption and thermoregulation in *Apis mellifera* workers and drones. *Journal of Comparative Physiology, A* 55: 355–357.

80. CORBET, S A; FUSSELL, M; AKE, R; FRASER, A; GUNSON, C; SAVAGE, A; SMITH, K (1993) Temperature and the pollinating activity of social bees. *Ecological Entomology* 18: 17–30.

81. STONE, G N; WILLMER, P G (1989) Warm-up rates and body temperatures in bees: the importance of body size, thermal regime and phylogeny. *Journal of Experimental Biology* 147: 303–328.

82. JUNGMANN, R; ROTHE, U; NACHT-IGALL, W (1989) Flight of the honey bee I. Thorax surface temperature and thermoregulation during tethered flight. *Journal of Comparative Physiology, B* 158: 711–718.

83. FELLER, P; NACHTIGALL, W (1989) Flight of the honeybee II. Inner- and surface thorax temperatures and energetic criteria, correlated to flight parameters. *Journal of Comparative Physiology, B* 158: 719–727.

84. HEINRICH, B (1980) Mechanisms of body temperature regulation in honeybees, *Apis mellifera* II. Regulation of thoracic temperature at high air temperatures. *Journal of Experimental Biology* 85: 73–87.

85. HEINRICH, B (1979) *Bumble bee economics.* Harvard University Press; Cambridge, MA, USA; 246 pp.

86. WOLF, T J; SCHMID-HEMPEL, P; ELLINGTON, C P; STEVENSON, R D (1989) Physiological correlates of foraging efforts in honey-bees: oxygen consumption and nectar load. *Functional Ecology* 3: 417–424.

87. HEINRICH, B (1975) Thermoregulation in bumble bees II. Energetics of warm-up and free flight. *Journal of Comparative Physiology, B* 96: 155–166.

88. MARDEN, J H (1987) Maximum lift production during take-off in flying animals. *Journal of Experimental Biology* 130: 235–258.

89. CHAI, P; CHEN, J S C; DUDLEY, R (1997) Transient hovering performance of hummingbirds under conditions of maximal loading. *Journal of Experimental Biology* 200: 921–929.

Chapter 8. Glands

1. ECKERT, R; RANDALL, D; AUGUSTINE, G (1988) *Animal physiology, mechanisms and adaptations.* W H Freeman and Co.; New York, USA; 683 pp (3rd edition).

2. GULLAN, P J; CRANSTON, P S (1994) *The insects: an outline of entomology.* Blackwell Science Ltd; Oxford, UK; 470 pp.

3. WINSTON, M L; SLESSOR, K N (1992) The essence of royalty: honey bee queen pheromone. *American Scientist* 80(4): 374–385.

4. NOIROT, C; QUENNEDEY, A; SMITH, R F (1974) Fine structure of insect epidermal glands. *Annual Review of Entomology* 19: 61–81.

5. BLUM, M S (1992) *Honey bee pheromones.* In Graham, J N (ed.) *The hive and the honeybee.* Dadant and Sons Hamilton, IL, USA; pp 373–400.

6. UEYAMA, Y; HASHIMOTO, S; NII, H; FURUKAWA, K (1990) The essential oil from the flowers of *Rosa rugosa* Thunb. var. *plena* Regel. *Flavour and Fragrance Journal* 5: 219–222.

7. HANSSON, B S (1995) Olfaction in Lepidoptera. *Experientia* 11: 1003–1027.

8. BOECKH, J; KAISSLING, K E; SCHNEIDER, D (1965) Insect olfactory receptors. *Coldspring Harbor Symposia on Quantitative Biology* 30: 263–280.

9. WILSON, E O (1963) Pheromones. *Scientific American* 208: 2–11.

10. WILSON, E O (1971) *The insect societies.* The Belknap Press of Harvard University Press; Cambridge, MA, USA; 584 pp.

11. BLUM, M S; BRAND, J M (1972) Social insect pheromones: their chemistry and function. *American Zoologist* 12(3): 553–576.

12. SCHMIDT, J O; SLESSOR, K N; WINSTON, M L (1993) Roles of Nasonov and queen pheromones in attraction of honeybee swarms. *Naturwissenschaften* 80: 573–575.

13. BRADSHAW, J W S; BAKER, R; HOWSE, P E (1975) Multicomponent alarm pheromones of the African weaver ant. *Nature, London* 258: 230–231.

14. BRADSHAW, J W S; BAKER, R; HOWSE, P E (1979) Multicomponent alarm pheromones in the mandibular glands of major workers of the African weaver ant, *Oecophylla longinoda. Physiological Entomology* 4(1): 15–25.

15. FREE, J B (1987) *Pheromones of social bees*. Chapman and Hall; London, UK; 218 pp.

16. SNODGRASS, R E (1956) *The anatomy of the honey bee*. Comstock Publishing Associates and Cornell University Press; Ithaca, NY, USA; 334 pp.

17. VECCHI, M A (1960) The scent gland of *Apis mellifica* L. *Bollettino dell'Istituto di Entomologia dell'Universita degli Studi Bologna* 24: 53–66.

18. CASSIER, P; LENSKY, Y (1994) The Nassanov gland of the workers of the honey bee (*Apis mellifera* L.): ultrastructure and behavioral function of the terpenoid and protein components. *Journal of Insect Physiology* 40(7): 577–584.

19. PICKETT, J A; WILLIAMS, I H; MARTIN, A P; SMITH, M C (1980) Nasonov pheromone of the honey bee *Apis mellifera* L. (Hymenoptera: Apidae) Part 1. Chemical characterization. *Journal of Chemical Ecology* 6(2): 425–434.

20. BOCH, R; SHEARER, D A (1963) Production of geraniol by honey bees of various ages. *Journal of Insect Physiology* 9: 431–434.

21. PICKETT, J A; WILLIAMS, I H; SMITH, M C; MARTIN, A P (1981) Nasonov pheromone of the honey bee *Apis mellifera* L. (Hymenoptera: Apidae) Part III. Regulation of pheromone composition and production. *Journal of Chemical Ecology* 7(3): 543–554.

22. WILLIAMS, I H; PICKETT, J A; MARTIN, A P (1981) Nasonov pheromone of the honey bee *Apis mellifera* L. (Hymenoptera: Apidae) Part II. Bioassay of the components using foragers. *Journal of Chemical Ecology* 7(2): 225–237.

23. WINSTON, M L (1987) *Biology of the honey bee*. Harvard University Press; Cambridge, MA, USA; 281 pp.

24. FERGUSON, A W; FREE, J B (1981) Factors determining the release of Nasonov pheromone by honeybees at the hive entrance. *Physiological Entomology* 6(1): 15–19.

25. FREE, J B; WILLIAMS, I H (1970) Exposure of Nasonov gland by honeybees (*Apis mellifera*) collecting water. *Behaviour* 37: 286–290.

26. FREE, J B; WILLIAMS, I H (1983) Scent-marking of flowers by honeybees. *Journal of Apicultural Research* 22(2): 86–90.

27. BUTLER, C G; FLETCHER, D J C; WALTER, D (1969) Nest-entrance marking with pheromones by the honeybee, *Apis mellifera* L., and a wasp, *Vespula vulgaris* L. *Animal Behaviour* 17: 142–147.

28. LENSKY, Y; SLABEZKI, Y (1981) The inhibiting effect of the queen bee (*Apis mellifera* L.) foot-print pheromone on the construction of swarming queen cups. *Journal of Insect Physiology* 27(5): 313–323.

29. ARNHART, L (1923) Das Krallenglied der Honigbiene. *Archiv für Bienenkunde* 5: 37–86.

30. LENSKY, Y; CASSIER, P; FINKEL, A; DELORME-JOULIE, C; LEVINSOHN, M (1985) The fine structure of the tarsal glands of the honeybee, *Apis mellifera* L. (Hymenoptera). *Cell and Tissue Research* 240: 153–158.

31. CHAUVIN, R (1962) Sur l'épagine et sur les glandes tarsales d'Arnhart. *Insectes Sociaux* 9: 1–5.

32. ZUPKO, K; SKLAN, D; LENSKY, Y (1993) Proteins of the honeybee (*Apis mellifera* L.) body surface and exocrine gland secretions. *Journal of Insect Physiology* 39(1): 41–46.

33. LENSKY, Y; CASSIER, P; FINKEL, A; TEESHBEE, A; SCHLENGER, R; DELORME-JOULIE, C; LEVINSOHN, M (1984) Les glandes tarsales de l'abeille mellifique (*Apis mellifera* L.) reines, ouvrières et faux-bourdons (Hymenoptera, Apidae). II. Role biologique. *Annales des Sciences Naturelles, Zoologie* 6: 165–175.

34. LENSKY, Y; FINKEL, A; CASSIER, P; TEESHBEE, A (1987) [The tarsal glands of honeybees (*Apis mellifera* L.) queens, workers and drones — chemical characterization of footprint secretions.] *Honeybee Science* 8(3): 97–102 (in Japanese).

35. GOULD, J L; GOULD, C G (1988) *The honey bee.* Scientific American Library; New York, USA; 239 pp.

36. HEPBURN, H R (1986) *Honeybees and wax.* Springer-Verlag; Heidelberg, Germany; 205 pp.

37. CASSIER, P; LENSKY, Y (1995) Ultrastructure of the wax gland complex and secretion of beeswax in the worker honey bee, *Apis mellifera* L. *Apidologie* 26(1): 17–26.

38. TULLOCH, A P (1980) Beeswax — composition and analysis. *Bee World* 61(2): 47–62.

39. LOCKE, M (1961) Pore canals and related structures in insect cuticle. *Journal of Biophysical and Biochemical Cytology* 10: 589–618.

40. KURSTJENS, S.P; McCLAIN, E; HEPBURN, H R (1990) The proteins of beeswax. *Naturwissenschaften* 77(1): 34–35.

41. SEELEY, T D (1995) *The wisdom of the hive: the social physiology of honey bee colonies.* Harvard University Press; Cambridge, MA, USA; 295 pp.

42. FRISCH, K VON (1974) *Animal architecture.* Harcourt Brace Jovanovich; New York, USA; 306 pp.

43. SEELEY, T D (1985) *Honeybee ecology. a study of adaptation in social life.* Princeton University Press; Princeton, NJ, USA; 210 pp.

44. GONTARSKI, H (1949) Über die Vertikalorientierung der Bienen beim Bau der Waben und bei der Anlage des Brutnestes. *Zeitscrift für Vergleichende Physiologie* 31: 652–670.

45. VANDENBERG, J D; MASSIE, D R; SHIMANUKI, H; PETERSON, J R; POSKEVICH, D M (1985) Survival, behaviour and comb construction by honey bees, *Apis mellifera*, in zero gravity aboard NASA Shuttle Mission, STS-13. *Apidologie* 16(4): 369–383.

46. MARTIN, H; LINDAUER, M (1966) Sinnesphysiogische Leistungen beim Wabenbau der Honigbiene. *Zeitschrift für Vergleichende Physiologie* 53: 372–404.

Chapter 9. Defending the colony

1. WINSTON, M L (1987) *The biology of the honey bee.* Harvard University Press; Cambridge, USA; 281 pp.

2. COLLINS, A M; RINDERER, T E; TUCKER, K W; SYLVESTER, H A; LOCKETT, J T (1980) A model of honeybee defensive behaviour. *Journal of Apicultural Research* 19(4): 224–231.

3. SHING, H; ERICKSON, E H (1982) Some ultrastructure of the honeybee (*Apis mellifera* L.) sting. *Apidologie* 13(3): 203–213.

4. SNODGRASS, R E (1956) *Anatomy of the honey bee.* Comstock Publishing Associates and Cornell University Press; Ithaca, NY, USA; 334 pp.

5. RIETSCHEL, F (1937) Bau und Funktion des Wehrstachels der staatendbilden Bienen und Wespen. *Zeitschrift für Morphologie und Ökologie der Tiere* 33: 313–357.

6. SCHMIDT, J O (1982) Biochemistry of insect venoms. *Annual Review Entomology* 27: 339–368.

7. RICHES, H R C (1982) Hypersensitivity to bee venom. *Bee World* 63(1): 7–22.

8. OGAWA, H; KAWAKAMI, Z; YAMAGUCHI, T (1995) Motor pattern of the stinging response in the honeybee *Apis mellifera*. *Journal of Experimental Biology* 198(1): 39–47.

9. HERMANN, H R; DOUGLAS, M E (1976) Comparative survey of the sensory structures on the sting and ovipositor of hymenopterous insects. *Journal of the Georgia Entomological Society* 11(3): 223–239.

10. BLUM, M S (1992) Honey bee pheromones. *In* Graham, J M (ed.) *The hive and the honey bee.* Dadant and Sons; Hamilton, IL, USA; pp 373–400.

11. BOCH, R; SHEARER, D A; STONE, B C (1962) Identification of iso-amyl acetate as an active component in the sting pheromone of the honey bee. *Nature, London* 195: 1018–1020.

12. GRANDPERRIN, D; CASSIER, P (1983) Anatomy and ultrastructure of the

Koschewnikow's gland of the honey bee, *Apis mellifera* L. (Hymenoptera: Apidae). *International Journal of Insect Morphology and Embryology* 12(1): 25–42.

13. CASSIER, P; TEL-ZUR, D; LENSKY, Y (1994) The sting sheaths of honey bee workers (*Apis mellifera* L.): structure and alarm pheromone secretion. *Journal of Insect Physiology* 40(1): 23–32.

14. PICKETT, J A; WILLIAMS, I H; MARTIN, A P (1982) (*Z*)-11-eicosen-1-ol, an important new pheromonal component from the sting of the honey bee, *Apis mellifera* L. (Hymenoptera, Apidae). *Journal of Chemical Ecology* 8(1): 163–175.

15. GHENT, R L; GARY, N E (1962) A chemical alarm releaser in honey bee stings (*Apis mellifera* L.). *Psyche (Cambridge, Mass.)* 69(1): 1–6.

16. COLLINS, A M; BLUM, M S (1982) Bioassay of compounds derived from the honeybee sting. *Journal of Chemical Ecology* 8(2): 463–470.

17. COLLINS, A M; BLUM, M S (1983) Alarm responses caused by newly identified compounds derived from the honeybee sting. *Journal of Chemical Ecology* 9(1): 57–65.

18. VALLET, A; CASSIER, P; LENSKY, Y (1991) Ontogeny of the fine structure of the mandibular glands of the honeybee (*Apis mellifera* L.) workers and the pheromonal activity of 2-heptanone. *Journal of Insect Physiology* 37(11): 789–804.

19. FREE, J B (1987) *Pheromones of social bees*. Chapman and Hall; London, UK; 218 pp.

When Dr Lesley Goodman finally lost her fight against lung cancer in 1998, she left behind her vision for an accessible, authoritative reference work for bee scientists, undergraduates and beekeepers. Form and Function in the Honey Bee is the fruition of her work — a posthumous tribute to her life and interests.

Containing over 340 diagrams, micrographs and colour illustrations, Form and Function works equally well as an expert guide to the physiology and anatomy of the honey bee, and as an introduction to this fascinating field for students and others.

The chapters take the reader through the major structures and activities of the honey bee — the antennae, compound eyes, dorsal ocelli, the bee's response to gravity, feeding, respiration, flight, glands and colony defence are all examined in detail to give the reader a comprehensive understanding of how and why the honey bee behaves as it does.

The book has been completed posthumously by Prof. Richard J Cooter, Chair of the L J Goodman Insect Physiology Trust, and Dr Pamela Munn, Deputy Director of the International Bee Research Association.

Chapters
1. The antennal sense organs: smelling, tasting, touching and hearing in the bee
2. Vision in the bee: the compound eye
3. The dorsal ocelli: the bee's second set of eyes
4. The bee's response to gravity: which way is up?
5. Feeding:
       1. Using the mouthparts
       2. Tasting the food
       3. Collecting the pollen
6. Respiration: how do bees breathe?
7. Flight: wings, aerodynamics, sensory control and metabolism
8. Glands: chemical communication and wax production
9. Defending the colony: the sting

www.ingramcontent.com/pod-product-compliance
Lightning Source LLC
Chambersburg PA
CBHW060957030426
42334CB00032B/3266